Ecological Research Monographs

Series Editor: Yoh Iwasa

For further volumes:
http://www.springer.com/series/8852

Nobukazu Nakagoshi • Jhonamie A. Mabuhay
Editors

Designing Low Carbon Societies in Landscapes

Springer

Editors
Nobukazu Nakagoshi
IDEC
Hiroshima University
Higashi-Hiroshima, Japan

Jhonamie A. Mabuhay
Biology Department
College of Natural Sciences and Mathematics
Mindanao State University
Marawi, Philippines

ISSN 2191-0707
ISBN 978-4-431-54818-8
DOI 10.1007/978-4-431-54819-5
Springer Tokyo Heidelberg New York Dordrecht London

ISSN 2191-0715 (electronic)
ISBN 978-4-431-54819-5 (eBook)

Library of Congress Control Number: 2014932982

© Springer Japan 2014
This work is subject to copyright. All rights are reserved by the Publisher, whether the whole or part of the material is concerned, specifically the rights of translation, reprinting, reuse of illustrations, recitation, broadcasting, reproduction on microfilms or in any other physical way, and transmission or information storage and retrieval, electronic adaptation, computer software, or by similar or dissimilar methodology now known or hereafter developed. Exempted from this legal reservation are brief excerpts in connection with reviews or scholarly analysis or material supplied specifically for the purpose of being entered and executed on a computer system, for exclusive use by the purchaser of the work. Duplication of this publication or parts thereof is permitted only under the provisions of the Copyright Law of the Publisher's location, in its current version, and permission for use must always be obtained from Springer. Permissions for use may be obtained through RightsLink at the Copyright Clearance Center. Violations are liable to prosecution under the respective Copyright Law.
The use of general descriptive names, registered names, trademarks, service marks, etc. in this publication does not imply, even in the absence of a specific statement, that such names are exempt from the relevant protective laws and regulations and therefore free for general use.
While the advice and information in this book are believed to be true and accurate at the date of publication, neither the authors nor the editors nor the publisher can accept any legal responsibility for any errors or omissions that may be made. The publisher makes no warranty, express or implied, with respect to the material contained herein.

Printed on acid-free paper

Springer is part of Springer Science+Business Media (www.springer.com)

Preface

Violent tropical low-pressure airs or typhoons emerge from the Pacific Ocean and attack Eastern Asia regularly. However, the advent of "super typhoons," which have caused unprecedented devastation in recent years, has increased fears among the people. Last year alone, several typhoons damaged many regions including Japan and other Asian countries. The Philippines, home country of co-editor Prof. Jhonamie A. Mabuhay, was not an exception. Destruction was enormous. Thousands of precious lives and properties were lost. In addition, huge tracts of farmland were ruined by flooding and high-tide phenomena. Crops almost ready for harvest, grown painstakingly by hardworking farmers, vanished, and their work was in vain. A few reliable scientific research outcomes support the hypothesis that super typhoons occur as a result of global warming. Accumulated meteorological data also appear to confirm that the current rise in seawater temperature is caused by the rise in air temperature. Aside from typhoons, abnormal weather patterns have also been observed more frequently compared to a decade, two decades, and even half a century ago.

The major cause of global warming is the high concentration of greenhouse gases present in the air. Greenhouse gases such as CO_2 and CH_4 are released through human activities. Among these harmful elements, CO_2 is known as the main culprit. Its presence in the atmosphere is primarily the result of biotic respiration and burning of fossil fuels. The rapid increase of CO_2, however, is largely caused by the latter. Fossil fuels such as coal, petroleum, and natural gas were formed though photosynthetic organisms from prehistoric eras. When burned, their carbon content, which had been stored for millions of years, is released into the atmosphere. But the problem goes beyond this. At present, the total area of major CO_2 absorbers such as forests, wetlands, and coral reefs has been shrinking. The ecological balance has deteriorated as a result of destructive human practices. If the reduction of CO_2 absorbers accelerates the increase of CO_2 in the atmosphere, it is common knowledge that controlling CO_2 emissions and increasing the amount of CO_2 absorbers are essential keys to total CO_2 reduction. Ecological research groups have already developed several effective measures such as limitation and optimization of land use, recovery of ecological systems, and development of bio-energy, to name a few. Still, use of a single method alone cannot eradicate the problem.

Low carbon societies can be established by restoring the CO_2 balance through integration of multiple tools as well as simultaneous application of various fields such as physics and chemistry, physiology and humanities, and education. This goal can be realized based upon a universally accepted philosophical way of thinking that the global environment can be protected if people will make certain sacrifices—abandoning daily conveniences and giving up profit-gains, among others. Achieving this low carbon society is our top priority, and landscape conservation is regarded as the first step in this ecological research.

The research introduced in this book focuses on three elements: conservation of ecosystem complexes, changes of arrangement, and creation as a means to achieve a low carbon society. A landscape is a collection of various ecosystems. Specifically, landscapes are classified into three categories: urban, agricultural, and natural, in relationship to humans. The landscape in urban areas is intensively influenced by citizens, whereas the agricultural landscape symbolizes the coexistence between farmers and the ecosystem. Further, the CO_2 balance in a natural vegetation landscape differs greatly from that in urban and agricultural landscapes. There are specific countermeasures for carbon absorption among these three types of landscapes because they have their respective carbon balances. Urban landscapes in East Asia, for example, which have been affected by heat island phenomena, are subject to fierce temperature fluctuation. School biotope projects in Japan, although still small in scale at the moment, are considered a promising solution because young people are encouraged to join environmental protection activities. Similar endeavors aim to rebuild damaged biodiversity in urban areas. Agricultural landscapes, on the other hand, have evolved moderately and thus have created biodiversity through long periods of years. Such landscapes have adjusted themselves to human activities. This book contains several research cases on tropical regions that had not been actively explored in the past. Studies on natural landscapes conducted in wildlife conservation areas consist of those held in nature reserves where local people live, as well as in their surrounding areas. In developed countries, natural areas have been regarded as conserved nature protection areas since early times. By contrast, abundant nature had remained in developing countries such as Indonesia and Malaysia. Later, nature protection areas were designated along with development. Now, developmental pressures have penetrated the nature protection areas. Under these circumstances, landscape ecological projects are needed. Therefore, research outcomes focused on tropical regions became the key in selecting the contents for this book.

Readers may freely choose their favorite chapters according to their individual interests in landscapes. Nineteen chapters chosen for the book were revised papers presented at the 8th World Congress of International Association for Landscape Ecology, Beijing 2011. All the papers were evaluated for technical and scientific quality by experts and accepted by Springer for publication. In fact, there were more than 19 candidates at the time of the proposal; but in the final phase, only those that qualified through peer review have been presented here.

I compiled this book as the main editor, but nothing would have been possible without the frequent discussions with my co-editor, Prof. Jhonamie A. Mabuhay.

Preface vii

My consultations with her were indeed relevant and valuable. Completing the book would have been impossible without the cooperation of many people from various sectors. The following professors and researchers generously and critically peer-reviewed all articles submitted for this publication: Saiful Arif Abdullah (Universiti Kebangsan Malaysia), Hadi Susilo Arifin (Bogor Agricultural University), Jürgen Breuste (University of Salzburg), Inocencio E. Buot, Jr. (University of the Philippines), Sun-Kee Hong and Jae-Eun Kim (Mokpo National University), Jhonamie A. Mabuhay (Mindanao State University), and Yun Pan (Capital Normal University, Beijing). Dr. R. L. Kaswanto of Bogor Agricultural University and two Ph.D. candidates, Rachmad Firdaus and Beni Raharjo of Hiroshima University, worked hard, and I am grateful for their assistance in editing this book. From at least 27 supporting agencies, there were eight large fund donors: the GELs Program of Hiroshima University; a Grant-in Aid for Scientific Research of JSPS; UNESCO; MAB Korea and Jeollanam-do Province; the Ministry of Science, Technology and Innovation Malaysia; the Ministry of Forestry Indonesia; BAPPENAS Indonesia; the UNREDD Program Indonesia; and the Joint Japan World Bank Graduate Scholarship Program.

Last, I thank my wife, Naomi Nakagoshi of JICA, because she has provided worthwhile advice to the young scholars studying at Hiroshima University and to the visiting professors working in the same university. Through her support, our team was able to discuss successfully the environmental and local situations in the developing countries involved in our studies. I believe many contributors to this book are thankful to her as well.

Higashi-Hiroshima, Japan   Nobukazu Nakagoshi

Contents

Part I Introduction

1 Landscape Ecological Approaches to a Low Carbon Society 3
Nobukazu Nakagoshi and Jhonamie A. Mabuhay

Part II Urban Landscape Ecology

2 Cooling Potential of Urban Green Spaces in Summer 15
Kochi Tonosaki, Shiro Kawai, and Koji Tokoro

3 A Study on the Restoration of Urban Ecology: Focus on the Concept of Home Place in Callenbach's *Ecotopia*—A Park Conservation and Community Networks 35
Masami Kato

4 Transpiration Characteristics of Chinese Pines (*Pinus tabulaeformis*) in an Urban Environment 57
Hua Wang, Zhiyun Ouyang, Weiping Chen, Xiaoke Wang, and Hua Zheng

5 Landscape Design for Urban Biodiversity and Ecological Education in Japan: Approach from Process Planning and Multifunctional Landscape Planning 73
Keitaro Ito, Ingunn Fjørtoft, Tohru Manabe, and Mahito Kamada

Part III Ecologies in Cultural Landscapes

6 Can Satoyama Offer a Realistic Solution for a Low Carbon Society? Public Perception and Challenges Arising 89
Yuuki Iwata, Takakazu Yumoto, and Yukihiro Morimoto

Contents

7 Effects of Sustainable Energy Facilities on Landscape: A Case Study of Slovakia 109
Katarina Pavlickova, Anna Miklosovicova, and Monika Vyskupova

8 Low Carbon Society Through *Pekarangan*, Traditional Agroforestry Practices in Java, Indonesia 129
Hadi Susilo Arifin, Regan Leonardus Kaswanto, and Nobukazu Nakagoshi

9 Challenges and Goal of the Sustainable Island: Case Study in UNESCO Shinan Dadohae Biosphere Reserve, Korea 145
Sun-Kee Hong, Heon-Jong Lee, Bong-Ryong Kang, Jae-Eun Kim, Kyoung-Ah Lee, Kyoung-Wan Kim, and Dae-Hoon Jang

10 The Neglect of Traditional Ecological Knowledge on Wild Elephant-Related Problems in Xishuangbanna, SW China 163
Zhao-lu Wu, Qing-cheng He, and Qiu-jun Wu

11 Effects of Tropical Successional Forests on Bird Feeding Guilds 177
Eurídice Leyequién, José Luis Hernández-Stefanoni, Waldemar Santamaría-Rivero, Juan Manuel Dupuy-Rada, and Juan Bautista Chable-Santos

Part IV Ecologies in Protected Areas

12 Understanding Development Trends and Landscape Changes of Protected Areas in Peninsular Malaysia: A Much Needed Component of Sustainable Conservation Planning 205
Saiful Arif Abdullah, Shukor Md. Nor, and Abdul Malek Mohd Yusof

13 Land Use Trends Analysis Using SPOT 5 Images and Its Effect on the Landscape of Cameron Highlands, Malaysia 223
Mohd Hasmadi Ismail, Che Ku Akmar Che Ku Othman, Ismail Adnan Abd Malek, and Saiful Arif Abdullah

14 The Relationship Between Land Use/Land Cover Change and Land Degradation of a Natural Protected Area in Batang Merao Watershed, Indonesia 239
Rachmad Firdaus, Nobukazu Nakagoshi, Aswandi Idris, and Beni Raharjo

15 Ecotourism Activities for Sustainability and Management of Forest Protected Areas: A Case of Camili Biosphere Reserve Area, Turkey 253
Mustafa Fehmi Türker, İnci Zeynep Aydın, and Türkan Aydın

Contents

16 Community Aspects of Forest Ecosystems in the Gunung Gede Pangrango National Park UNESCO Biosphere Reserve, Indonesia ... 271
Nobukazu Nakagoshi, Heri Suheri, and Rizki Amelgia

17 Landscape Ecology-Based Approach for Assessing Pekarangan Condition to Preserve Protected Area in West Java ... 289
Regan Leonardus Kaswanto and Nobukazu Nakagoshi

18 Integrating the Aerial Photos and DTM to Estimate the Area and Niche of *Arundo formosana* in Jiou-Jiou Peaks Natural Reserve of Taiwan ... 313
Jeng-I Tsai and Fong-Long Feng

19 REDD+ Readiness Through Selected Project Activities, Financial Mechanisms, and Provincial Perspectives in Indonesia ... 327
Ima Yudin Rayaningtyas and Nobukazu Nakagoshi

Index ... 349

Part I
Introduction

Chapter 1
Landscape Ecological Approaches to a Low Carbon Society

Nobukazu Nakagoshi and Jhonamie A. Mabuhay

Abstract The first Global Environmental Leaders (GELs) education programs were commenced in 2008 as a 5-year national program financed by the Special Coordination Funds for Promotion of Science and Technology under the Ministry of Education, Culture, Sports, Science and Technology, Japan. The GELs program for designing a low carbon society of Hiroshima University has successfully developed human resources selected from graduate schools and government institutions throughout the world. Capable young researches and experts have been fostered to the level that they are now able to participate in the international scientific societies such as the International Association for Landscape Ecology (IALE), the International Association for Vegetation Science (IAVS), and the East Asian Federation of Ecological Societies (EAFES). As a result, issues relating to the low carbon society are actively discussed in those conferences and the academic quality is upgraded, which contribution was undoubtedly made by GELs researchers. Along with the implementation of this program, numerous new scientific findings have emerged, and various technological fields are remarkably advanced thanks to the constructive debates conducted at both international and local meetings. Our research at Hiroshima University particularly highlights the following areas: land use and land cover change, population dynamics and ecosystem function, multistage sensing from urban to natural landscapes, new energy from biomass resources, and development of wise managing philosophy in developing countries. It is concluded that continuous collaborations among experts are

N. Nakagoshi (✉)
Graduate School for International Development and Cooperation, Hiroshima University,
1-5-1 Kagamiyama, Higashi-Hiroshima 739-8529, Japan
e-mail: nobu@hiroshima-u.ac.jp

J.A. Mabuhay
College of Natural Sciences and Mathematics, Mindanao State University,
Malawi City 9700, Philippines
e-mail: jhonamie@gmail.com

N. Nakagoshi and J.A. Mabuhay (eds.), *Designing Low Carbon Societies in Landscapes*, Ecological Research Monographs, DOI 10.1007/978-4-431-54819-5_1, © Springer Japan 2014

indispensable for designing an expected low carbon society that effectively and efficiently uses biomass and services in landscapes, so that sustainability can be realized on the global level.

Keywords Biomass use • Ecosystem sustainability • Global environmental leaders • Holistic approach • Landscape type • Systematic education

1.1 Introduction

Sustainable use and management of ecosystems are essential to preserve carbon sinks and to further develop alternative energy sources instead of relying on fossil fuels. It is our hope that the realization of low carbon societies will forestall the acceleration of global warming. The ongoing degradation of ecosystems continues to be a serious global problem. The Wise Use of Biomass Resources (WUBR) research group of the Global Environmental Leaders (GELs) program in Hiroshima University is mainly engaged in education and research activities on biomass/ecosystem resources for building a low carbon society (Nakagoshi 2011). As a result, the group published numerous scientific papers in the 5-year period (FY2008–2012) of the project. During this period, the first author contributed by editing two related textbooks entitled *Landscape Ecological Applications in Man-Influenced Areas* (Hong et al. 2008) and *Landscape Ecology in Asian Cultures* (Hong et al. 2011). This book is a textbook succeeding the previous two publications. The major difference of this book compared with the previous two books is consideration for the realization of sustainable landscapes in natural and man-made environments. For this book to be excellent, we invited outstanding papers on designing low carbon societies presented during the 8th World Congress of the International Association for Landscape Ecology (IALE) in Beijing, 2011. In addition, two related papers were collected with those peer-reviewed from the 5th East Asian Federation of Ecological Societies (EAFES) International Congress in Otsu, 2012, and the 55th Symposium of the International Association for Vegetation Science (IAVS) in Mokpo, 2012.

1.2 Framework of Research on Wise Use of Biomass Resources

The rate of deforestation has remained at 14 million ha annually since 1990. Most (78.6 %) of this decrease has occurred in tropical rainforests, which have large CO_2 absorption capabilities. The tropics have high socioeconomic dependency on natural resources (Amelgia et al. 2009), and land-use management schemes and guidelines are lenient or poorly designed (Caesariantica et al. 2011; Firdaus et al. 2011; Raharjo and Nakagoshi 2012). In fact, deforestation and cultivation in

1 Landscape Ecological Approaches to a Low Carbon Society

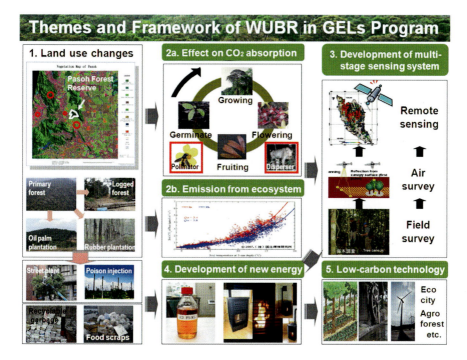

Fig. 1.1 Themes and framework of the Wise Use of Biomass Resources (WUBR) research group of the Global Environmental Leaders (GELs) program of Hiroshima University. The major research results are shown in the references. Concerning the chapters in this book, we included papers that are related to two or more themes among the five themes in this figure. All articles conformed to the low carbon technologies and societies. (Redrawn from HICEC 2009)

Southeast Asia account for about 75 % of the total release of carbon to the atmosphere. Such ecosystem disturbances brought about by human activities not only decrease the standing carbon stocks in ecosystems but also fragment and inhibit the regeneration of ecosystems (Abdullah and Nakagoshi 2008; Kaswanto et al. 2010; Lubis and Nakagoshi 2011). Given this situation, the introduction of ecosystem management to stimulate sustainable natural resource usage and carbon sink protection in developing counties is an emergency issue. Within this framework, the WUBR group has set five areas as priorities for research and education (Fig. 1.1). This book was designed to help in the landscape ecological training of graduate students and young researchers in universities and allied institutions. The synthesized ideas and technologies through these 19 chapters are easy to understand, and the vast list of references can be useful.

1.2.1 Landscape Degradation Caused by Land-Use Changes

Landscape degradation is caused by land-use and land cover changes. We hope to discuss ecosystem management strategies that balance sustainable natural resources usage and carbon sink protection among different types of ecosystems. For example, the Wise Use of Biomass Resources (WUBR) group investigated the losses of ecosystem services in devastated ecosystems and landscapes such as forests (Raharjo and Nakagoshi 2012; Mabuhay and Nakagoshi 2012), bamboo thickets (Suzuki and Nakagoshi 2008; Someya et al. 2009), savannahs (Caesariantica et al. 2011), grasslands (Lee and Nakagoshi 2010), and arable lands (Tokuoka et al. 2011). Appropriate management is also needed for rivers (Kohri et al. 2011) and wetlands (Byomkesh et al. 2009), including groundwater regulation (Pan and Nakagoshi 2008; Pan et al. 2008, 2011).

1.2.2 Effects of Ecosystem Degradation on CO_2 Absorption and Emission

Isolation of habitat causes genetic differentiation in temperate plants (Kaneko et al. 2008; Kondo et al. 2009). Extinction often occurs among smaller populations, particularly in endemic plants (Abe et al. 2008). Similarly, in the tropics fragmentation of ecosystems and decreasing population density as a result of both land-use changes and natural resource usage may inhibit the regeneration of a plant species through isolation among its local populations. Therefore, such human activities depress the future ability of ecosystems to serve as carbon sinks and decrease the standing carbon pools in the ecosystems. It is necessary to investigate the effects of isolation using field surveys and genetic marker analysis in the ecosystems of developing countries. In addition, soil respiration and greenhouse gas emission from ecosystems can be investigated by assessing temporal changes in the microorganisms and fauna linked with land use and ecosystem management (Mabuhay and Nakagoshi 2012).

1.2.3 Development of Multistage Sensing System

For regional valuation of ecosystems as carbon sinks, current technology must be scaled up so that ecological data can be extrapolated to wider areas including territories in developing countries. One potential technology is quantitative classification through remote sensing, a technology that allows continuous collection of data from across a landscape (Hong et al. 2008). Remote sensing and geographic information systems (GIS) are increasingly being used as tools to make an inventory of ecosystem resources, integrate data, and support decision making through analysis, modeling (Wang and Nakagoshi 2010), and forecasting (Johaerudin and Nakagoshi 2011).

Under this research theme, we focus on the following three technical research aims. The first is improvement in multistage sensing systems that can provide timely and accurate information at different scales for concurrent prediction of biomass in ecosystems. The second is the development of methods for monitoring human impact on ecosystems. The third aim is the proposal of precision ecosystem management schemes that can achieve successful results which reflect plants (Rotaquio et al. 2008; Kaneko et al. 2008; Kohri et al. 2011), animals (Wicaksono et al. 2011), landscape (Lubis and Nakagoshi 2011), and economic responses (Diep et al. 2012a, b).

1.2.4 Development of New Energy from Residual Biomass Resources

To utilize biomass at a high efficiency as a renewable energy source and to contribute to the mitigation of global warming by a great reduction of CO_2 emission are fields of interest with urgent consequences. Especially, if economical technologies can be established to utilize residual wood biomass from ecosystems, a large amount of woody biomass such as unused trees (Nakagoshi et al. 2011), lumber scraps, and construction waste can be utilized (Taib and Nakagoshi 2010). The first step in constructing a truly sustainable society is to establish methods on how to utilize renewable biomass resources (Taib et al. 2010). In this research theme, innovative biofuel production technologies from various biomass stocks (Takur and Nakagoshi 2011), focusing on plant biomass including the inedible part of harvests (Diep et al. 2011, 2012a, b; Wicaksono et al. 2010a) with larger CO_2 fixation capability, are being developed.

1.2.5 Development of Low Carbon Technologies

Land-use change, such as developing agricultural crops on farmlands (Abdullah and Nakagoshi 2008) and oil palm plantations (Johaerudin and Nakagoshi 2011), is one of the primary causes of forest loss in Southeast Asia. These human impacts also affect global warming through the loss of carbon sinks and subsequently increased CO_2 emission. The relationship between these two factors, however, has not been quantified. Quantification is vital to improve the planning and management of land use. Given this situation, landscape ecological approaches and GIS are increasingly being used. Under this research theme, we proposed land-use planning and design strategies that balance natural resources usage and carbon sink protection in landscapes by analyzing the patterns of land-use change from the aspect of realizing a low carbon society.

The idea of significant landscape is acceptable, and the following three types can be recognized: urban/technical, cultural, and natural landscapes. We proposed effective management on all the three types of landscapes: urban areas (Pham and Nakagoshi 2008; Kong et al. 2010; Arifin and Nakagoshi 2011; Byomkesh et al. 2012), forestry and agriculture areas (Abdullah and Nakagoshi 2008; Wicaksono and Nakagoshi 2009; Wicaksono et al. 2010a, b; Hakim and Nakagoshi 2010; Hong et al. 2011; Kaswanto and Nakagoshi 2012), national parks (Amelgia et al. 2009; Caesariantica et al. 2011; Firdaus and Nakagoshi 2013), and natural areas (Lin et al. 2009).

Biomass energy technology and the development of sustainable agriculture are essential for present and future humankind. We are currently conducting research studies in many fields such as biotechnology, plant protection, weed management, utilization of allelopathy, analytical analysis, chemical synthesis, crop breeding, development of crops tolerant to climatic changes, and biomass energy technologies. We have proposed some technologies for development of biomass energy (Xuan et al. 2009, 2012; Khanh et al. 2013) and reviewed the current trend of biomass technology to support policy planning of the agricultural sector in developing countries (Diep et al. 2012b).

In the area of biomass energy polices, we try to develop the social systems required to promote the effective use of biomass resources in developing countries. We proposed a systematic approach to design a social system in the Yellow River Basin, China (Higashi and Shirakawa 2011). We also focused on the roles of stakeholders to contribute effective institutional design for sustainable forest management in developing countries (Higashi 2011; Higashi and Shirakawa 2011; Higashi et al. 2012).

1.3 Symposia at the 8th World Congress of IALE in Beijing

In August 18–23, 2011, IALE organized the 8th World Congress in Beijing, China, a highlight event for us to summarize our project in intermediary terms. It was good timing to develop ideas on designing low carbon societies. The first author and his collaborators organized three symposia in the congress. Their titles and organizers are "Urban Green Spaces, Human Health, and Eco-environment Quality" (Nakagoshi N, Kong F), "Landscape Ecological Approaches to Develop the Low Carbon Societies in East and Southeast Asia" (Nakagoshi N, Fu B, Hong SK), and "Sustainability of Protected Area Landscapes in Asian Tropical Regions" (Abdullah SA, Nakagoshi N).

The 16 peer-reviewed papers among 29 presented in three symposia and one paper in a poster session were directly related to designing low carbon societies. We invited these 17 papers as chapter articles in this book. These 19 papers including two papers in EAFES and IAVS were rearranged and distributed into four parts, namely, one introductory article of Part I; four papers in Part II: Urban Landscape Ecology; six papers in Part III: Ecologies on Cultural Landscapes; and eight papers in Part IV: Ecologies on Protected Areas.

1.4 Summary

At first, we introduced the majority of the results from the WUBR group of GELs program in Hiroshima University, which was a research and educational project from FY2008 to FY2012. After the 8th Congress of IALE 2011 in Beijing, we invited almost all papers in the three symposia to be published in the Ecological Research Monographs series by Springer. Finally, 19 papers were selected through the peer-review process. These successful papers are presented as chapter articles in the four different categories.

It is our primary goal to help young scholars gain sufficient knowledge and skills to be able to work as a specialist in his/her area of expertise, especially in landscape ecology. This book also enables learning about areas other than your own area of ecological expertise to become an expert in that field. We hope that collaborations among experts and practitioners in science and technology will continue to work toward designing a low carbon society that uses biomass and services in landscapes. Because science and technology are constantly evolving, this kind of textbook on landscape ecology has a limited lifespan. However, we urge you to take full advantage of this book. To facilitate access to relevant research results, all articles of the book have been carefully discussed among the author(s), reviewers, and editors.

References

Abdullah SA, Nakagoshi N (2008) Change in agricultural landscape pattern and its spatial relationship with forestland in the state of Selangor, peninsular Malaysia. Landsc Urban Plan 87:147–155

Abe T, Wada K, Nakagoshi N (2008) Extinction threats of a narrowly endemic shrub, *Stachyurus macrocarpus* in the Ogasawara Islands. Plant Ecol 198:169–183

Amelgia R, Wicaksono KP, Nakagoshi N (2009) Forest product dependency excluded timber in Gede Pangrango National Park in West Java. Hikobia 15:331–338

Arifin HS, Nakagoshi N (2011) Landscape ecology and urban biodiversity in tropical Indonesian cities. Landsc Ecol Eng 7:33–43

Byomkesh T, Nakagoshi N, Shadedur RM (2009) State and management of wetland in Bangladesh. Landsc Ecol Eng 5:81–90

Byomkesh T, Nakagoshi N, Dewan AM (2012) Urbanization and green space dynamics in Greater Dhaka, Bangladesh. Landsc Ecol Eng 8:45–58

Caesariantica E, Kondo T, Nakagoshi N (2011) Impact of *Acacia nilotica* (L.) Willd. Ex Del invasion on plant species diversity in the Bekol Savanna, Baluran National Park, East Java, Indonesia. Tropics 20:45–53

Diep NQ, Nakagoshi N, Fujimoto S, Minowa T, Sakaguchi K (2011) Potential for fuel ethanol production from rice straw in Vietnam. The 8th Biomass-Asia Workshop, Hanoi, 2011. http://www.ibt.ac.vn/8th _biomassasia_ws/fulltext

Diep NQ, Fujimoto S, Minowa T, Sakanishi K, Nakagoshi N (2012a) Estimation of the potential of rice straw for ethanol production and the optimum facility size for different regions in Vietnam. Appl Energy 93:205–211

Diep NQ, Fujimoto S, Yanagida T, Minowa T, Sakanishi K, Nakagoshi N, Tran XD (2012b) Comparison of the potential for ethanol production from rice straw in Vietnam and Japan via techno-economic evaluation. Int Energy J 13:113–122

Firdaus R, Nakagoshi N (2013) Assessment of the relationship between land use land cover and water quality status of the tropical watershed: a case of Batang Merao watershed, Indonesia. J Biodivers Environ Sci 3(11):21–30

Firdaus R, Nakagoshi N, Raharjo B (2011) Changes in land use/land cover and priority determination on handling land degradation in Cirasea sub-watershed, West Java. Hikobia 16:9–20

Hakim L, Nakagoshi N (2010) Ecotourism in Asian tropical countries: planning a destination's site-plan to meet education objectives. J Int Dev Coop 16(1):13–21

HICEC (2009) Global environmental leader education program for designing a low carbon society. http://hicec.hiroshima-u.ac.jp

Higashi O (2011) Summer course 2010 group work development within a low carbon world: preparing professionals for participatory approaches in planning and implementing climate change policies. J Int Dev Coop 18(2):1–6

Higashi O, Shirakawa H (2011) International cooperation for building low-carbon and water-saving society: case study of Japan and China. J Int Dev Coop 17(3):71–81

Higashi O, Abdullah SA, Nakagoshi N, Shirakawa H, Patricia SM (2012) Evaluating the capacity of forest governance system for effective and efficient REDD-plus policies in the state of Pahang, peninsular Malaysia. EAAERE 2012

Hong SK, Nakagoshi N, Fu B, Morimoto Y (eds) (2008) Landscape ecological applications in man-influenced areas. Springer, Dordrecht [Paperback]

Hong SK, Wu J, Kim JE, Nakagoshi N (eds) (2011) Landscape ecology in Asian cultures. Springer, Tokyo

Johaerudin, Nakagoshi N (2011) GIS-based land suitability assessment for oil palm production in Landak Regency, West Kalimantan. Hikobia 16:21–31

Kaneko S, Nakagoshi N, Isagi Y (2008) Origin of the endangered tetraploid *Adonis ramose* assessed with chloroplast and nuclear DNA sequence data. Acta Phytotax Geobot 59(2):165–174

Kaswanto, Nakagoshi N (2012) Revitalizing Pekarangan home gardens, a small agroforestry landscape for a low carbon society. Hikobia 16:161–171

Kaswanto, Nakagoshi N, Arifin HS (2010) Impact of land use changes on spatial pattern of landscape during two decades (1989–2009) in West Java region. Hikobia 15:363–376

Khanh TD, Anh TQ, Buu BC, Xuan TD (2013) Applying molecular breeding to improve soybean rust resistance in Vietnamese elite soybean. Am J Plant Sci 4:1–6

Kohri M, Kamada M, Nakagoshi N (2011) Spatial-temporal distribution of ornithochorous seeds from an *Elaeagnus umbellata* community dominating a riparian habitat. Plant Species Biol 26:174–185

Kondo T, Nakagoshi N, Isagi Y (2009) Shaping of genetic structure along Pleistocene and modern river systems in the hydrochorous riparian azalea, *Rhododenderon ripense*. Am J Bot 96:1532–1543

Kong F, Yim H, Nakagoshi N, Zong Y (2010) Urban green space network development for biodiversity conservation: identification based on graph theory and gravity modeling. Landsc Urban Plan 95:16–27

Lee HJ, Nakagoshi N (2010) Grassland vegetation change after riverbank restoration in Jungnang River, Seoul, Korea. Hikobia 15:377–384

Lin H, Wang L, Yang J, Nakagoshi N, Liang C, Wang W, Lu Y (2009) Predictive modeling of the potential natural vegetation pattern in northeast China. Ecol Res 24:1313–1321

Lubis JPG, Nakagoshi N (2011) Land use and land cover change detection using remote sensing and geographic information system in Bodri watershed, central Java, Indonesia. J Int Dev Coop 18(1):1–12

Mabuhay JA, Nakagoshi N (2012) Response of soil microbial communities to changes in a forest ecosystem brought about by pine wilt disease. Landsc Ecol Eng 8:189–196

Nakagoshi N (2011) Strategic education programme for designing low carbon societies in Asia. In: IALE (ed) Proceedings of the 8th World Congress of IALE. IALE, Beijing (CD-ROM)

Nakagoshi N, Kuti FO, Yamaba A, Watanabe S, Saito I (2011) Biodiversity and fuel potential of wood species from unmanaged Satoyama forest in Higashi-Hiroshima. The 8th Biomass-Asia Workshop, Hanoi, 2011. http://www.ibt.ac.vn/8th_biomassasia_ws/fulltext

Pan Y, Nakagoshi N (2008) Effects of groundwater disturbance on vegetation implicated by eco-hydrology. Hikobia 15:177–184

Pan Y, Nakagoshi N, Gong H (2008) Using three-dimensional visualization to represent hydrological influence on wetland plants. Ecohydrol Hydrobiol 8(2–4):317–329

Pan Y, Gong H, Zhou D, Li X, Nakagoshi N (2011) Impact of land use change on groundwater recharge in Guishui River Basin, China. Chin Geogr Sci 21:734–743

Pham UD, Nakagoshi N (2008) Application of land suitability analysis and landscape ecology to urban greenspace planning in Hanoi, Vietnam. Urban For Urban Greening 7:25–40

Raharjo B, Nakagoshi N (2012) Tree resource diversity in the biological preservation block of the Sultan Adam Forest Park, South Kalimantan. Hikobia 16:151–160

Rotaquio EL Jr, Nakagoshi N, Rotaquio RL (2008) Does mangrove *Kandelia candel* (L.) Druce follow a mangrove zonation, soil salinity and substrate for survival? Hikobia 15:165–176

Someya T, Suzuki S, Nakagoshi N (2009) Regional comparison of bamboo forest expansion during the last 20 years (1980s–2000s) in Hiroshima Prefecture, Japan. Hikobia 15:299–310

Suzuki S, Nakagoshi N (2008) Expansion of bamboo forests caused by reduced bamboo-shoot harvest under different natural and artificial conditions. Ecol Res 23:641–647

Taib SM, Nakagoshi N (2010) Sustainable waste management through international cooperation: review of comprehensive waste management technique 2 training course. J Int Dev Coop 16 (1):23–34

Taib SM, Amagaki K, Nakagoshi N (2010) Assessing applicability of technologies for waste to energy in developing Asian cities. J Int Dev Coop 16(1):5–12

Takur IS, Nakagoshi N (2011) Production of biofuels from lignocellulosic biomass in pulp and paper mill effluents for low carbon society. J Int Dev Coop 18:1–12

Tokuoka Y, Ohigashi K, Nakagoshi N (2011) Limitations on tree seedling establishment across ecotones between abandoned fields and adjacent broad-leaved forests in eastern Japan. Plant Ecol 212:923–944

Wang R, Nakagoshi N (2010) Urban heat island change in the Beijing-Tianjin-Hebei metropolitan area driven by land cover change. Jpn J Biometeorol 47(2):77–89

Wicaksono KP, Nakagoshi N (2009) Agriculture profile and sustainability in Okinawa Prefecture Japan and East Java Province of Indonesia. ISTECS J 9(Special issue):1–7

Wicaksono KP, Murniyanto E, Nakagoshi N (2010a) Distribution of edible wild taro on the different altitude. Agrivita 32(3):225–233

Wicaksono KP, Prasetyo A, Nakagoshi N (2010b) Performance of Bondoyudo Mayang irrigation system in East Java, Indonesia. J Int Dev Coop 16(2):69–80

Wicaksono KP, Suryanto A, Nugroho A, Nakagoshi N, Karniawan N (2011) Insect as biological indicator from protected to the disturbed landscapes in central Java, Indonesia. Agrivita 33 (1):75–84

Xuan TD, Toyama T, Fukuta M, Khanh TD, Tawata S (2009) Chemical interaction in the invasiveness of cogongrass (*Imperata cylindrica* (L.) Beauv.). J Agric Food Chem 57:9448–9453

Xuan TD, Toyama T, Khanh TD, Tawata S, Nakagoshi N (2012) Allelopathic interference of sweet potato with cogongrass and relevant species. Plant Ecol 213:1955–1961

Part II
Urban Landscape Ecology

Chapter 2
Cooling Potential of Urban Green Spaces in Summer

Kochi Tonosaki, Shiro Kawai, and Koji Tokoro

Abstract The urban heat island phenomenon has recently become a serious subject of public concern. This chapter aims to clarify the cooling potential of urban green spaces in summer. At first, it shows that green space and anthropogenic heat emission have a great effect on the temperature in downtown areas from the various data collected from 27 observation points in Minato-ku, Tokyo. Then, it clarifies the cooling potential of green spaces. The results of multiple regression analysis, using the mean daily maximum temperatures in August as dependent variables and the size of woodland area and the amount of anthropogenic heat emissions as explanatory variables, showed that an increase in trees contributes to a reduction in temperature in urban areas, and that an increase in the amount of anthropogenic heat emissions causes a rise in temperature. A multiple regression model in equations and a coefficient of correlation among mean daily maximum temperature, woodland area, and amount of anthropogenic heat emissions were obtained as follows: $Y = 32.0011 - 0.001(X_1) + 0.0033(X_2)$, $r = 0.7276$, where $Y =$ the mean daily maximum temperature, $X_1 =$ woodland area, and $X_2 =$ the amount of anthropogenic heat emissions. From this regression analysis, it can be said that the cooling influence by green spaces of 22,500 m^2 is equivalent to the heating influence by the anthropogenic heat released from 70 offices of average size in Minato-ku, having a total floor area of about 211,726 m^2. Furthermore, the cooling potential of a green space of 22,500 m^2 during July to September can be expected to reduce about 236 times as much carbon dioxide as the same green space absorbs for 1 year. In conclusion, green spaces in urban downtown areas have the function of air conditioning provided by nature.

K. Tonosaki (✉)
Organization for Landscape and Urban Green Infrastructure, Kandajimbou-cho 3-2-4,
Tamura Bld 2F Chiyoda-ku, Tokyo 101-0051, Japan
e-mail: tonosaki@urbangreen.or.jp

S. Kawai • K. Tokoro
Soken, Inc., Nagoya, Japan
e-mail: kawai@soken.co.jp; tokoro@soken.co.jp

N. Nakagoshi and J.A. Mabuhay (eds.), *Designing Low Carbon
Societies in Landscapes*, Ecological Research Monographs,
DOI 10.1007/978-4-431-54819-5_2, © Springer Japan 2014

Keywords Anthropogenic heat emission • Carbon dioxide • Cooling potential • Green coverage ratio • Green space • Heat island phenomena • Minimum air temperature • Tokyo • Woodland

2.1 Introduction

The heat island effect is acknowledged by the Japanese government as a pollution issue: heat islands were identified in 2003 by the Ministry of the Environment as causing thermal pollution of urban air and were noted in the Third Basic Environment Plan drawn up in April 2006 as an air pollution problem in Japanese cities. With the emergence of the heat island phenomenon during the past few years as a major social issue affecting urban areas, there has been a growing move to promote urban greening in view of its potential benefits in a variety of areas—not only in heat island mitigation but also in absorbing the greenhouse gases that cause global warming and in enhancing the cityscape. In particular, from the perspective of heat island mitigation, interest in the cooling effects of urban greening is mounting, and a number of studies have been conducted in this area.

From the correlation between green coverage ratio and the minimum air temperature during days on which "tropical nights" occur, Owada et al. (2007) showed that the air temperature tends to be lower the higher the green coverage ratio. By analyzing land-use data derived from Landsat data, and air temperatures based on AMeDAS data, Irie (2003) demonstrated that the effect of green space is most apparent at times when the air temperature is lowest. Ando et al. (2008) examined actual measurements of the cool air drifting out of the Imperial Palace gardens as a green effect during the hottest times of day. Yamada and Maruta (1991) took the highest and lowest thermometer readings at 57 sites in Tokyo's Suginami-ku and found that wooded areas had an effect in reducing both maximum and minimum temperatures.

On the other hand, in April 2007, the U.S. Supreme Court handed down a ruling requiring regulation of CO_2 and other greenhouse gases that cause global warming to the federal Environmental Protection Agency. In March 2010, the Climate Action Reserve (2010) provided guidance for calculating, reporting, and verifying greenhouse gas (GHG) emission reductions associated with a planned set of tree planting and maintenance activities to permanently increase carbon storage in trees. The appendix, which was included in the Protocol, introduced previous studies on the effect of air conditioning: for example, McPherson and Simpson (2003), EIA (2002), CARB (2007), Nowak and Crane (2002), Sampson et al. (1992), Trexler (1991), Moulton and Richards (1990), and McHale et al. (2007). Akibari (2002) noted, "Urban tree planting can reduce net cooling and heating energy consumption in urban landscapes by 25 %."

All these previous studies evaluated the effects on air temperature of the different types of ground cover that make up green space (woodland, grassland, bare ground, etc.). However, there appear to be no previous studies that evaluate the

2 Cooling Potential of Urban Green Spaces in Summer

effects of the floor area of buildings, amount of vegetation cover, or amount of anthropogenic heat emissions on air temperature.

Against this background, our study considers at the effect of woodlands in reducing air temperature, based on air temperature data obtained in August 2007 from digital weather stations set up at 27 locations in elementary schools and parks in Minato-ku, Tokyo. In addition, by showing that the temperature at the observation points can largely be explained in terms of the size of the woodland and the amount of anthropogenic heat emissions, we ascertained the ratio of these two factors (negative and positive, respectively) in contributing to air temperature. Furthermore, we tried converting the reduction in air temperature per unit area of woodland into the cooling potential of air conditioners and other such equipment.

2.2 Methods

Minato-ku is the ward located in the southeastern part of Tokyo that faces Tokyo Bay to the east (Fig. 2.1). The total area of Minato-ku is 20.34 km^2 as of October 1, 2008. As of April 1, 2010, the population of Minato-ku, according to the Basic Resident Registry, was 202,505 residents.

The diagram in Fig. 2.2 shows the procedure used in this study.

The focus of this study was Minato-ku, where the heat island effect was pronounced, as Bureau of the Environment, Tokyo Metropolitan Government (2005), and Minato-ku (2006a, b) pointed out. First, we obtained temperature data in summertime from digital weather stations set up at 1.5 m above ground level at 27 sites in Minato-ku.

We calculated the mean daily maximum temperature during August 2007 at each site and used these figures as temperature data. Mean values were taken to eliminate confounding factors in the temperature data caused by wind or building orientation, for instance. The observation points at the 27 locations in Minato-ku are shown in Table 2.1 and Fig. 2.3. To plot the temperature data in a 50 × 50 m mesh, we took the mean daily maximum temperature at each of the 27 sites as the data for each grid cell containing an observation point. We then performed surface interpolation of the gridded data using GIS and estimated the temperature distribution over the entire Minato-ku area.

Based on findings noted in the literature about the heat island effect by Ojima (2002), Tokyo Heat Island Mitigation Council (2003), and Moriyama (2004), we collected and compiled data about factors we considered to be relevant to temperature formation: the size of woodlands and green spaces, amount of anthropogenic heat emissions, building floor area, and site elevation. The actual methods used to calculate this factor data and grid it on a mesh are described next.

To calculate green space, we used available GIS data that show green areas. These data were created by identifying vegetation cover from color digital orthophoto data and infrared orthophoto data obtained from the aerial photographic data taken by a digital aerial camera at the time of the 7th Minato-ku Green Space Survey (2006).

Fig. 2.1 Location of Minato-ku in 23 wards in Tokyo

In the available data, locations with at least 200 m² continuous green space had been specifically extracted as "woodlands." We decided to utilize this existing information. Thus, the green-related factors in temperature formation were classified and organized under two separate categories: "woodlands" and "green space" (comprising areas of vegetation or tree cover, grasslands, green roofs, bare ground, and water features).

In calculating the amount of anthropogenic heat emission, we focused in this study on the emission stage, which has a direct influence on urban air temperatures. We considered the heat released from commercial and residential buildings, and from vehicles and trains, which together account for 90 % of the total anthropogenic heat emissions in Minato-ku. To derive emissions from these sources, we followed the heat emission intensities and calculation procedures described in the report by Ministry of Land, Infrastructure, Transport and Tourism and Ministry of the Environment (2004). Using the appropriate heat emission intensities given in that report, anthropogenic heat emissions were calculated By Building Size and Purpose in the case of commercial and residential buildings, By Vehicle Type, Speed, and Time of Day in the case of vehicles, and By Railway Line in the case of railway heat emissions.

2 Cooling Potential of Urban Green Spaces in Summer

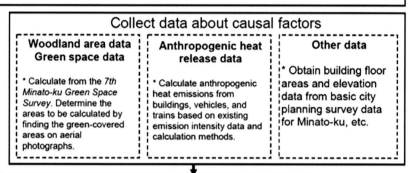

Fig. 2.2 Procedural flow

To derive anthropogenic heat emissions from buildings on this basis, first of all we identified the purpose of each building using the basic city planning survey data for Minato-ku given in the Tokyo City Planning Geographical Information System. The area of each building was found from building polygon data, and this figure was multiplied by the number of floors in the building to give the total floor area. The building floor area was then multiplied by the relevant heat emission intensity unit (building size and purpose) to give the amount of anthropogenic heat released from that particular building.

Table 2.1 Observation points at 27 locations

Number	Name	North latitude	East longitude
1	Aoyama junior high school	35°40′13″	139°43′34″
2	Akasaka elementary school	35°40′22″	139°44′16″
3	Seinan elementary school	35°39′37″	139°43′10″
4	Nogizaka tunnel	35°39′45″	139°43′38″
5	Akasaka elementary school	35°39′58″	139°44′03″
6	Kouryo junior high school	35°39′12″	139°43′34″
7	Azabu elementary school	35°39′32″	139°44′30″
8	Roppongi junior high school	35°39′29″	139°44′05″
9	Eco-plaza	35°39′44″	139°44′58″
10	Sakurada park	35°39′43″	139°45′35″
11	Motomura junior high school	35°38′47″	139°43′58″
12	Akabane junior high school	35°39′00″	139°44′43″
13	Shiba junior high school	35°38′52″	139°45′12″
14	Minato city hall	35°39′18″	139°43′14″
15	Sibarikyu children's park	35°39′10″	139°45′43″
16	Shirokanenomori nursing home	35°38′02″	139°43′30″
17	Asahi junior high school	35°38′25″	139°43′47″
18	Onda elementary school	35°38′23″	139°44′28″
19	Shibaura elementary school	35°38′31″	139°43′08″
20	Shibaura monitoring station	35°38′49″	139°45′42″
21	Takanawa park	35°37′46″	139°44′19″
22	Shibaura central park	35°37′57″	139°44′49″
23	Wharf park	35°38′18″	139°45′30″
24	Takanawa-minamimachi park	35°37′18″	139°44′19″
25	Konan junior high school	35°37′40″	139°45′08″
26	Konan park	35°37′15″	139°45′00″
27	Nijinohashi kindergarten	35°37′50″	139°46′51″

Anthropogenic heat emissions from vehicles were calculated from traffic volume and speed data, categorized by time of day, area, and vehicle type in the Road Traffic Census 2005, multiplied by the relevant heat emission intensity unit (vehicle type, speed, and time of day). To derive anthropogenic heat emissions from railways, we calculated the number of trains traveling on each of the railway lines between the stations in Minato-ku, based on the 2006 JTB timetable, and then multiplied those figures by the heat emission intensity unit per railway line to determine the amount of anthropogenic heat emitted on each stretch of railway.

Next, we describe how the various factor data were plotted on a 50-m mesh. Heat emissions from buildings were calculated per structure and plotted as attribute values on a building polygon in the GIS map. By summing the anthropogenic heat emissions from individual buildings, we calculated the total anthropogenic heat emissions from buildings in each grid cell. Where a building crossed the grid cell boundary, the building area was apportioned between the two cells. Anthropogenic heat emissions from vehicles and railways were calculated by area (census area for vehicles; station-to-station for railways) and were assigned in the GIS as

Fig. 2.3 Observation points at 27 locations

polyline attributes representing roads and railways recorded on Digital Map 2500 published by the Geospatial Information Authority of Japan. We divided these data over a 50-m mesh, and then calculated the amount of anthropogenic heat emission from the vehicles and railways in each grid cell.

The floor area of each building was calculated by multiplying the building area by the number of floors, using the basic urban planning survey data for Minato-ku given in the Tokyo City Planning Geographical Information System. The calculations were summed by grid cell.

For the site elevation data, we used data from digital map 50-m mesh elevations (Geospatial Information Authority of Japan).

In analyzing the effects of the various factors on air temperature in this study, because the smallest mesh size used in our study was 50 × 50 m, we decided for convenience's sake to set three cell sizes (50, 150, and 250 m) encompassing a central temperature measurement point. As shown in Fig. 2.4, if we take the 150-m cell, its factor data are determined by adding the factor data for the basic 50-m cell containing the temperature measurement point (basic cell) to the factor data summed for the eight cells that surround it. The factor data for the 250-m cell are determined by adding the factor data for the 150-m cell to the factor data summed for the 16 cells that surround it in the same manner.

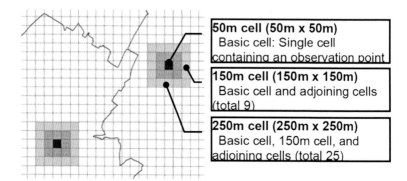

Fig. 2.4 Conceptualization of cell sizes

Next, among the varying temperatures at the 27 observation points, we determined those points where readings were identical and summed the various factor data contained in the 50-m basic cells, 150-m cells, and 250-m cells related to those observation points. In each case we found the mean values by cell size and temperature. That is, we calculated the amount of green space, the amount of anthropogenic heat emissions, building floor area, and site elevation, and performed correlation analysis on this dataset. From the results we determined which factors have the greatest effect on temperature formation, and performed multiple regression analysis on those data.

As the temperature data used for regression analysis in this study, for all cell sizes we used only the readings at the 27 locations in the 50 × 50 m basic cells.

2.3 Results and Discussion

2.3.1 Collected Temperature and Factor Data, and Mesh Data Results

The mean temperatures in Minato-ku in August are shown in Fig. 2.5. Some of the temperature data and factor data used in this study are described as follows. Temperature data were obtained in August 2007 from readings taken by digital weather stations set up at 1.5 m above ground level at 27 sites in Minato-ku, including schools, parks, and other public facilities. The mean daily maximum temperature was calculated for each site, and the values between the observation points were found by surface interpolation (Fig. 2.6).

From Fig. 2.6 it can be seen that the hottest region of 33 °C or higher occurs primarily around Roppongi in the northern Azabu area, and extends from the

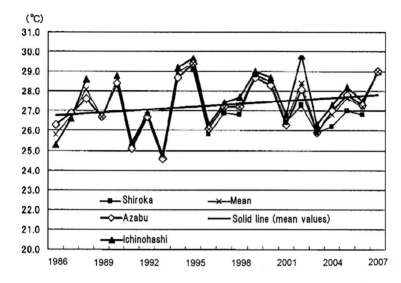

Fig. 2.5 Mean temperatures in Minato-ku in August

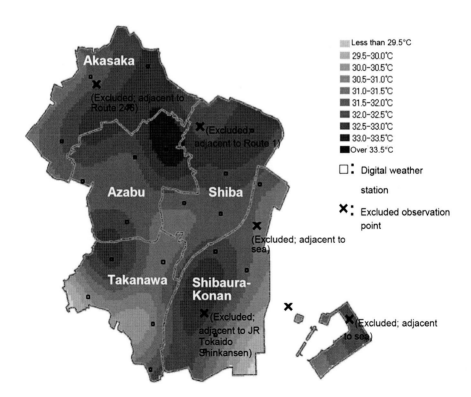

Fig. 2.6 Temperature distribution on a 50-m mesh

Fig. 2.7 Green space distribution

northern Shiba area to Akasaka. The next warmest regions are Shirokane in the Takanawa area and Takanawa in the Shibaura-Konan area.

Figure 2.7 shows the distribution of green space in Minato-ku. Those areas with the largest vegetation cover are the Akasaka Imperial Gardens and Aoyama Cemetery to the north, the Institute of Nature Study attached to the National Museum of Nature and Science in the west, and Shiba Park and its environs in the center of Minato-ku. Large parks account for the main green spaces. Other areas lack green space and have less than 500 m^2 vegetation per 2,500 m^2 (less than 20 % green coverage ratio).

Figure 2.8 shows the distribution of anthropogenic heat emissions in Minato-ku. Emissions are highest in the area reaching from Shimbashi and Roppongi to Hamamatsu-cho, which has a high concentration of commercial buildings, and crowded arterial routes such as metropolitan expressways, national highways, and areas adjacent to those routes.

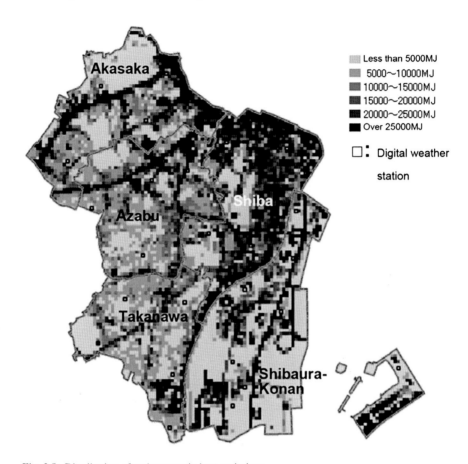

Fig. 2.8 Distribution of anthropogenic heat emissions

2.3.2 Results of Categorizing Factor Data by Cell Size and Temperature

Tables 2.2 and 2.3 list factor data by cell size and temperature, calculated in accordance with the procedure described in the research methodology. However, because of the presence of sea breezes and transportation systems (cars and railways), abnormal values that would always affect the temperature are very likely to be recorded at some of the 27 observation points in Minato-ku. For this reason, a total of 22 observation points are given in Tables 2.2 and 2.3. The five excluded sites are two sites near the sea, two bordering major arterial highways (Routes 246 and 1), and one adjacent to a JR railway line.

The factor data in the table are mesh values compiled separately for each of the cell sizes that encompass an observation point (50, 150, and 250 m).

Table 2.2 Factor data at observation points for 50- and 150-m cells

Mean daily maximum temperature in August (°C)	Factor data for 50-m cells					Factor data for 150-m cells				
	Woodland area (m²)	Green space (m²)	Anthropo-genic heat emissions (MJ/day)	Building floor area (m²)	Site elevation (m)	Woodland area (m²)	Green space (m²)	Anthropo-genic heat emissions (MJ/day)	Building floor area (m²)	Site elevation (m)
30.44	397	581	23,834	6,205	27	10,152	14,009	54,497	16,780	28
30.74	1,635	1,932	2,682	834	13	10,705	12,365	35,824	12,319	13
31.33	2	397	863	2,882	21	4,325	7,309	14,944	15,614	21
31.63	480	495	867	3,076	10	5,596	7,268	55,538	27,131	10
31.65	639	1,185	1,438	5,120	30	4,936	13,199	44,007	25,294	30
31.67	0	173	0	32	2	900	4,183	133,835	33,607	2
31.69	0	344	4,636	5,747	4	211	1,995	174,042	61,315	4
31.73	337	2,498	0	0	26	4,608	8,705	28,462	17,863	26
31.76	0	44	3,674	5,937	2	405	4,982	77,545	50,675	2
31.77	299	1,385	2,000	90	2	2,425	8,847	44,123	3,788	2
31.80	673	807	5,174	1,889	17	3,951	6,276	52,445	22,837	17
31.82	538	1,034	3,268	3,328	2	3,020	11,312	17,436	11,489	2
32.02	209	446	2,042	3,697	31	2,087	4,792	92,147	34,277	31
32.02	197	1,199	930	3,300	20	2,352	7,013	98,751	35,883	21
32.12	240	393	2,315	4,429	20	3,207	5,237	40,076	24,618	21
32.36	0	148	159	563	3	1,264	7,069	80,941	27,410	3
32.44	14	334	59,751	13,278	4	4,634	7,087	293,277	68,344	4
32.63	380	568	5,654	4,218	20	951	2,487	190,569	56,438	19
32.69	346	766	776	2,753	19	6,859	10,057	24,523	17,733	20
32.74	0	156	4,207	5,902	19	538	2,758	99,179	39,974	19
32.87	166	1,306	1,674	321	3	318	2,958	220,554	40,785	3
33.41	0	200	14,473	5,229	24	1,390	4,281	298,063	35,568	25
Total	6,552	16,391	140,417	78,612	–	74,834	154,189	2,170,778	679,742	–

2 Cooling Potential of Urban Green Spaces in Summer 27

Table 2.3 Factor data at observation points for 250-m cells

| Mean daily maximum temperature in August (°C) | Factor data for 250-m cells | | | | |
	Woodland area (m²)	Green space (m²)	Anthropogenic heat emissions (MJ/day)	Building floor area (m²)	Site elevation (m)
30.44	29,118	36,496	168,866	57,145	28
30.74	19,907	24,599	363,105	67,824	14
31.33	7,169	16,443	123,686	61,447	19
31.63	8,184	17,138	287,245	91,278	9
31.65	15,502	33,419	181,596	93,276	30
31.67	900	8,194	441,768	119,949	2
31.69	959	7,210	906,243	258,285	4
31.73	10,141	20,481	137,615	84,101	25
31.76	408	15,451	369,268	262,256	2
31.77	5,629	19,859	320,065	60,120	2
31.80	13,182	17,925	137,197	57,860	16
31.82	8,534	29,767	57,258	30,277	2
32.02	3,888	11,210	286,905	115,340	30
32.02	8,593	16,499	470,440	126,267	22
32.12	7,218	13,931	146,852	72,258	21
32.36	2,366	17,991	338,481	93,573	3
32.44	13,815	20,273	647,000	149,346	4
32.63	1,789	6,260	649,964	172,644	19
32.69	14,626	21,122	103,139	67,061	19
32.74	5,387	12,012	264,938	124,193	20
32.87	399	6,184	728,528	134,066	3
33.41	2,458	10,739	577,586	115,158	25
Total	180,172	365,219	7,369,399	2,320,187	–

2.3.3 Results of Correlation Analysis of Temperature Data and Factor Data

Table 2.4 lists the results of correlation analysis of temperature data and factor data. Negative correlations were present between the mean daily maximum temperatures in August and woodland area and between the mean daily maximum temperatures and green spaces. Positive correlations were present between the mean daily maximum temperatures and the amount of anthropogenic heat emissions and between the mean daily maximum temperatures and building floor areas. Site elevation was little related to the mean daily maximum temperatures. The factor data that correlated most highly with the mean daily maximum temperatures in August were woodland area followed by green spaces; this applied for both the 150- and 250-m grid cells. Correlation coefficients were generally low for the 50-m grid cells. In the 150-m cells, the highest

Table 2.4 Results of correlation analysis of temperature data and factor data by cell size

	Maximum temperature	Woodland area	Green space	Anthropogenic heat emissions	Floor area	Elevation
50-m cells						
Maximum temperature	1					
Woodland area	−0.492	1				
Green space	−0.281	0.609	1			
Anthropogenic heat emissions	0.072	−0.158	−0.223	1		
Floor area	0.141	−0.288	−0.473	0.804	1	
Elevation	−0.057	0.160	0.165	−0.084	−0.082	1
150-m cells						
Maximum temperature	1					
Woodland area	−0.627	1				
Green space	−0.603	0.821	1			
Anthropogenic heat emissions	0.602	−0.429	−0.565	1		
Floor area	0.425	−0.514	−0.665	0.761	1	
Elevation	−0.051	0.349	0.239	−0.186	−0.217	1
250-m cells						
Maximum temperature	1					
Woodland area	−0.583	1				
Green space	−0.573	0.831	1			
Anthropogenic heat emissions	0.354	−0.464	−0.615	1		
Floor area	0.228	−0.529	−0.549	0.731	1	
Elevation	−0.047	0.414	0.236	−0.367	−0.290	1

correlation was between the mean daily maximum temperatures in August and the size of woodland area; correlation with the amount of anthropogenic heat emissions was the next highest value after correlation with green space size.

2.3.4 Results of Multiple Regression Analysis Using Temperature as the Dependent Variable

Based on the results of correlation analysis of temperature data and factor data, we performed multiple regression analysis using the mean maximum daily temperature in August as the dependent variable. The size of woodland area and amount of anthropogenic heat emissions, which correlated highly with mean maximum daily temperatures in August, were used as explanatory variables.

2 Cooling Potential of Urban Green Spaces in Summer

Table 2.5 Results of multiple regression analysis by cell size

| | Range of collected data | | | | | |
| | A: 50-m cells | | B: 150-m cells | | C: 250-m cells | |
Explanatory variables	Coefficient	P value	Coefficient	P value	Coefficient	P value
Woodland area (m^2)	−0.0009	0.025	−0.0001	0.018	−0.0001	0.013
Anthropogenic heat emissions (GJ/day)	−0.0003	0.979	0.0033	0.03	0.0002	0.678
Intercept	32.2398	–	32.0011	–	32.3125	–
Multiple correlation coefficient	0.4924		0.7276		0.5885	
P value	0.0715		0.0008		0.0176	
Number of readings	22		22		22	

Although the amount of green space also correlates highly with the mean maximum daily temperatures in August, a high correlation coefficient was found between the amount of woodland and amount of green space, which suggests a likelihood of multicollinearity. For this reason we chose to use data from woodland areas rather than green space, because the former has a higher correlation with mean maximum daily temperatures in August.

The most significant result of multiple regression analysis was found for B (150-m cells), which gave good P values and multiple correlation coefficients in Table 2.5. From this result we determined the multiple regression model in Eq. (2.1), where Y is the mean daily maximum temperature in August (°C), X_1 is the woodland area (m^2), and X_2 is the amount of anthropogenic heat emissions (GJ/day):

$$Y = 32.0011 - 0.0001(X_1) + 0.0033(X_2) \tag{2.1}$$

It should be noted that the multiple regression model obtained in this study is applicable within a mean daily maximum temperature range in August between 30.4 ° and 33.4 °C when the data were actually collected.

The effects on temperature formation given in Eq. (2.1), as obtained from the correlation analysis results, show that the woodland area coefficient has a negative effect whereas the anthropogenic heat emissions coefficient has a positive effect. This result can be said to validate the hypothesis given in the introduction.

2.3.5 Evaluation of Green Space in Terms of Its Effect in Reducing Temperature

To derive the relative contribution of woodland area and amount of anthropogenic heat emissions to temperature formation from Eq. (2.1), we obtained Eq. (2.2), which applies when the Y-value is identical to the intercept. Here X_1 is the woodland area (m^2) and X_2 is the amount of anthropogenic heat emission (GJ/day):

$$0.0001\ (X_1) = 0.0033\ (X_2) \tag{2.2}$$

From Eq. (2.2), the cooling effect of trees per grid cell ($150 \times 150\,m = 22{,}500\,m^2$) would be equivalent to offsetting 681.8 GJ of anthropogenic heat emissions per day. To explore the implications, let us convert this 681.8 GJ/day into the anthropogenic heat emissions from commercial buildings, which are typical of the type and size of buildings in Minato-ku, and consider how many square meters of floor area this emission intensity would equate to. Referring to the heat emission intensities of commercial buildings (August estimates) given in the Ministry of Land, Infrastructure, Transport and Tourism and Ministry of the Environment (2004), the intensity of anthropogenic heat released per day from a commercial building ranked as having from 2,000 m^2 to less than 5,000 m^2 of floor area is stated as being 3,220.29 kJ/m^2/day in total (sensible heat, 2,925.66 kJ/m^2/day; latent heat, 270.87 kJ/m^2/day; sewage heat, 23.76 kJ/m^2/day). Using these emission intensities, and calculating back from the calculated cooling effect (681.8 GJ/day), we derive a figure of 211,726 m^2 as the total floor area of multiple commercial buildings of the type typical in Minato-ku. That is, 681.8 GJ/day of heat would be equivalent to the anthropogenic heat emissions per day from approximately 70 commercial buildings each with a floor area of about 3,000 m^2. Note that this figure of 681.8 GJ anthropogenic heat released per day, as described in detail in the *2003 Investigative Report*, is the total amount of heat released to the atmosphere, primarily from external air conditioning units and ventilation systems in an office environment that has heat sources such as lights, computers, air conditioners, ventilation systems, kitchen equipment, water heaters, and various other equipment in operation, together with the waste heat released through sewage systems.

Therefore, from Eq. (2.2), the presence of trees covering an area of 22,500 m^2 would have the effect of moderating or offsetting a temperature rise of 2.25 °C caused by 681.8 GJ of heat released per day from all the various systems and equipment installed in approximately 70 office buildings each having a floor area of approximately 3,000 m^2.

2.3.6 Evaluation of Green Space in Terms of Its Effect in Reducing CO_2

Next, we investigated the cooling potential of a 22,500 m^2 area of trees in terms of its effect in reducing the greenhouse gas CO_2. In the 2006 Forest Sink Measures to Prevent Global Warming (Forestry Agency of the Japan Ministry of the Environment), 50-year-old cedars are given as a guideline for assessing the amount of CO_2 absorbed by trees. The same publication states that the average amount of CO_2 absorbed by one 50-year-old cedar is approximately 14 kg/year (Ministry of the Environment and Forestry Agency 2006), and the average area occupied by one tree in a cedar forest is 12 m^2. As approximately 12 m^2 is required to grow one Japanese cedar, 1,875 cedars can be planted in an area of 22,500 m^2. Looking at the amount of CO_2 absorbed by these trees, because one cedar absorbs 14 kg CO_2 annually, 1,875 cedars would absorb and sequester 26,250 kg CO_2 in 1 year.

In a typical office building equipped with air conditioning, setting aside the waste heat in sewage, anthropogenic heat emissions for the most part would equate to the heat discharged outdoors from external air conditioning units, which are run on electric power to lower the room temperature in an office environment that is warmed by thermal radiation from the sun and is further heated by the use of heat sources such as lights, computers, and kitchen equipment (Tokyo Metropolitan Government 2005). Based on this premise, if we subtract sewage waste heat from the 681.8 GJ of heat released per day from approximately 70 commercial buildings, each with a floor area of about 3,000 m^2, and then calculate the heat emissions from those buildings, we get 681.8 GJ/day minus the sewage heat waste (sewage heat intensity, 23.76 kJ/m^2/day $\times$ 3,000 m^2 $\times$ 70 buildings = 5 GJ/day), which comes to 676.8 GJ/day.

In contrast to air conditioners, which run on supplied power whose generation produces CO_2, trees and woodlands cool the atmosphere through shade and evapotranspiration and can exhibit cooling potential without producing any CO_2. It would seem essential to take this point into account in assessing the functions of green space.

With this idea in mind, we tried converting the 676.8 GJ/day heat emissions into the power consumption of air conditioning equipment.

The basic approach used to calculate the heat discharged from external air conditioning units is taken from the Ministry of Land, Infrastructure, Transport and Tourism and Ministry of the Environment (2004). Using Eq. (2.3) given in that report, we calculated the amount of fuel (calorific value) used in running an air conditioner as follows:

$$\begin{aligned} \text{Air conditioner heat discharge} = {} & \text{fuel used(calorific value) to run the air conditioner} \\ & \times \text{energy efficiency} \\ & \times \text{air conditioner COP} \end{aligned}$$

$$(2.3)$$

where COP (coefficient of performance) is the value of the cooling capacity (W) divided by the power consumption (W) while the air conditioner is in cooling operation.

In regard to the energy efficiency in Eq. (2.3), we referred to data from the Energy Conservation Center, Japan (ECCJ), which states the average all-day efficiency as 36.9 %, assuming the installation site is some distance from the power station.

For air conditioner COP, we used the value (COP = 4.159) given in the ECCJ (2006). It lists the cooling capacity (kW), power consumption (kW), cooling COP, and other characteristics of ten medium-size air conditioner models (cooling capacity, 10.0 kW class; four-way cassette) for offices and stores being marketed by leading manufacturers as of March 2006. In this study, we took the mean value (4.159) of the cooling COPs of these ten models (2.87–4.90). Assigning these values to the variables in Eq. (2.3) above, and calculating the fuel used (calorific value) during air conditioner operation, we obtain the value 441 GJ/day. Because the conversion factor is given as 1 kW h = 3.6 MJ17, converting this 441 GJ/day calorific value into power consumption gives 122,600 kW h/day. From this result, if 676.8 GJ of anthropogenic heat is discharged per day from a commercial building, the power station needs to generate 122,600 kW h of power per day.

According to the ordinance issued by the Ministry of the Environment on March 23, 2006, which amends part of the Act on Promotion of Global Warming Countermeasures, the emission factor for calculating how much CO_2 is produced in generating 1 kW h of electricity is given as 0.555 kg CO_2. Therefore, the amount of CO_2 produced in generating 122,600 kW h/day of electricity will be 68,043 kg CO_2/day. Because this value equates to the daily average for the month of August, if we take the cumulative effect during the period July to September (90 days) when coolers are in frequent use, the presence of woodlands would reduce the amount of CO_2 in the atmosphere by approximately 6,124 t CO_2.

As already discussed, the amount of carbon dioxide absorbed by planting a 22,500 m^2 area of cedars would be of the order of 26 t CO_2 (26,250 kg CO_2) per year. However, because tree species are not necessarily planted at a uniform density, if we estimate the cooling potential of a 22,500 m^2 plantation using conservatively estimated GIS data, we find that the expected reduction in CO_2 from the trees' cooling potential would be as much as 236 times the amount of CO_2 that the trees absorb during the months of July to September (6,124 t CO_2/26 t CO_2).

From these findings, any future assessment of woodlands would require a more multifaceted approach: their role needs to be examined not only from the well-canvassed perspective of CO_2 absorption and sequestration, but must also take into account the potential of woodlands to moderate the amount of CO_2 in the atmosphere, their capacity in summertime in particular to reduce power consumption by air conditioning equipment, effectively helping to reduce the amount of CO_2 released from power stations and related facilities.

2.4 Summary and Conclusion

The results obtained through our investigation and analysis are summarized as follows.

- The results of multiple regression analysis, using the mean daily maximum temperatures in August as dependent variables and the size of woodland area and amount of anthropogenic heat emissions as explanatory variables, showed that an increase in trees contributes to a reduction in temperature in urban areas, and that an increase in the amount of anthropogenic heat emissions causes a rise in temperature.

- The cooling potential of 22,500 m^2 of trees in Minato-ku is equivalent to moderating and offsetting the total anthropogenic heat released per day from approximately 70 commercial buildings of a typical type and size for Minato-ku, each having a floor area of approximately 3,000 m^2. In other words, wooded areas in the densely and increasingly overpopulated commercial heart of the city have the effect of moderating and offsetting the rise in temperature caused by anthropogenic heat emissions from air conditioning systems, which inevitably arise in downtown Tokyo. That is, green spaces in urban areas have an air conditioning function provided by nature.
- The value of green space consisting of trees and other vegetation has generally been assessed in terms of how much CO_2 the trees absorb. This study demonstrates that green space can be evaluated from a new perspective, the cooling potential of green space, particularly in urban areas, in offsetting the waste heat from air conditioners (anthropogenic heat emissions) that spreads throughout the neighborhood, and contributing to a reduction in the amount of CO_2 in the atmosphere released by power stations and related facilities. Comparing the effects per 22,500 m^2 of trees in Minato-ku, we found that trees have the potential to reduce 236 times as much CO_2 as the amount of CO_2 they absorb.

References

Akibari H (2002) Shade trees reduce building energy use and CO_2 emissions from power plants. Environ Pollut 116:S119–S126

Ando K et al. (2008) Observations on the heat island mitigation effects of large green space (Parts 1–3), Summaries of technical papers of Annual Meeting Architectural Institute of Japan, pp 983–987 (in Japanese)

Bureau of the Environment, Tokyo Metropolitan Government (2005) Guidelines for heat island control measures [summary edition], pp 1–7. http://www.kankyo.metro.tokyo.jp/en/attachement/heat_island.pdf. Accessed 21 Apr 2013

California Air Resources Board (2007) California 1990 greenhouse gas emissions level and 2020 emissions limit. California Air Resources Board, Sacramento, p 29

Climate Action Reserve (2010) Urban forest project protocol. http://www.climateactionreserve.org/how/protocols/urban-forest/. Accessed 21 Apr 2013

Energy Conservation Center JAPAN (ECCJ) (2006) Energy conservation performance of air conditioners for offices and stores, spring (in Japanese)

Energy Information Administration (EIA) (2002) Updated state-level greenhouse gas emission coefficients for electricity generation 1998–2000. Energy Information Administration, Washington, DC, p 9

Irie T (2003) Study on effect of open space in reducing heat island by presuming the temperature. J Jpn Inst Landsc Architect 66(5):889–892 (in Japanese)

McHale MR, McPherson EG, Burke IC (2007) The potential of urban tree plantings to be cost effective in carbon credit markets. Urban For Urban Green 6:49–60

McPherson EG, Simpson JR (2003) Potential energy savings in buildings by an urban tree planting programme in California. Urban For Urban Green 2:73–86

Minato-ku (2006a) FY 2007 Heat island effect analysis report (in Japanese). http://www.city.minato.tokyo.jp/chikyukankyou/kankyo-machi/kankyo/chosa/h17/index.html. Accessed 21 Apr 2013

Minato-ku (2006b) The 7th Minato-ku green space survey (in Japanese)

Ministry of Land, Infrastructure, Transport and Tourism and Ministry of the Environment (2004) FY 2003 investigative report on urban heat island mitigation by curbing anthropogenic heat emissions (in Japanese). http://www.env.go.jp/air/report/h16-05/. Accessed 21 Apr 2013

Ministry of the Environment and Forestry Agency (2006) Sink measures by forest to prevent global warming (in Japanese). Ministry of the Environment and Forestry Agency, http://www.rinya.maff.go.jp/kanto/aizu/knowledge/breathing.html/. Accessed 21 May 2014

Moriyama M (2004) Heat island countermeasures and technologies. Gakugei Shuppansha, pp 1–206 (in Japanese)

Moulton RJ, Richards KR (1990) Costs of sequestering carbon through tree planting and forest management in the United States. Gen. Tech. Rep. WO-GTR-58. Forest Service, U.S. Department of Agriculture, Washington, DC

Nowak DJ, Crane DE (2002) Carbon storage and sequestration by urban trees in the USA. Environ Pollut 116:381–389

Ojima T (2002) Heat islands. Toyo Keizai, Tokyo, pp 1–157 (in Japanese)

Owada M, Nakagawa Y, Iwata M, Sakurai M, Umeda Y (2007) Effect of green space and distribution of hot summer night in Nagoya City, Japan. Bulletin of Aichi University of Education (Natural Science), pp 19–24 (in Japanese)

Sampson RN, Moll GA, Kielbaso JJ (1992) Opportunities to increase urban forests and the potential impacts on carbon storage and conservation. In: Sampson RN, Hair D (eds) Forests and global change: opportunities for increasing forest cover, vol 1. American Forests, Washington, DC, pp 51–72

Tokyo Heat Island Mitigation Council (2003) The principle of heat island mitigation. Tokyo Heat Island Mitigation Council, pp 1–43 (in Japanese). https://www.kankyo.metro.tokyo.jp/climate/attachement/heathousin.pdf/. Accessed 21 May 2014

Tokyo Metropolitan Government (2005) Manual for the Tokyo building environment planning system: description of assessment standards and methods in the green building design guidelines. Tokyo Metropolitan Government (in Japanese). http://www7.kankyo.metro.tokyo.jp/building/doc/old/s_3-2.pdf/. Accessed 21 May 2014

Trexler MC (1991) Minding the carbon store: weighing U.S. strategies to slow global warming. World Resources Institute, Washington, DC, p 81

Yamada Y, Maruta Y (1991) A quantitative analysis on the mitigation of city temperature by the open spaces in urban area. J Jpn Inst Landsc Architect 54(5):299–304 (in Japanese)

Chapter 3
A Study on the Restoration of Urban Ecology: Focus on the Concept of Home Place in Callenbach's *Ecotopia*—A Park Conservation and Community Networks

Masami Kato

Abstract More than ever before, after the nuclear power plant disaster caused by the Great Tohoku Earthquake on March 11, 2011, people living in urban regions are beginning to consider a more eco-friendly and sustainable lifestyle. To present a practical vision that citizens can translate into action, this study examines North American writer Ernest Callenbach's important concept of "home place" in his novel *Ecotopia* as a sustainable vision to apply to the restoration of urban ecology. As a case study, we review a citizen movement to preserve a municipal park in Bunkyo Ward, Tokyo, Japan. This study illustrates how diverse citizen activities can organically form multiple layered and intertwined networks that strengthen local communities and function effectively even in difficult situations—a small but inspiring example that Ecotopian principles can emerge and thrive even within the context of a large city such as Tokyo.

Keywords Bunkyo Ward • Citizen movement • Eco-friendly • Sustainable • Tokyo • Urban region

3.1 Introduction

The purpose of this study is to consider whether the realization of Ecotopian values and lifestyle could be possible in large cities such as Tokyo. For the purpose of this chapter, *Ecotopia* represents the title of the 1974 novel written by North American writer Ernest Callenbach (1929–2012) and Ecotopian represents how Callenbach's thoughts could be realized in an ecologically sustainable society.

M. Kato (✉)
Department of Value and Decision Science, Tokyo Institute of Technology,
Tokyo 152-8552, Japan
e-mail: mkato@valdes.titech.ac.jp

N. Nakagoshi and J.A. Mabuhay (eds.), *Designing Low Carbon Societies in Landscapes*, Ecological Research Monographs,
DOI 10.1007/978-4-431-54819-5_3, © Springer Japan 2014

How to bring about and live in a society underpinned by principles of sustainability is one of the most important challenges for the twenty-first century, and mitigation of global warming has been a major environmental topic. As Mander and Callenbach point out in their prologue to Gar Smith's *Nuclear Roulette*, nuclear energy has been considered as a "grand 'green' climate-friendly energy system that can successfully replace fossil fuel and continue to sustain our industrial society at its present level" since the beginning of the twenty-first century (Mander and Callenbach 2011). However, the nuclear power plant accident in Fukushima caused by the Tohoku earthquake and tsunami in 2011 has clearly revealed that nuclear energy can also be extremely hazardous and uncontrollable when accidents happen.

Environmental philosopher Toshio Kuwako views the nuclear power disaster in 2011 not only as the collapse of the local commons, but also the contamination and destruction of the global commons as well, and believes that from now on Japan "must bear a burden to restore" both the disrupted local as well as the global commons (Kuwako 2011).

The unprecedented nuclear power catastrophe in Japan has brought into public awareness the social and environmental issues surrounding the use of nuclear energy. Tokyo's residents experienced a severe reduction of their power supply that restricted the functioning of the city and negatively impacted their lifestyle. As a result, more people than ever have begun to join the discussion around achieving sustainability, which alters the entire social system as well as affecting individual lifestyles.

In addition, it is a critical challenge to achieve sustainability in urban regions, where nature has been forced to yield so much to human-made environments. The urban population has been expanding all over the world, and many urban dwellers that who were born, grew up, or have lived in urban areas for a long time have not had the opportunity to experience close contact with their local ecosystems.

3.1.1 Objectives

This study focuses on an urban community and on individual lifestyles drawn from Ernest Callenbach's concepts in his novel *Ecotopia* regarding what defines a sustainable society (Capra 1995; Callenbach 1995) and how the quality of urban ecology can be improved (Callenbach 2004). Many of his holistic ideas introduced in 1974 about sustainable policies, technologies, social systems, and ecological ways of living have become reality; we can find many Ecotopian ideas put into practice in cities and regions around the world (windmill power stations, photovoltaic power generation, organic recycle systems, and so on). Eisaku Tsuruta, who translated Callenbach's works into Japanese, points out that the major characteristic of Callenbach's thoughts is to propose a practical vision that people can take into action (Capra 1995; Callenbach 1995; Tsuruta 1995). Many readers and critics have focused on Callenbach's ideas of a nation in which nature-friendly practices are accomplished through the application of science and technology. This study focuses

3 A Study on the Restoration of Urban Ecology: Focus on the Concept of Home... 37

on the importance of people's feeling a sense of living in a "home place." The first half of this chapter examines two key concepts of Ecotopian principles, "home place" and "stable state" and discusses two major criticisms leveled against Ecotopia. The second half considers whether certain Ecotopian concepts are applicable to real urban life in Japan through a case study of a citizen movement to stop the local government's City Planning Agency plan to replace a small municipal park, the Motomachi Park in Bunkyo Ward, Tokyo, by a high-rise building.

3.2 Review of ECOTOPIA

3.2.1 Is ECOTOPIA an Ecological Utopia or a Dystopia?

Many readers have the impression that Ecotopia is a story of ecological utopia. For instance, the Amazon Readers Evaluation: ECOTOPIA introduces 49 readers' book reviews of Ecotopia (Amazon Book Review) (Table 3.1). The readers' evaluation is divided into two group: readers who assess the book with five stars, the highest evaluation, view the novel as based on Callenbach's scientific research. On the other hand, five of seven readers who rated the book "one star" regard the story as being a "utopia" and denounce the novel as "nonsense."

Hiroshi Shioda in his review "Ecotopia: its contradiction and violence—in reading Callenbach's dystopia" maintains that the word "ecotopia" means an ecological utopia and states that in Callenbach's ecological civilization, the life of human beings somehow becomes inhuman and miserable. And he sees the country as a "dystopia" that brainwashes its people to force an environmentally coexistent lifestyle upon them (Shioda 2008).

His criticism resembles the early concerns of William Weston, the main character of the novel. Weston shows strong suspicion of and antagonism toward the environmentally friendly co-existing society in Ecotopia. Similar to Weston, people who are accustomed to life in the industrial society tend to see eco-centered practices as inconvenient and unpleasant and requiring much physical labor.

Scott Slovic, a professor at the University of Nevada, Reno (USA) and a growing leader of a literature-and-the-environmental-movement, edited *Ecotopia and the Environmental Justice Reader* and views Ecotopia as "the ideal topos of ecology" for which people long (Slovic 2008). Slovic lists Callenbach's *Ecotopia*, which is reviewed by Shioda, as one of 22 ecotopia titles in the book.

Table 3.1 Amazon Readers Evaluation: ECOTOPIA. www.amazon.com/Ecotopia-Ernest-Callenbach/product-Reviews/0553348477(2011.07.19)

Evaluation	Lower higher					Total/average
Number of stars	1	2	3	4	5	49
Numbers of readers	7	2	5	13	22	3.83

Takaaki Okuda, a researcher of the Ecotopia Science Institute, University of Nagoya, points out that the most difficult thing is peoples' longing for a much more inconvenient Ecotopian society rather than benefiting from the present condition of free society (Okuda 2009). Okuda sees Ecotopia as a restrictive society.

3.2.2 Is Ecotopia a Story of the Pacific Northwest?

According to Callenbach, *Ecotopia* has been criticized for the setting of the novel, a narrow strip of coastal region of the Pacific Northwest in North America. The critics point out that the emergence of this ecologically oriented country was facilitated by the region's situation in one of the best parts of the United States: land rich in nature, clean water, fertile soil, and tree-covered mountains (Callenbach interviewed 2003).

This region is often called Ecotopia because of the title of Callenbach's novel. For example in his nonfiction book *Nine Nations of North America*, Joel Garreau, a Canadian journalist calls the region Ecotopia after the novel Ecotopia as Ecological Utopia and introduces cultural and geographic aspects characteristic of the region (Garreau 1981).

Scott Timberg, a journalist, also introduces ECOTOPIA in the New York Times as "the novel that predicted Portland" in a feature article about Callenbach (Timberg 2008), stating that the cities in the novel remind him of Portland.

Callenbach counters this argument: "Ecotopia is not a story of people who enjoy living in good environments" (Callenbach interviewed 2003). He believes that stable state is achieved when people live by a lifestyle suitable to the bioregion, where "characteristic plants, animals, birds, insects, fish, and other inhabitants live, adapted to the region's climate, landforms, and soils" (Callenbach 2008). In his nonfiction book, *Bring Back the Buffalo*, he shows the Great Plains could be a sustainable region if people practice a sustainable lifestyle within its bioregion. The region is so dry to allow a smaller population to live that Garreau calls it the Empty Quarter (Garreau 1981). The lifestyle in the Great Plains, where people and buffalo coexist, is obviously different from the one in the Pacific Northwest (Callenbach 2000).

3.2.3 Two Ecotopia Writers in Japan

In *Ecotopia; a Proposal for a Recycle-Oriented Society*, Masaaki Naito traces back the cause of greenhouse effects from the beginning of the civilization to the recent industrialization and describes how human beings have behaved toward natural environments. Then, he proposes the necessity of change from a heavy industrial society to an agriculture-based recycle-oriented society (Naito 1992).

"News from Ecotopia, Kyoto, Arashiyama, An invitation to lifestyle based on the Natural Circulation Systems" shows what author Takayuki Mori has been doing for 50 years in quest of an ecologically coexisting lifestyle within the natural environment (Mori 2009). His ecological garden seems paradise; however, it requires local ecological knowledge, adequate care, and daily hard work. The combination of the ancient wisdom and allocation of advanced science and technology makes his natural circulation system possible. Mori believes that an Ecotopian lifestyle is a suitable way to adjust to an ecological environment (Mori 2011)

3.3 What is Callenbach's Ecotopia?

3.3.1 *Meanings of the Word ECOTOPIA*

The word Ecotopia was not originated by Callenbach. He heard the word "ecotopia" on the radio while he was writing the novel (Callenbach interviewed 2003). Callenbach indicates in the preface of the book that "ecotopia" is rooted in the Greek components "eco" and "topo," giving it the meaning of "Home Place" in English. In addition, he states in his afterword in ECOTOPIA's 30th anniversary edition that "Ecotopia does not portray a utopia—an imaginary country where everything is entirely perfect forever (…we know that homes are not perfect forever)" (Callenbach 2004).

Utopia is also based on Greek with the words "U" and "Topo" meaning "No" and "Place." So utopia means "a place which does not exist in reality" and it evolves into an "ideal place." Both Ecotopia and utopia have the same suffix, "topia," and it might be natural for some people to understand Ecotopia as an ecological utopia (Table 3.2). Accordingly, Callenbach's *Ecotopia* should be differentiated from a utopia.

3.3.2 *The Concepts of ECOTOPIA by Ernest Callenbach*

The novel tells the story of William Weston, a journalist from New York, a large city in the industrial United States, who visits Ecotopia, an ecologically sustainable and independent country that seceded from the United States 20 years earlier. But the more he investigates the country the more he learns the meaning of "stable state." Also, the more he knows about the people in Ecotopia, the more he

Table 3.2 Greek components of Ecotopia and Utopia (The Shorter Oxford 1973)

Component	Greek	English	Japanese
Eco	οἶκος	Home	家
Topo	τόπος	Place	場所
U	οὐ	No	無

understands the meaning of "home place." He decides to stay in the country at the end of the story. These two words are the main concepts of the novel.

3.3.2.1 The Concept of Stable State

The concept of "stable state" is explained by an Ecotopian in the novel as follows:

> But, we've actually achieved something like stability. Our system meanders on its peaceful way, while yours has constant convulsions. I think of ours as like a meadow in the sun. There's a lot of change going on—plants growing, other plants dying, bacteria decomposing them, mice eating seeds, hawks eating mice, a tree or two beginning to grow up and shade the grasses. But the meadow sustains itself on a steady-state basis—unless men come along and mess it up (Callenbach 2004).

The meadow changes slowly over time, but it is the place for production, consumption, and decomposition. An organic recycling system keeps the meadow ecosystem in a stable state. However, the meadow ecosystem is affected when men come in. In Ecotopia, people seek the point of stability within its ecosystem, while people from the industrial society tend to see the meadow as their resource of exploitation and "mess it up."

3.3.2.2 The Concept of Home Place

Another theme of ECOTOPIA is "Home Place." Weston learns that it is important for people to support each other in Ecotopian society. Meanwhile, he realizes what a lonely and stressful life he used to live in the industrial society. The story ends when he finds his home in Ecotopia. Home is a niche where individuals restore themselves and feel secure. At the same time, home is the Earth for all the living beings. That is how F. Capra, a natural philosopher, describes ecology: a study to connect all the living beings with one another on the earth as home (Capra 1995; Callenbach 1995).

3.3.3 Defining Ecotopia

Two concepts of Ecotopia have been reviewed: Callenbach believes that a "stable state" is ecologically essential for all living beings, including humans. In addition, the feeling of living in and sharing "home place" provides essential security for people and their community. At the end of this section, we would like to define "ecotopia" as follows: "Ecotopia is a place where human beings try to live harmoniously within their existing ecosystem."

Callenbach has written about "urban ecology" in *Ecology: A Pocket Guide* and closes the article as follows:

3 A Study on the Restoration of Urban Ecology: Focus on the Concept of Home... 41

Modern eco-cities based on these strategies will offer comfortable human habitats. They have more variety of entertainments, more lively streets, and more delightful accidental encounters. They will also offer "green cracks" in the urban concrete where some wild species can coexist with us as visible companions reminding us that even in the hearts of cities we are still part of the vast ecological web of planetary life (Callenbach 2008).

The next section, as an example of the possibility of an Ecotopia style living in large cities, we examine a citizen movement to preserve a small municipal park from urban redevelopment in Bunkyo Ward, Tokyo.

3.4 Ecotopia in the Cities: Layers of Citizen Activities Networks

3.4.1 About Motomachi Park

Bunkyo Ward is located in the central part of Tokyo Special district (Fig. 3.1). Tablelands, valleys, and slopes form the distinguished landscapes of Bunkyo Ward. There are many historic buildings and gardens of the Edo, Meiji, Taisho, and the early Showa era; more than 300 educational and cultural institutions, including the University of Tokyo, call it green.

Motomachi Park is located on 1-1 Hongo, Bunkyo Ward. The area had been called Hongo, but it was changed to "Hongo Motomachi" in the Edo era (1696; in Fig. 3.2). Motomachi means "Old-Town." "Motomachi" was used for the town name between 1906 and 1965, and the park was constructed during this period.

At that time in the Edo era, a tip of the Tableland was a viewpoint for Mt. Fuji. The scenery from there was drawn in "Ukinoyo-e," colored woodblock prints (Tatemono Ouendan Various timelime 2006).

Motomachi Park (opened in 1930 and consisting of 3,520.44 m^2) was "build after the Great Kanto Earthquake Disaster in 1923 by the City of Tokyo, one of 52 "Commemoration of Reconstruction" small parks which were designed at that time as attachment to the elementary schools, although it now is the only park maintaining the original design (Kouichi Tonuma 2007; in Fig. 3.3). Adjacent to Motomachi Park is the closed Motomachi Elementary School (built in 1927, with ground area of 4,146.76 m^2). The school, originally founded in 1910, burnt down in the Great Kanto Earthquake Fire and was relocated to its present site by the readjustment of various town lots (Tatemono Ouendan Various timelime 2006)."

According to Yoko Kano, a researcher of Landscape Architecture, the "Commemoration of Reconstruction Project" indicates the modernization of Japanese society. The reconstruction of small parks and elementary schools next to each other in Tokyo was recognized as "Merkmal" (a feature) as Modern City Planning in Japan (Kano 2007)." Moreover, the historical value of Motomachi Park is enhanced by the fact that the closed elementary school's spatial arrangement and main structures are kept in good original conditions. The other parks and

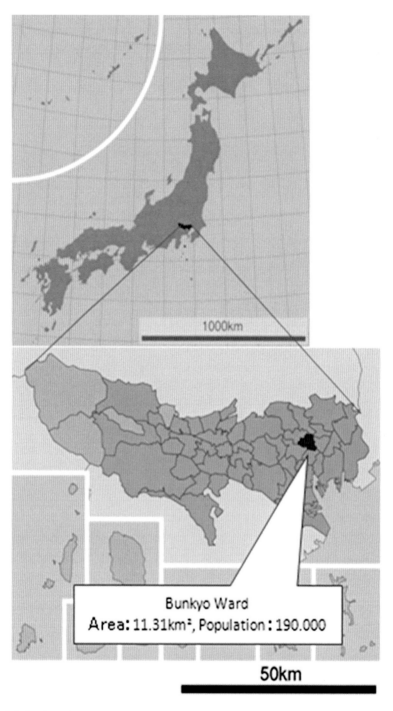

Fig. 3.1 Bunkyo Ward Maps Japan. http://ja.wikioedea.org/wikiTokyohttp://ja.wikipedia.org/wiki

3 A Study on the Restoration of Urban Ecology: Focus on the Concept of Home...

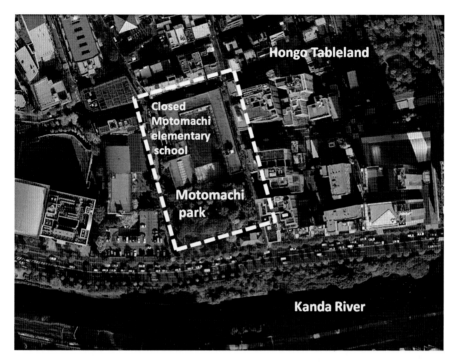

Fig. 3.2 The site of Motomachi Park, located in the middle of southernmost part of the ward, situated on the cliff line between the north side of Hongo Tableland and the south side of Kanda River. The park faces south to No.2 Rink; further to south runs the Japan Rail Soubu line (Google map 20110719)

elementary schools built in the same project have already disappeared, or were reformed by war damage or by other fluctuations through times (Kano and Kawanishi 2008).

Motomachi Park and the elementary school building bear witness to the changes of time; a bomb shelter was dug during the World War II; after the War, a baseball team was organized in the park in 1950; and a temporary day care facility for children operated between 1950 and 1967.

Motomachi Elementary School was closed on March 31, 1998 in accordance with Bunkyo Ward's public school closure and integration policy, the so-called "Domino Toppling Plan." Because of the trend toward fewer children in the district, the Ward planned to close 7 of 20 elementary schools consolidating the students and reassigning them to 13 schools (Bunkyo Ward hp). The school building was used by Hongo Elementary School for 2 years during the process, then it was rented to two private schools for 4 years from 2002 to 2006 (Tatemono Ouendan Various timelime 2006).

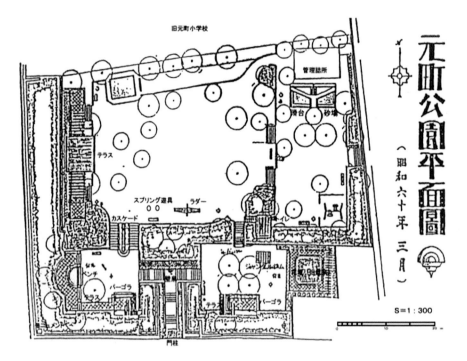

Fig. 3.3 Plan of Motomachi Park. The park restoration was done in 1981. http://www.geocities.jp/zouenkasyudan/52parks/pa23.html

3.4.2 Preserve Motomachi Park Movement

3.4.2.1 The Ward's Initial Proposal to Close Motomachi Park

Motomachi Park became a center of an argument over preservation or closure when Bunkyo Ward first announced an initial proposal to its Ward Council (assembly) on March 6, 2006 (Tatemono Ouendan Various timelime 2006). The initial proposal included replacement of the park to the north side location of the closed Motomachi Elementary School (Kano 2007). The Ward has three goals after the closure of the park.

1. To create a larger lot by clearing both the park and the elementary school.
2. To construct a high rise building following the proposal of a private company.
3. To use a part of the existing building for a new public gymnasium and other municipal facilities for residents' wellness. An old gymnasium in another district needed to be replaced (Tatemono ouendan). Kano sees the city planning of the ward "on the basic idea, public land where 500 % of a floor-area ratio, should be heavily utilized the commercially other than as a park" (Kano 2007).

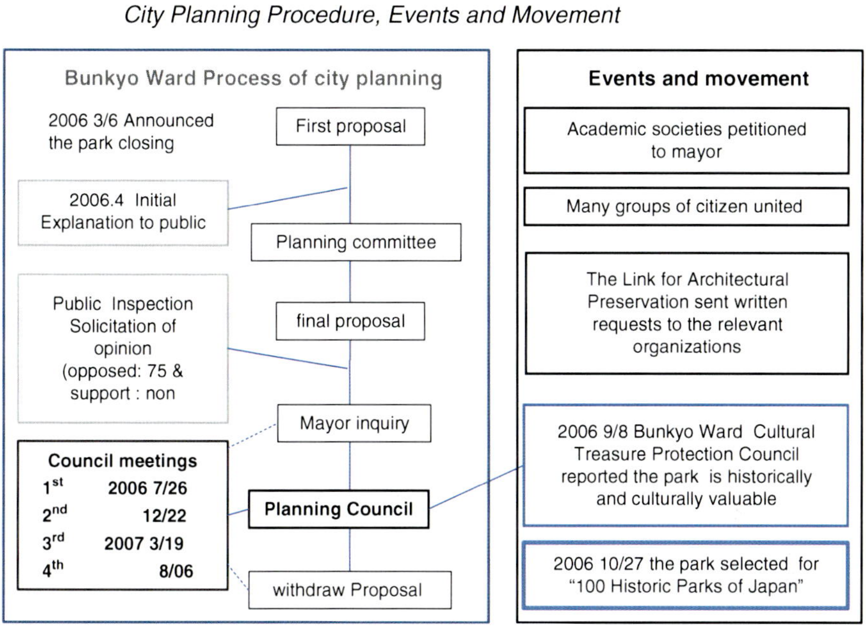

City Planning Procedure, Events and Movement

Bunkyo Ward Process of city planning		Events and movement
2006 3/6 Announced the park closing	First proposal	Academic societies petitioned to mayor
2006.4 Initial Explanation to public	Planning committee	Many groups of citizen united
Public Inspection Solicitation of opinion (opposed: 75 & support : non	final proposal	The Link for Architectural Preservation sent written requests to the relevant organizations
	Mayor inquiry	2006 9/8 Bunkyo Ward Cultural Treasure Protection Council reported the park is historically and culturally valuable
Council meetings 1st 2006 7/26 2nd 12/22 3rd 2007 3/19 4th 8/06	Planning Council	
	withdraw Proposal	2006 10/27 the park selected for "100 Historic Parks of Japan"

Fig. 3.4 The decision-making process of city planning of the closure of Motomachi Park. The right column of the *left frame* is the simplified city planning decision-making procedure. The left column of the *left frame* shows the critical events in the process. The *right frame* shows the essence of events and citizen movement (Bunkyo Ward: Tonuma 2007: tatemono ouendan http://www.toshima.ne.jp/~tatemono/page047.html)

People concerned about Motomachi Park took actions. Accordingly the preservation movement started, and it was resolved when Bunkyo Ward City Planning Council decided not to endorse the final proposal.

3.4.2.2 City Planning Procedure of the Closing of Motomachi Park

Planning Procedure

Bunkyo Ward, as a municipality, conforms to the City Planning Law, Article 15–24, the Procedure for Decision Making for Plans and Changes of Plan. As per Fig. 3.4 (left column of left frame), the following are the procedures of "Changes of Tokyo Metropolis City Park" to Motomachi Park:

1. When the initial proposal was made, the Ward held three explanatory meetings in April 2006 (Bunkyo Word 2006). Local communities, in the vicinity of the park, requested the Ward to hold an explanatory meeting, but the request was not met (Tatemono Ouendan Various timelime 2006).

2. The Ward explained the initial proposal, reported citizen comments and opinions at the explanatory meetings, and consulted with the City Planning Council to make the final proposal.
3. The Ward rendered the final proposal open to public inspection and received 75 opinions from citizens. All the opinions were in opposition, and none endorsed the proposal.
4. The mayor inquired the Planning Council for deliberation and advice of the final proposal to which was attached the sign of consent by the governor of Tokyo Metropolis.

When the procedure had reached this point, the Ward and also Tonuma, the Chairperson of the City Planning Council, received many requests from citizen groups. He reported the procedure of closing Motomachi Park and reminisced "as a Chairman of the City Planning Council, consider the issue (receptively or) without any bias (Tonuma 2007)."

Citizens Actions and Events

The citizen groups did as many things as they could after the announcement of the initial proposal. As per Fig. 3.4 (right frame):

1. The Japanese Institute of Landscape Architecture and four other academic societies wrote petitions to the mayor.
2. Many citizen groups held two joint symposiums in the latter half of 2006. "Pappato@Kaigi@Motomachi Park (2006)" tried to reach out to a wide range of people to emphasize the importance of preservation of the Park. Also, the home page of "23 Motomachi Park Commemoration of Earthquake Disaster" provided basic information about the park.
3. "The Bunkyo Link for Architectural Preservation" submitted five requests to the Mayor and four other requests to the related organizations. They opened their home pages as "Tatemono Ouendan," which gave the history of the park, and they wrote many petitions and requests to the Mayor of the ward and related organizations.
4. "The Society for Bunkyo's Cultural Heritage" sent Tokyo Metropolitan Park Association a recommendation that Motomachi Park be listed on the 100 historical parks of Japan, a program to celebrate the 50th anniversary of enactment of the Urban Park Act. As a result, the park was selected as one of the historical parks in Japan (Tatemono Ouendan Preservation Action 2006).

The Conclusion of the City Planning Council

The City Planning Council was not able to make final decision until the fourth meeting on August 6, 2007. It took more than 1 year before they reached their conclusion. During that time, the Ward inquired of experts, presented much detailed

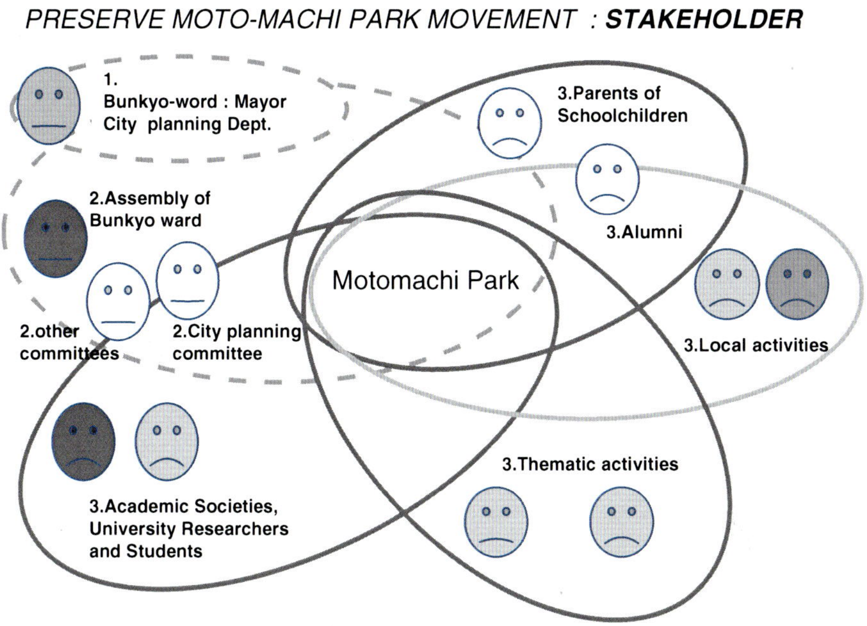

3 A Study on the Restoration of Urban Ecology: Focus on the Concept of Home... 47

PRESERVE MOTO-MACHI PARK MOVEMENT : *STAKEHOLDER*

Fig. 3.5 The network of stakeholders

information, and made efforts to improve the final proposal. Some of the related committees and councils of the Ward suggested the mayor consider the preservation of the park. The City Planning Council invited experts of landscape architecture, cultural landscape,s and preservation of the cultural property fields as temporary council members. The meetings proceeded in a tense atmosphere full of public observers.

At the end, the City Planning Council concluded not to endorse the final proposal unanimously (Tonuma 2007). This was a rare case for a municipal City Planning Council to return a negative report to the mayor's inquiry. As a result, the final proposal was withdrawn. Aforementioned are the circumstances and chronological sequences about who and how Motomachi Park was saved.

3.4.2.3 Stakeholders of the Preservation Movement

There were three kinds of stakeholders (Fig. 3.5).

1. A decision-making body: the mayor and the department of city planning.
2. Groups of neutral position: such as the City Planning council, other committees, councils, and the Assembly of Bunkyo Ward.
3. Groups of people who wanted to preserve Motomachi Park.

There were many different groups of people with different interests involved in the movement, including

1. Academic societies, university researchers, and students.
2. Citizen activity groups, who had thematic interests and goals. Most of them were already introduced in "Citizens Actions and Events." These groups were interested in history, culture, preserving historic architectures, gardening, and environmental issues. Some of the members were active and joined several groups. Some of the groups had link to many other groups outside the ward. "PapattoKaigi @ Motomachi Park" was a group organized for the Movement.
3. Associations of local communities, residents, and shopping districts.
4. Alumni Association of Motomachi elementary school.
5. Parents of Public Schoolchildren. Bunkyo Ward's "Domino Toppling Plan," referred to in Sect. 3.1, was not welcome by the parents of public schoolchildren. The parents, who were discontented with the plan, joined the movement (Fujiwara interviewed 2011).

There were groups with different backgrounds and interests united for the same goal. It was one of the key factors to successful movement, although it was not easy to form a network with various groups of people with different interests. In the case of the Motomachi Park movement, key persons knew how to communicate with others and formed organic networks through their experiences. They had learned these things through participating in events and daily activities for a long time.

3.4.3 Kougenn-ji: A Center for Local Activities

3.4.3.1 Events in Kougenn-ji

In Kougenn-ji, various citizen groups participate in events and easily create networks.

The Annual "Houzuki Sennari Ichi"

One of the most awaited events in Kougenn-ji is the annual Houzuki Sennari Ichi, a Lantern Plant Market, on July 8 and 9 (Fig. 3.6). The market is held on the day of "46,000 Day Buddhist Service." This market was discontinued once before, but Kougenn-ji invited citizen groups to resume it in 2001. The event is organized and executed by participants from the planning stage, and the cozy atmosphere is very popular for both visitors and participants.

According to the program in 2011, there are about 50 individuals and groups participating in performances and in the market. For example, there are ethnic dances, electric piano concerts, local drums, and other performances. Also, a variety of participants sell their handmade products in a kind of Flea Market

Fig. 3.6 Houzuki Sennari Ichi, known as "Komagome, Okannonn-sama, Kougenn-ji," is located on the northern part of the ward, where the old downtown atmosphere is still preserved. (Photograph by Kato 2011.7.8)

(Kougenn-ji 2011). In 2011, the Houzuki Sennari Ichi had the special purpose of supporting the victims of the Great Tohoku Earthquake and Nuclear Disaster. They have special programs for "disaster drills," "cheerleading for the disaster regions," and "A concert for Prayer for the Disaster-stricken Area" was performed; also, many participants sold their homemade goods and donated the revenue for charity to support victims of the disaster at the market corner.

Support for the Victims of the Great Tohoku Earthquake

Kougenn-ji with the local activity groups started voluntary activities from March 12, the very next day after the disaster. Kougenn-ji became an emergency kitchen; some prepared foods, and others visited the disaster-stricken area to distribute the food to the victims.

One of the participants was grateful that she was able to join the volunteer activity through Kougenn-ji network; participating in daily activities and events led her to volunteer through the network (Matsumoto 2011).

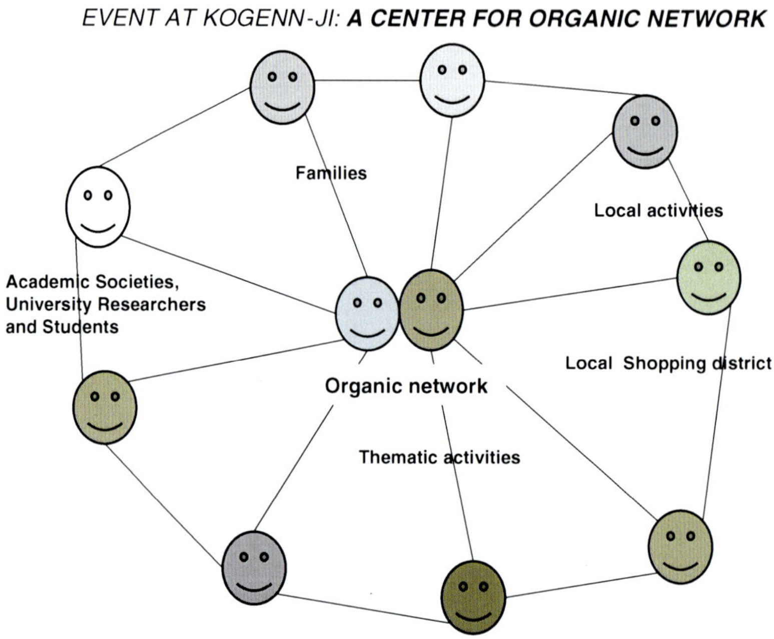

EVENT AT KOGENN-JI: **A CENTER FOR ORGANIC NETWORK**

Fig. 3.7 A network centered on events in Kougenn-ji

Kougenn-ji, the Center for an Organic Network

Centered on Kougenn-ji, various groups of people form an organic network (Fig. 3.7). It is a core for the local people, citizen activities, events, and problem solving. People interact, exchange information, and feel as if they are a part of the organic network. People usually have different interests and participate in different activities. Some people get together when their interests coincide, such as for the preservation of Motomachi Park, or support for the victims of the disaster, which eventually become a movement.

The participants in the Annual Houzuki Sennari Ichi are drawn from a diverse group: local communities, local shopping districts, town watchers, local magazines, theaters, people with special interest in preservation of historic architecture, composting and gardening, providing welfare for handicapped: in addition, participants include university researchers and students, the International Friendship and Refugees, makers of handcrafts, cafés, and more... Some of the people participating in multiple group activities were key persons in the Preservation of Motomachi Park Movement (Kougenn-ji 2011).

In the next part, we look into daily activities of a NPO, a part of the organic networks of Kougenn-ji.

Fig. 3.8 Composting site under the Ochanomizu Bridge. (Photograph by Kato 2011.6.15)

3.4.4 Daily Activities of NPO Bank of Green Resources

3.4.4.1 Daily Activities

Centered on the goal "Let's make city greener with garbage recycling," main activities of NPO Bank of Green Resources are to compost organic materials (such as kitchen garbage, rice bran, and dead leaves) and to revitalize the soil with their compost and take care of the flower beds in public places. Because Bunkyo Ward is located in the center of the congested city, the NPO has experienced difficulty in finding a site for their activities: despite the difficulty, they have established their organic recycling system.

Place for Composting

Since 2005, the NPO has made compost under Ochanomizu-bridge where the NPO and Bunkyo Ward entered into partnership for a compost-making experiment (Fig. 3.8). The Ward offers three wooden composting frames (3 × 5 m) under the bridge and collect leaves and grasses from parks and public green areas. The NPO brings about 50 kg of garbage from members' kitchens, which was kept in a fermented mixture of rice bran to prevent decay and odor, and adds it to the compost frames. Several NOP members and some workers from a private company perform the work once a month as their environmental social contribution.

Fig. 3.9 Soil revitalizing and gardening at the Kasuga Intersection. (Photograph by Kato 2011.6.13)

Place for Soil Revitalizing and Gardening

Another activity of the NPO is to revitalize the soil and grow flowers: they care for flower beds on the pavement along the streets in public places, and a pocket park at one corner of the Kasuga Intersection (Fig. 3.9). They take care of the flower beds once a week all year around, revitalizing soil with compost from under the bridge, planting out seedlings, raising plants, and clearing plant debris after the growing season (Matsumoto interviewed 2011).

Support Daily Activity Networking

Centered on their goal of the NPO, Bunkyo Ward, private companies, schools, Kougen-ji and local communities are in the network (Fig. 3.10).

The Chief of NPO recently expressed her satisfaction with NPO activities: she personally works with good companies, maintaining her health by riding a bicycle instead of driving, associates with committed citizens, enjoys composting (an activity that helps her feel close to organic networks), and receives much positive feedback as a result of the lectures she presents all over Japan (Matsumoto interviewed 2011; Matsumoto 2008).

3 A Study on the Restoration of Urban Ecology: Focus on the Concept of Home... 53

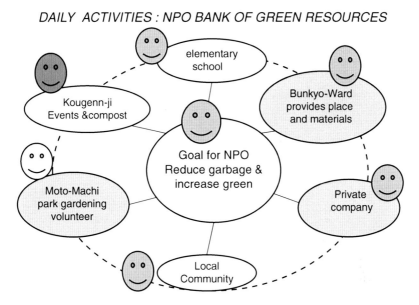

Fig. 3.10 A network centered with the goal of an NPO

3.5 Conclusion

In this study, we have reviewed two major concepts underlying the North American writer E. Callenbach's thoughts expressed in his novel *Ecotopia* to discover ways to achieve a sustainable society in a specific urban region of Tokyo.

First, we examined two key concepts of Ecotopia, "stable state" and "home place." The concept of "stable state" embodies the ecological condition of sustainability, whereas "home place" means Ecotopia, the fictional country, and represents a place where people support one another and practice a sustainable lifestyle. Thus, we define Ecotopia by reviewing Callenbach's work and the work of two Japanese books written about the book *Ecotopia* by Naito and Mori: "Ecotopia is a place where human beings try to live harmoniously within an ecosystem."

In the latter half of this study, we examined citizen activities in the central part of Tokyo to see if there was a possibility of living with a sense of "home place" in large cities. We studied the case of the Movement for the Preservation of Motomachi Park (Fig. 3.11) in the Bunkyo Ward. The Ward's City Planning Division almost replaced Motomachi Park, overlooking the values that citizens considered important, yet the park was saved for the following reasons:

1. Motomachi Park and the closed Motomachi Elementary School represented historical, cultural, and ecological values. The park and school building constituted valuable "commons" for citizens involved in the preservation movement and who shared a "sense of place" in the park.

Fig. 3.11 Motomachi Park. (Photograph by M. Kato 2011)

2. Motomachi Park was saved by many stakeholders, many different groups of people united to achieve a common goal: to accomplish this, Bunkyo Ward made the final decision to withdraw the proposal; the City Planning Council deliberated fully before coming to their conclusion; experts in many fields clearly emphasized the value and necessity of preserving the park; and citizen groups with different interests united to accomplish a common goal.
3. The Motomachi Park Preservation Movement was supported by multiple layers of citizen activity networks: movements, events, and daily activities. Each network had a unique goal and formation. People became acquainted through participation in daily activities and events. Participants became part of a network and bonded that network to other networks, which gradually formed multiple layers of networks with intertwined structures, energizing and strengthening communities.

The saving of Motomachi Park demonstrates that a citizen movement, organized on an as-needed basis when problems arise (such as the 2011 Earthquake and Nuclear Disaster), can strongly bring its influence from outside to bear on the decision-making process of a public agency. The saving of Motomachi Park may be a small, local issue, but it is one that has a significant meaning for our society. Yet the preservation of the Park is a rare case in Japan, and as Callenbach suggests (Callenbach 2008), we should take it as the appearance of "green cracks" in the congested urban environment. An important challenge to urban ecology could start with such small green cracks.

3 A Study on the Restoration of Urban Ecology: Focus on the Concept of Home... 55

To conclude, we would like to consider whether it is possible for Ecotopian principles to emerge in large cities. An Ecotopian living style is possible if people try to live harmoniously and sustainably within their existing ecosystem. To do so, people need to experience an awareness of themselves as part of a local organic community. In an urbanized area, people need to find their "home place" by exploring the local ecology, geography, history, culture and community. Nothing sensational, but it must have been what Callenbach hoped to show through the explorations of Weston, his journalist from industrial society, who came home to Ecotopia.

References

Bunkyo Word (2006) Report about the citizen opinions and requests and responses at the explanatory meetings about "City Planning Changes (Motomachi Park)" (in Japanese) http://www.city.bunkyo.lg.jp/library/sosiki_busyo/keikakutyousei/keikaku/setumeikai/motomachikouen_kaitou.pdf. Accessed 19 July 2011

Callenbach E (1995) A model society in the 21 century: "ECOTOPIA". In: Capra F, Callenbach E (eds) Thought in deep ecology—for a sustainable future. Kousei Shuppan, Tokyo (in Japanese)

Callenbach E (2000) Bring back the buffalo. University of California Press, Berkeley

Callenbach E (2003) Interviews conducted by M. Kato at Callenbach's home in Berkeley, California, USA

Callenbach E (2004) Ecotopia: the 30th anniversary edition. Banyan Tree Books, Berkeley

Callenbach E (2008) Ecology: a pocket guide, revised and expanded. University of California Press, Berkeley

Capra F (1995) The origin of word ecology. In: Capra F, Callenbach E (eds) Thought in deep ecology—for the sustainable future. Kousei Shuppan, Tokyo (in Japanese)

Fujiwara M (2011) Interviews conducted by M. Kato at office of the Bunkyo Ward Assembly. Bunkyo Ward City Hall

Garreau J (1981) The nine nations of North America. Avon Books, New York

Kano Y (2007) Small parks for commemoration of the great earthquake, consideration of Motomachi Park. Nihon Landscape Architecture Society Kanto Division Conference Case study, series of research and report No. 25 (in Japanese)

Kano Y, Kawanishi T (2008) Utilization and reconstruction elementary schools after the 1923 Great Kanto Earthquake Disaster in Tokyo. An abstract for an academic lecture presented at the Conference of Architecture Institute of Japan (Chugoku) 2008, 09 (in Japanese)

Kougenn-ji (2011) Houzuki Sennari Ichi leaflet in Kougenn-ji, Komagome, Bunkyo-Ward, Accessed 8–9 July 2011

Kuwako T (2011) Consensus building for reconstruction from the disaster of earthquake, tsunami, large scale nuclear contamination. ERI Review, vol 9, pp 5–6 (in Japanese)

Mander J, Callenbach E (2011) Prologue: False Solutions. In: Smith G, Callenbach E (eds) Nuclear Roulette: the case against a "Nuclear Renaissance". No. 5 in the International Forum on Globalization series focused on False Solutions to the Global Climate Crisis, p 4

Matsumoto M (2008) Let's make Tokyo greener with waste management. NPO Bank of Green Resources, lively NPOs in City Planning, City. Plann Rev 269 (in Japanese)

Matsumoto M (2011) Interviews conducted by M. Kato at Bunkyo Ward City Hall, Kasuga, Bunkyo Ward, Tokyo

Mori T (2009) Kyoto Arasiyama, a news from Ecotopia—introduction of ecological life style. Shougakukann, Tokyo (in Japanese)

Mori T (2011) Interviews conducted by M. Kato at Mori's Ecotopian Garden. Arashiyama, Kyoto

Naito M (1992) Ecotopia: a proposal for a recycle-oriented society. Nikkann Kougyo-sha (in Japanese)

Okuda T (2009) A city aim to bring about Ecotopia: numerical values for simulation of regions (in Japanese) (2011). http://www.esi.nagoya-u.ac.jp/pdf/jp__t-okuda__20091017081824__1207.pdf

Pappato@Kaigi@Motomachi Park (2006) (in Japanese) http://www.nporprogram.jp/motomachi/blog/ieieaiecoei/. Accessed 21 June 2011

Shioda H (2008) "Ecotopia:" its contradictions and violence—reading in dystopia from Callenbach's 'Ecotopia'. In: Slovic S (ed) Ecotopia and the environmental justice reader. Koyo Shobo, Kyoto (in Japanese)

Slovic S (2008) Preface. In: Slovic S (ed) Ecotopia and the environmental justice reader. Koyo Shobo, Kyoto (in Japanese)

Tatemono Ouendan Various timeline (2006) (in Japanese) http://www.toshima.ne.jp/~tatemono/page0.47.html. Accessed 19 July 2011

Tatemono Ouendan Preservation Action (2006) (in Japanese) http://www.toshima.ne.jp/~tatemono/page0.25.html. Accessed 19 July 2011

Timberg S (2008) A 70s cult novel is relevant again. New York Times (2011.09.). http://www.nytimes.com/2008/12/14/fashion/14ecotopia.html

Tonuma K (2007) A significant role for a small park—the issue of preservation of the park for the commemoration of the great earthquake. The frontier of Machidukuri City. Plann Rev 270:70–73 (in Japanese)

Tsuruta E (1995) A commentary. In: Capra F, Callenbach E (eds) Thought in deep ecology—for the sustainable future. Kousei Shuppan, Tokyo (in Japanese)

Chapter 4
Transpiration Characteristics of Chinese Pines (*Pinus tabulaeformis*) in an Urban Environment

**Hua Wang, Zhiyun Ouyang, Weiping Chen,
Xiaoke Wang, and Hua Zheng**

Abstract Urban environments can significantly influence the transpiration of isolated plants. Therefore, optimal green space design, tree species selection, and tree maintenance require that the water use patterns of urban plants be quantified. In this study, the transpiration from individual Chinese pines (*Pinus tabulaeformis*) in the center of Beijing, China was measured continuously over a 2-year period. The response of whole-tree transpiration (E_t) to environmental factors was investigated in multiple time scales. Maximum sap flux density (J_s) ranged from 3.34E-05 to 8.2E-03 cm/s. E_t was much higher in summer (32.93 kg/day) than in winter (6.22 kg/day). E_t in the urban environment was much higher than that reported for Chinese pines with similar diameters at breast height (DBH) during 2000–2005 in suburban Beijing. Great differences were observed in the response of E_t to environmental factors at different time scales. At the diurnal scale, hourly mean J_s was linearly related to photosynthetically active radiation (PAR) and vapor pressure deficit (D), whereas at the daily scale, daily mean E_t was linearly related to PAR, air temperature (T_a), and soil water content (SWC), and was curvilinearly related to D. At the annual scale, E_t was similar in the growing seasons of 2008 (a wet year) and 2009 (a dry year), even though the annual precipitation (P) and irrigation times were significantly different (724.8 vs. 432.8 mm; 2 vs. 12).

H. Wang
State Key Laboratory of Urban and Regional Ecology, Research Center for Eco-Environmental Sciences, Chinese Academy of Sciences, Beijing 100085, China

Institute of Forestry and Pomology, Beijing Academy of Agriculture
and Forestry Sciences, Beijing 100093, China
e-mail: wanghuaphd@gmail.com

Z. Ouyang (⊠) • W. Chen • X. Wang • H. Zheng
State Key Laboratory of Urban and Regional Ecology, Research Center for Eco-Environmental Sciences, Chinese Academy of Sciences, Beijing 100085, China
e-mail: zyouyang@rcees.ac.cn; wpchen@rcees.ac.cn; wangxk@rcees.ac.cn;
zhenghua@rcees.ac.cn

N. Nakagoshi and J.A. Mabuhay (eds.), *Designing Low Carbon
Societies in Landscapes*, Ecological Research Monographs,
DOI 10.1007/978-4-431-54819-5_4, © Springer Japan 2014

From this result, it can be concluded that urban soil water conditions affected by both P and irrigation practice were a major cause of interannual E_t variation.

Keywords Green space • Isolated tree transpiration • Sap flux density • Soil water content • Urban environment • Urban soil water

4.1 Introduction

Urbanization significantly influences local and regional climate, water resources, the atmosphere, and land use (Wu 2008). Urban green spaces including parks, street trees, gardens, agricultural areas, rehabilitated areas, fragmented natural areas in a city, natural areas surrounding a city, and other open areas are providers of urban ecosystem service (Niemelä et al. 2010). An urban green ecosystem can mitigate many of the environmental impacts of urban development by moderating climate, reducing atmospheric carbon dioxide, improving air quality, lowering rainfall runoff, and reducing noise levels (Nowak and Dwyer 2007). We see urban green areas as the most effective environmental protection tool and the foundation of the urban ecological framework (skeleton) or green infrastructure. Tree transpiration, in particular, cools the air and reduces storm water runoff (McPherson et al. 2005; Nowak and Dwyer 2007). At the same time, the transpiration pattern of trees in cities and surrounding areas may be significantly influenced by urban environmental changes (Gregg et al. 2003). Many metropolitan cities, including Beijing, expend great effort to improve tree cover. However, inappropriate green space design, tree species selection, and tree maintenance can increase water consumption. Furthermore, isolated urban plants are thought to be susceptible to "the clothesline effect," which causes high rates of evapotranspiration (Hagishima et al. 2007; van Bavel et al. 1962). Therefore, it is imperative to study the transpiration patterns and factors affecting transpiration in the urban environment.

Many studies have examined the factors affecting urban plant transpiration, which factors include plant density, irrigation, energy exchange with building walls, and pollutant concentrations (Hagishima et al. 2007; Heilman et al. 1989; Martin and Stabler 2002; Montague and Kjelgren 2004; Neighbour et al. 1988). These studies have focused primarily on responses of potted plants to a single feature of the urban environment. However, the effects of changes in the urban environment on the water use of trees in situ have seldom been studied.

Time scale is an important component in evaluating the factors influencing transpiration. Photosynthetically active radiation and vapor pressure deficit and soil moisture content affect transpiration on a daily time scale, whereas leaf area varies within and among species on seasonal and interannual time scales (Ohta et al. 2008; Phillips and Oren 2001). Diurnal and seasonal variations in tree transpiration have been broadly studied. However, knowledge regarding interannual variation is insufficient, especially about trees in urban environments with irrigation. Soil moisture changes are also a major cause of interannual variation in the transpiration period (Yoshifuji et al. 2007). Garden management techniques such as irrigation are being adopted in urban environments, especially in arid and semiarid areas or under dry weather conditions. Thus, in irrigated systems the interannual variations in soil water content do not correspond to the amount of precipitation. The effects of irrigation on interannual variations in transpiration are poorly understood.

Basic data of urban green characteristics in Beijing were listed as follows: 658,914.07 ha forest area, 61,695 ha green area, 44.4 % urban green coverage, 22 parks, 315 public gardens, 100 boulevards, 100,000 m^2 roof green area, and 210 ha community green area (Statistical Yearbook of Beijing 2010; www.bjyl. gov.cn). During the past decades, both urban green coverage (22.3–44.4 %) and green area (26,680–61,695 ha) have notably increased (Statistical Yearbook of Beijing 2010). Trees are essential components of all urban green spaces. More than 2,056 species of vascular plants (He et al. 1993) and more than 61 million trees (Beijing Municipal Bureau of Landscape and Forestry 2005) are planted across the city. Chinese pine (*Pinus tabulaeformis*), a species endemic to China, has been planted widely in northern Chinese cities because of its wide adaptability and aesthetic value. It is one of the top five evergreen tree species in Beijing in terms of number of individuals and ecological importance value (Beijing Gardening and Greening Bureau 2005; Meng 2004). In this study, we monitored the diurnal, daily, seasonal, and annual patterns of transpiration of Chinese pines in the center of Beijing with the help of the thermal dissipation probe method (TDP). The influences of urban environmental changes on the transpiration of Chinese pine were evaluated at different time scales.

4.2 Materials and Methods

4.2.1 Site Description

Beijing, the capital of China, is one of the oldest and fastest developing capital cities in the world. It has a population of 16.33 million and a built-up area of 873 km^2 (Beijing Statistics Yearbook 2007). This study was conducted in the Beijing Teaching Botanical Garden, which covers an area of $11.65 \times 10^4 \, m^2$ and is in the Chongwen District in the center of Beijing. Beijing, situated in a warm temperate zone, has a typical continental monsoon climate. Mean annual precipitation is about 585.8 mm with more than 70 % of the annual total occurring between July and August. Mean annual temperature is about 11.8 °C, varying between 11 and 12 °C.

Table 4.1 Characteristics of the trees sampled for sap flow measurements

Year	Tree number	DBH (cm)	Height (m)	Sapwood area (cm^2)	LAI	Orientation of sensor	Number of sensors
2008	No. 1	17	5.9	163.56	2.07	South, north	2TDP30
	No. 2	16.2	5.7	147.52	2.36	South, north	2TDP30
	No. 3	18.7	5.8	200.60	2.82	South, north	2TDP30
2009	No. 1	17.1	5.9	165.63	1.79	South, north	2TDP30
	No. 2	17.55	5.7	175.10	2.08	South, north	2TDP30
	No. 3	20.3	5.8	239.17	2.23	South, north	2TDP30

LAI, leaf area index

4.2.2 Estimation of Sapwood Cross-Sectional Area, Leaf Area Index, and Transpiration

Sapwood cross-sectional area (A_s, cm^2) was estimated from sapwood cross-sectional area and diameter at breast height (DBH) data collected in the Beijing Teaching Botanical Garden and Jiu Feng mountain in suburban Beijing using the relationship:

$$A_s = 0.3786 \times (\text{DBH})^{2.1419} \quad \left(n = 22; R^2 = 0.9772; p < 0.0001 \right). \tag{4.1}$$

Whole-tree leaf area index (LAI) was measured using a plant canopy analyzer (LAI2000; USA), and it was measured every 2 or 3 days during leaf expansion and defoliation and once a week during other periods under diffuse light conditions on cloudy days or at dusk.

Three Chinese pine trees of uniform size that were 45 years old were selected for sap flow measurements. The characteristics of the sampled trees are summarized in Table 4.1. Thermal dissipation probes (Dynamax, USA) were horizontally inserted in the sapwood of the trunk at breast height on both the north and south side of every sampled tree. Based on an empirical relationship (Granier 1987), sap flux density was derived from the temperature difference between the upper, constant-heated probe and the lower, unheated probe, which acted as a reference. Measurements of sap flow were taken every 10 s, and 10-min averages were stored in a datalogger (CR1000; Campbell Scientific, UK). Sap flux density measurements made in stems were scaled to each individual tree by its A_s (Granier et al. 1992):

$$E_t = \sum_t \left(\left(\left(J_{s-\text{north}} + J_{s-\text{south}} \right) \times A_s \times 3600 \right) / (2 \times 1000) \right). \tag{4.2}$$

4.2.3 Environmental Monitoring

An automated weather station located in the Teaching Botanical Garden was used to measure meteorological parameters, using an air temperature (T_a) and relative humidity (RH) probe (HMP45C; Vaisala, Helsinki, Finland), a wind (w) sensor

4 Transpiration Characteristics of Chinese Pines (*Pinus tabulaeformis*)... 61

(034B Met One Windset; Campbell Scientific, Logan, UT, USA), a quantum sensor (PAR Lite; Kipp and Zonen, Delft, Netherlands), a soil temperature (T_s) probe at a depth of 10 cm (109; Campbell Scientific), and a rainfall (P) gauge (TE525MM; Campbell Scientific). Soil moisture content was measured at depths of 10 and 30 cm (SWC_{10}, SWC_{30}) using soil moisture sensors (ECH$_2$O; Decagon Devices, Pullman, WA, USA) in the center of the sampled Chinese pine trees. All these meteorological data were sampled and recorded at the same frequency as the sap flow measurements. Vapor pressure deficit (D) was calculated using the 10-min averages of temperature and relative humidity as follows:

$$D = a \times \exp(bT_a/(T_a + c)) \times (1 - RH) \tag{4.3}$$

where a, b, and c are fixed parameters equal to 0.611 kPa, 17.502, and 240.97 °C, respectively (Campbell and Norman 1998).

4.2.4 Statistical Analyses

Statistical analyses were performed using SPSS 11.5 (SPSS, Chicago, IL, USA) and Sigmaplot 10.0 (Systat Software, San Jose, CA, USA). A paired-samples t test was performed in SPSS with a significance level of $p = 0.05$ for the comparison of mean E_t. The relationships between E_t and T_a, PAR, D, and SWC on both the diurnal and daily scale were investigated using curve estimation analyses performed with Sigmaplot. Linear regression analyses were conducted to study the influences of climate variables on E_t at multiple time scales using the stepwise procedure in SPSS with a significance level of $p = 0.05$.

4.3 Results and Discussion

4.3.1 Urban Environmental Conditions

Table 4.2 summarizes the daily variation in environmental parameters during 2008 and 2009. There was no significant difference between the air temperature (T_a), wind speed (w), soil temperature (T_{s10}), or photosynthetically active radiation (PAR) in 2008 and 2009. The annual average T_a, w, T_{s10}, and PAR were approximately 13.6 °C, 1.05 m/s, 14.2 °C, and 231.7 mol/m^2/s, respectively. The vapor pressure deficit (D) varied from 0.11 to 3.34 kPa and showed a seasonal trend (Fig. 4.1). D increased from around 0.57 kPa in March to a maximum of around 3.3 kPa in June before falling to less than 0.5 kPa at the end of the season in November (Fig. 4.1). The atmosphere was very dry in spring, and D often exceeded 2.0 kPa. Moreover, D during May and June in 2008 (1.37 and 1.02 kPa) was

Table 4.2 Average daily values of urban environmental factors and irrigation times in Beijing in 2008 and 2009

Parameter average value (range)	2008 ($n = 366$)	2009 ($n = 365$)
T_a (°C)	13.79 (−8.61 to 30.34)	13.51 (−9.36 to 31.18)
T_{s10} (°C)	13.85 (−1.15 to 28.21)	14.62 (−1.55 to 29.57)
w (m/s)	1.12 (0.22–3.11)	0.99 (0.19–4.07)
PAR (μmol/m^2/s)	242.86 (17.10–574.08)	220.61 (4.02–554.12)
D (kPa)	0.86 (0.11–2.55)	0.98 (0.11–3.34)
P (mm)	1.98 (0.00–52.70)	1.19 (0.00–55.60)
SWC$_{10}$ (%)	17.72 (7.68–29.42)	19.16 (8.50–34.19)
SWC$_{30}$ (%)	26.43 (19.36–36.07)	30.96 (22.32–38.76)
Frequency of irrigation (times)	2	12

significantly lower than in 2009 (1.83 and 2.10 kPa), which was the result of the lower air relative humidity (RH) and higher T_a during that period in 2009.

The year 2008 was the wettest year since records began to be kept in 1999, whereas 2009 was a dry year in which rainfall was below average (Statistical Bureau, 1999–2009). Rainfall in 2008 was 67.5 % higher than in 2009 (724.8 vs. 432.8 mm), but the time of local irrigation was only one-sixth of that in 2009 (2 vs. 12 times). As a consequence, the soil layer was 1.4 % wetter in 2009. In correspondence with the precipitation and irrigation, the soil moisture conditions varied greatly (Fig. 4.1). Moreover, soil water content at a depth of 10 cm was always lower and fluctuated more than that at a depth of 30 cm.

4.3.2 Transpiration Pattern of Chinese Pines at Multiple Time Scales

The diurnal patterns of sap flow density (J_s) in Chinese pines under sunny weather over four seasons are illustrated in Fig. 4.2a. Maximum J_s varied considerably among seasons, ranging from 3.34E-05 to 8.2E-03 cm/s. Diurnal J_s exhibited a broad peak pattern in summer and a narrow peak pattern in both spring and autumn, whereas the diurnal pattern was not pronounced in winter. In comparison with the J_s pattern in summer, in spring and autumn the timing of the onset and peak of J_s were delayed, whereas that of its decline was advanced. The diurnal pattern of sap flow indicated that sap flux density in the Chinese pine had no significant "noon depression" phenomenon. Noticeable sap flow in Chinese pine was evident in the nighttime, which may alleviate plant water stress.

The total transpiration from May 1, 2008 to April 30, 2009 amounted to 6,547.87 kg. Figure 4.2b illustrates the annual variation in E_t. The annual pattern

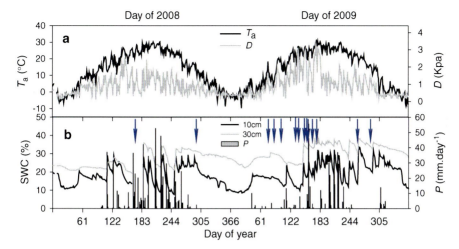

Fig. 4.1 Climatic data from 2008 to 2009. Reading from the top: daily air temperature (T_a; *black line*), mean vapor pressure deficit (D; *grey line*), precipitation (P; *bars*, right scale), soil water content (SWC) in the 0–10 cm soil layer (SWC_{10}; *black line*) and SWC in the 10–30 cm soil layer (SWC_{30}; *grey line*). Some data are missing because of power failure. *Arrows* indicate times of irrigation. Note that the Chinese pines received much less irrigation in 2008 than in 2009

of E_t was similar in 2008 and 2009. As shown in Fig. 4.2c, E_t increased rapidly from March 7 (8.46 kg/day), reached its peak on July 18 (54.94 kg/day), and then gradually decreased. Figure 4.2d illustrates the seasonal patterns of daily mean whole-tree transpiration. Strong seasonality was observed with E_t, which was much greater in the summer (June to August; 32.93 kg/day) than in winter (December to February; 2.78 kg/day), which may be attributed to the lower soil moisture content, vapor pressure deficit, and reduced radiation received in winter. Such seasonality in tree water use is generally observed in temperate and tropical systems (Melanie et al. 2006). Furthermore, E_t was much higher during the growing season (April to October) than during the nongrowing season (November to March). The nongrowing season E_t of Chinese pine cannot be neglected, as it accounts for about 8 % of the annual E_t and maintains the existing living cells. This finding is consistent with the results of Ceschiaa et al. (2002), which revealed that, during the nongrowing season, maintenance respiration of adult beech (*Fagus sylvatica*) trees ranged between 7.2 and 528 µmol/m^3/s at breast height and in the upper crown, respectively.

Statistical analysis showed that both the monthly E_t and average E_t (33.16 vs 31.61 kg/day) during the growing season (May to October) did not differ significantly in 2008 and 2009 (Fig. 4.3). Such maintenance of interannual variation of E_t was also observed in a pine forest and in an eastern Siberian larch forest (Ohta et al. 2008; Phillips and Oren 2001). However, there were obvious annual variations in tree water use observed in an open woodland of two co-occurring

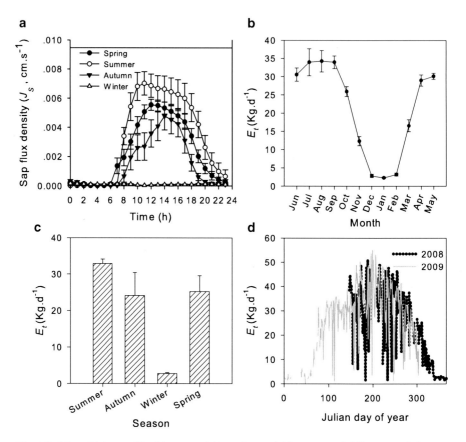

Fig. 4.2 Diurnal (**a**), monthly (**b**), seasonal (**c**), and annual (**d**) patterns of daily mean whole-tree transpiration (E_t) in Chinese pines. All points represent mean values of three sampled trees

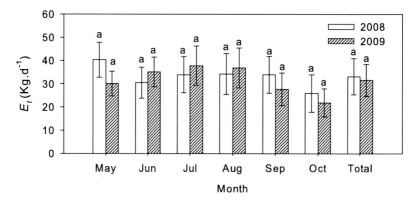

Fig. 4.3 Mean E_t (kg/day) for each month and the whole growing season in 2008 and 2009. Letters indicate significant differences between years ($p < 0.05$). *Vertical bars*, standard error

4 Transpiration Characteristics of Chinese Pines (*Pinus tabulaeformis*)... 65

Table 4.3 Chinese pine tree transpiration in suburban and central Beijing

	Results		Biological factors		
Time/soil/site	J_{s-max} (cm/s)	E_t (kg/day)	DBH (cm)	Height (m)	References
2005 cinnamon mixed forest	0.0025–0.0040	4.58–9.88	13.31–19.31	10.5–13.8	Liu (2008)
2002 cinnamon mixed forest	0.0001–0.0015		11.3	9	Nie et al. (2005)
2004 loess mixed forest	0.0030		21.5	12.8	Ma et al. (2006)
2001 loess mixed forest	0.00041–0.00080		18.1, 20.4	12.2,14.5	Ma and Wang (2002)
2001 brown mixed forest	0.0011	19.59	20.5	12.8	Ma et al. (2003)
2005 cinnamon sparse woods		4.0	14.8–15.4	6.7–7.2	Wang et al. (2008)
2008 cinnamon sparse woods	0.0082	33.17	16.2–18.7	5.7–5.9	Our results
2009 cinnamon sparse woods	0.010	31.60	17.1–20.3	5.7–5.9	Our results

species (a coniferous *Callitris* species and a broad-leaved *Eucalyptus* species), in a 19-year-old *Acacia mangium* plantation and in a Mediterranean *Quercus ilex* forest (Limousin et al. 2009; Ma et al. 2008; Melanie et al. 2006).

4.3.3 Comparison of Chinese Pine Transpiration in a Suburban Versus an Urban Environment

Table 4.3 summarizes transpiration in Chinese pine trees in a suburban area and in the center of Beijing (Liu 2008; Ma et al. 2006; Ma and Wang 2002; Ma et al. 2003; Nie et al. 2005; Wang et al. 2008). Both daily E_t and the maximum J_s observed in this study were much higher than those reported for Chinese pines with similar DBH in the suburban area of Beijing. The results suggest that the urban environment may promote plant water loss. The significant differences in transpiration between Chinese pine in the suburban and urban area may be attributable to the following factors: (1) increased Beijing urban air temperatures (Xiao et al. 2007) may enhance plant water loss (Wang et al. 2005); (2) low plant density could increase transpiration in both potted plants (Hagishima et al. 2007) and trees (Jimenez et al. 2008) because of the "clothesline effect" (van Bavel et al. 1962); (3) SWC is a very important factor for Chinese pine transpiration, particularly when the soil water is in deficit (Liu 2008), and the increased transpiration in urban areas could be attributable to a large increase in E_t in response to higher SWC; and (4) the

Table 4.4 Multiple linear regression analysis of the relationships between transpiration and meteorological conditions (T_a, RH, PAR, D, and w) at multiple time scales

Growing stage	Time scale	Main driving factors	R^2	p
Whole period	10 min	$y = -1.02\text{e-}06 + 3.379\text{e-}06\text{PAR} + 0.001D$	0.731	0.000
	1 h	$y = -2.60\text{e-}05 + 3.349\text{E-}06\text{PAR} + 0.001D$	0.740	0.000
	1 day	$y = -0.940 + 0.056\text{PAR} + 0.752T_a$	0.817	0.000
	1 month	$y = 4.891 + 1.212T_a$	0.941	0.000
Growing period	10 min	$y = -0.001 + 3.927\text{e-}06\text{PAR} + 0.000T_a$	0.741	0.000
	1 h	$y = -0.001 + 3.955\text{E-}06\text{PAR} + 0.000T_a$	0.747	0.000
	1 day	$y = 3.783 + 0.057\text{PAR} + 0.524T_a$	0.665	0.000
	1 month	$y = 11.603 + 0.922T_a$	0.651	0.001
Dormant period 11/2008 to 3/2009	10 min	$y = 0.000 + 0.001D + 1.339\text{e-}06\text{PAR}$	0.531	0.000
	1 h	$y = 0.000 + 0.002D + 1.378\text{e-}06\text{PAR}$	0.544	0.000
	1 day	$y = -1.377 + 1.012T_a + 0.046\text{PAR}$	0.791	0.000
	1 month	$y = -12.004 + 36.825D$	0.955	0.004

trees sampled in this study were shorter than the trees reported in the suburban area. A previous study reported that taller trees had lower mean canopy stomatal conductance (G_s, ~ 60 mmol/m^2 leaf area s^{-1}) than shorter trees (G_s, ~320 mmol/m^2 leaf area s^{-1}) (Schäfer et al. 2000).

4.3.4 Relationships Between the Transpiration of Chinese Pine and Urban Environmental Factors

Table 4.4 summarizes the relationships between the transpiration of Chinese pines and various climatic factors, including T_a, RH, PAR, D, and w, on multiple time scales. The sap flow rate in Chinese pines was highly related to the meteorological factors, but varied substantially with both growing season and time scale. For example, the trunk sap flow of the whole period was significantly correlated with PAR and D on both the 10-min scale and the 1-h scale, whereas the sap flow mainly depended on PAR and T_a on the daily scale and on T_a on the monthly scale. Similarly, the trunk sap flow of both the growth period and the dormant period was dependent on different climatic parameters at different time scales. The results suggest that the driving factors of E_t on the 10-min scale were the same as those on the 1-h scale. The E_t values at the 10-min, 1-h, and daily scales were highly correlated with PAR, whereas on the monthly scale, E_t was mainly affected by T_a (whole period and growth period) or D (dormant period).

The diurnal variation pattern of J_s closely matched that of PAR and D, and the correlation between J_s and PAR was generally better than that between J_s and D. For instance, on Julian day 118, the correlation coefficient (R^2) between J_s and PAR and J_s and D was 86 % and 71 %, respectively. PAR and D together explained 94.5 % of the variation in J_s. The diurnal pattern of J_s in the dormant period was very different from that in the growing period in that J_s was negatively related to

4 Transpiration Characteristics of Chinese Pines (*Pinus tabulaeformis*)... 67

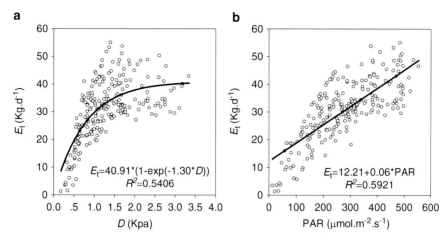

Fig. 4.4 Relationship between E_t and (**a**) vapor pressure deficit (*D*) and (**b**) photosynthetically active radiation (*PAR*) on the daily scale

PAR and *D* (e.g., $J_s = 0.0001$–$1.5083e{-}007\text{PAR}$, $R^2 = 0.56$, $p < 0.0001$; $J_s = 0.0002$–$0.0002D$, $R^2 = 0.56, p < 0.0001$ on Julian day 8). The results suggest that the shape of the diurnal flux in trees was mainly controlled by PAR, which might be caused by stomatal sensitivity to PAR. The research results are consistent with these on xylem water fluxes in ten tree species and two liana species (Phillips et al. 1999).

The day-to-day variation in E_t of Chinese pine was largely a function of daily differences in PAR, *D*, and SWC during the growing season. Increases in average daily *D* led to plateau-like increases in E_t [$E_t = 40.91(1 - \exp(-1.30 \times D)$, $R^2 = 0.54, p < 0.0001$] (Fig. 4.4a), indicating stomatal closure, which is consistent with studies on individual 30-year-old *Pinus sylvestris* L. trees (Wang et al. 2005), ten tree species (Phillips et al. 1999), and in a tropical rainforest (Granier et al. 1996), whereas these results differ from a study on large red maple trees (Wullschleger et al. 2000). Increases in the average daily PAR led to near-linear increases in E_t ($E_t = 12.21 + 0.06\text{PAR}, R^2 = 0.59, p < 0.0001$) (Fig. 4.4b). Daily PAR, however, had to exceed 100 μmol/m^2/s before rapidly increasing rates of E_t. A similar relationship was found by Wullschleger et al. in large red maple trees and by Zimmermann et al. in ten tree species (Phillips et al. 1999; Wullschleger et al. 2000).

When the rainy days were excluded, E_t was significantly correlated with SWC_{10} with medium explanatory power and with SWC_{30} with less explanatory power (Fig. 4.5a). Specifically, three representative patterns between E_t and SWC_{10} were identified (Fig. 4.5b–d). First, between days 195 and 214, it rained at an interval of a few days and SWC remained high (>19.5 %). During this time, the correlation between E_t and SWC was not significant. This result suggests that with sufficient soil water, E_t might be more sensitive to other variables, such as vapor pressure deficit, *D*, rather than SWC. Second, between days 164 and 183, after a first occurrence of 55 mm of rainfall, SWC increased to 20 % and then dropped again.

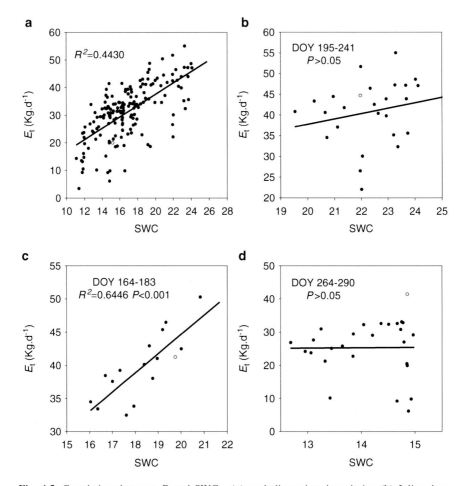

Fig. 4.5 Correlations between E_t and SWC_{10} (**a**) excluding rainy days during (**b**) Julian days (DOY) 195–241, (**c**) Julian days 164–183, and (**d**) Julian days 264–290 on the daily scale

The correlation between E_t and SWC was significant, indicating that the transpiration was affected greatly by SWC. Third, between days 264 and 290, no rainfall occurred and SWC remained low (<15 %). The correlation between E_t and SWC was not significant, which suggests that soil moisture was not the key factor regulating transpiration during this period. PAR and SWC together explained 74.3 % of the variation in E_t.

Meteorological factors affect instantaneous variability, whereas SWC determines the general level of tree sap flow (Huang et al. 2009; Liu 2008; Phillips and Oren 2001). SWC was a very important factor for E_t in the Chinese pine, particular after a rainfall followed by no rain for a few days. However, in a related study in a mixed forest of Chinese pines in suburban Beijing, sap flow showed a trend of accelerated growth when SWC increased from 4 % to 14 %, whereas the

4 Transpiration Characteristics of Chinese Pines (*Pinus tabulaeformis*)... 69

Table 4.5 t test of characteristics of three sampled trees between 2008 and 2009

Parameter	May to October in 2008	May to October in 2009	t	df	Sig. (two-tailed)
A_s (cm^2)	170.56	193.30	-2.103	2	0.170
A_c (m^2)	21.77	22.78	-2.831	2	0.105
LAI	2.42	2.03	3.657	2	0.067

A_c indicates canopy area

increase slowed down when SWC was greater than 14 % (Liu 2008). This slowing trend in sap flow was not observed in this study, which may be attributable to the soil water availability of the study site.

According to previous studies, the factors affecting interannual variability in transpiration are mainly growing season length, soil drought (Yoshifuji et al. 2007), leaf area dynamics (Phillips and Oren 2001), rainfall (Limousin et al. 2009), and compensatory mechanisms linking annual rainfall, leaf area index, and tree water use (Melanie et al. 2006). In our study, the sampled tree characteristics, including A_s, A_c, and LAI, and most environmental factors were similar in the two different years (Tables 4.2 and 4.5), whereas precipitation differed significantly in 2008 and 2009. Because of irrigation, soil water content was not exclusively affected by the amount of precipitation, and irrigation effectively offset soil drought (Fig. 4.1). Therefore, there was no significant decrease in soil moisture content in the dry year of 2009. These results indicate that soil moisture content was the most important variable among the factors determining E_t of Chinese pine at an interannual scale. Similarly to this, the annual evapotranspiration of eastern Siberian forests is relatively steady as a result of inflow from the deeper thawing layer affecting the soil moisture content (Ohta et al. 2008).

4.4 Conclusions

In this study, the water use patterns of Chinese pines in the center of Beijing were studied at multiple time scales. The results showed that water use of Chinese pines in the urban environment is mainly driven by photosynthetically active radiation and vapor pressure deficit at the diurnal scale and by soil water content at the seasonal and annual scale. Despite the drastic interannual variability in rainfall, E_t was almost the same in 2008 and 2009, suggesting that Chinese pine may be an appropriate species for the urban landscape of Beijing in the predicted future climate of reduced rainfall and higher air temperature. In comparison with suburban Beijing, the urban environment may significantly promote water loss in Chinese pines. Because of the significant correlation between transpiration and soil water content, it is appropriate to plant Chinese pines along roadsides or in plazas rather than on turf with regular irrigation.

Acknowledgments This study was supported by the Project of Knowledge Innovation of the Chinese Academy of Sciences for research into the urban ecosystem mechanisms of Beijing (KZCX2-YW-422). It was also supported by the "11th Five-Year Plan" to support science and technology projects (2007BAC28B01) and the Beijing Special Finance Investment on the Construction of a Public Education Platform for the Security of the Environment and the Ecosystem of the Capital (2008-0178). We thank the editor and two anonymous reviewers for their constructive comments and suggestions. We also thank all the members of Beijing Urban Ecosystem Research Station and Beijing Teaching Botanical Garden for their assistance in the field.

References

Beijing Gardening and Greening Bureau (2005) Collection of Beijing Urban Gardening and Afforestation Survey (monograph). Beijing Publishing House, Beijing (in Chinese)

Campbell GS, Norman JM (1998) An introduction to environmental biophysics. Springer, New York

Ceschiaa É, Damesina C, Lebaubeb S, Pontaillera JY, Dufrênea É (2002) Spatial and seasonal variations in stem respiration of beech trees (*Fagus sylvatica*). Ann For Sci 59:801–812

Granier A (1987) Evaluation of transpiration in a Douglas-fir stand by means of sap flow measurements. Tree Physiol 3:309–320

Granier A, Huc R, Colin F (1992) Transpiration and stomatal conductance of two rain forest species growing in plantations (*Simarouba amara* and *Goupia glabra*) in French Guyana. Ann For Sci 49:17–24

Granier A, Huc R, Barigah ST (1996) Transpiration of natural rain forest and its dependence on climatic factors. Agric For Meteorol 78:19–29

Gregg JW, Jones CG, Dawson TE (2003) Urbanization effects on tree growth in the vicinity of New York City. Nature (Lond) 424:183–187

Hagishima A, Narita K, Tanimoto J (2007) Field experiment on transpiration from isolated urban plants. Hydrol Process 21:1217–1222

Heilman JL, Brittin CL, Zajicek JM (1989) Water use by shrubs as affected by energy exchange with building walls. Agric For Meteorol 48:345–357

He SY, Xing QH, Yin ZT, Jiang XF (1993) Flora of Beijing. Beijing Publishing House, Beijing

Huang YQ, Zhao P, Zhang ZF, Li XK, He CX, Zhang RQ (2009) Transpiration of *Cyclobalanopsis glauca* (syn. *Quercus glauca*) stand measured by sap-flow method in a karst rocky terrain during dry season. Ecol Res 24:791–801

Jimenez E, Vega JA, Perez-Gorostiaga P, Cuinas P, Fonturbel T, Fernandez C, Madrigal J, Hernando C, Guijarro M (2008) Effects of pre-commercial thinning on transpiration in young post-fire maritime pine stands. Forestry 81:543–557

Limousin JM, Rambal S, Ourcival JM, Rocheteau A, Joffre R, Rodriguez-Cortina R (2009) Long-term transpiration change with rainfall decline in a Mediterranean *Quercus ilex* forest. Global Change Biol 15:2163–2175

Liu DL (2008) Spatial variation of sap flow of *Pinus tabulaeform*. J Northeast For Univ 36:15–18 (in Chinese with English abstract)

Ma LY, Wang HT (2002) Spatial and chronic fluctuation of sapwood flow and its relevant variables of Chinese pine. J Beijing For Univ 24:23–27 (in Chinese with English abstract)

Ma LY, Wang HT, Lin P (2003) Comparison of water consumption of some afforestation species in Beijing area. J Beijing For Univ 25:1–7 (in Chinese with English abstract)

Ma D, Li JY, Lin P (2006) Primary studies on water consumption of man-made forest in mountain area in Beijing. J Shanxi Agric Univ 26:48–51 (in Chinese with English abstract)

Ma L, Zhao P, Rao XQ, Cai XA, Zeng XP (2008) Diurnal, daily, seasonal and annual patterns of sap-flux-scaled transpiration from an *Acacia mangium* plantation in South China. Ann For Sci 65:402

Martin CA, Stabler LB (2002) Plant gas exchange and water status in urban desert landscapes. J Arid Environ 51:235–254

McPherson G, Simpson JR, Peper PJ, Maco SE, Xiao Q (2005) Municipal forest benefits and costs in five US cities. J For 103:411–416

Melanie JBZ, Isa AMY, Derek E (2006) Daily, seasonal and annual patterns of transpiration from a stand of remnant vegetation dominated by a coniferous *Callitris* species and a broad-leaved *Eucalyptus* species. Physiol Plant 127:413–422

Meng XS (2004) Composition of plant species and their distribution patterns in Beijing urban ecosystem. Dissertation, Beijing Forestry University (in Chinese with English abstract)

Montague T, Kjelgren R (2004) Energy balance of six common landscape surfaces and the influence of surface properties on gas exchange of four containerized tree species. Sci Hortic 100:229–249

Neighbour EA, Cottam DA, Mansfield TA (1988) Effects of sulphur dioxide and nitrogen dioxide on the control of water loss by birch (*Betula* spp.). New Phytol 108:149–157

Nie LS, Li JY, Zhai HB (2005) Study of the rate of stem sap flow in *Pinus tabulaeformis* and *Quercus variabilis* by using the TDP method. Acta Ecol Sin 25:1934–1940 (in Chinese with English abstract)

Niemelä J, Saarela S-R, Söderman T, Kopperoinen L, Yli-Pelkonen V, Väre S, Kotze DJ (2010) Using the ecosystem services approach for better planning and conservation of urban green spaces: a Finland case study. Biodivers Conserv 19:3225–3243

Nowak DJ, Dwyer JF (2007) Understanding the benefits and costs of urban forest ecosystems. In: Kuser JE (ed) Urban and community forestry in the northeast. Springer, New York, pp 25–46

Ohta T, Maximov TC, Dolman AJ, Nakai T, van der Molen MK, Kononov AV, Maximov AP, Hiyama T, Iijima Y, Moors EJ, Tanaka H, Toba T, Yabuki H (2008) Interannual variation of water balance and summer evapotranspiration in an eastern Siberian larch forest over a 7-year period (1998–2006). Agric For Meteorol 148:1941–1953

Phillips N, Oren R (2001) Intra- and inter-annual variation in transpiration of a pine forest. Ecol Appl 11:385–396

Phillips N, Oren R, Zimmermann R, Wright SJ (1999) Temporal patterns of water flux in trees and lianas in a Panamanian moist forest. Trees Struct Funct 14:116–123

Schäfer KVR, Oren R, Tenhunen JD (2000) The effect of tree height on crown level stomatal conductance. Plant Cell Environ 23:365–375

State Statistical Bureau (2007) Statistical yearbook of Beijing 2007. China Statistics, Beijing (in Chinese)

State Statistical Bureau (2010) Statistical Yearbook of Beijing 2010. China Statistics Press, Beijing (in Chinese)

Van Bavel CHM, Fritschen LJ, Reeves WE (1962) Transpiration by Sudangrass as an externally controlled process. Science 141:269–270

Wang KY, Kellomäki S, Zha T, Peltola H (2005) Annual and seasonal variation of sap flow and conductance of pine trees grown in elevated carbon dioxide and temperature. J Exp Bot 56:155–165

Wang RH, Ma LY, Xi RC, Li LP, Fan M (2008) Estimates of water consumption of seven kinds of garden plants and typical configuration in Beijing. Sci Silv Sin 44:63–68 (in Chinese)

Wu JJ (2008) Making the case for landscape ecology: an effective approach to urban sustainability. Landsc J 27:41–50

Wullschleger S, Wison KB, Hanson PJ (2000) Environmental control of whole-plant transpiration, canopy conductance and estimates of the decoupling coefficient for large red maple trees. Agric For Meteorol 104:157–168

Xiao RB, Ouyang ZY, Zheng H, Li WF, Schienke EW, Wang XK (2007) Spatial pattern of impervious surfaces and their impacts on land surface temperature in Beijing, China. J Environ Sci 19:250–256

Yoshifuji N, Tanaka N, Tantasirin C, Suzuki M (2007) Factors affecting interannual variability in transpiration in a tropical seasonal forest in northern Thailand: growing season length and soil drought. In: Sawada H, Araki M, Chappell NA, LaFrankie JV, Shimizu A (eds) Forest environments in the Mekong River basin. Springer, Tokyo, pp 56–66

Chapter 5
Landscape Design for Urban Biodiversity and Ecological Education in Japan: Approach from Process Planning and Multifunctional Landscape Planning

Keitaro Ito, Ingunn Fjørtoft, Tohru Manabe, and Mahito Kamada

Abstract The design of open and seminatural spaces in urban areas for urban biodiversity and ecological education is an important issue. There has been rapid decrease in the amount of open or natural space, in especially in urban areas in Japan, because of the development of housing areas. Thus, preserving these areas as wildlife habitats and spaces where children can play is a very important issue nowadays. This project, to design a park in Kitakyushu-City in the south of Japan, started in 2008. The aim of this project is to create an area for children's play and ecological education that can simultaneously form part of an ecological network in an urban area. The present projects have illustrated the importance of introducing natural environments into an urban park and thus enriching the learning environment for the children. Process planning and multifunctional landscape planning (MFLP) has been used for this project. Process planning would appear to be well suited for a long-term project such as a city park and school biotope. MFLP is thus considered suitable for the planning of a project such as a children's playground, which takes a long time to become established. The project mutually will serve as an example for future planning and development of children's

---

K. Ito (✉)
Laboratory of Environmental Design, Department of Civil Engineering, Kyushu Institute of Technology, 1-1 Sensui, Tobata, Kitakyushu 804-8550, Japan
e-mail: keitaro.ito.webmail@me.com

I. Fjørtoft
Telemark University College, 3679 Notodden, Norway
e-mail: Ingunn.Fjortoft@hit.no

T. Manabe
Kitakyushu Museum of Natural History and Human History,
2-4-1 Higashida yahatahigashi, Kitakyushu, Japan
e-mail: manabe@kmnh.jp

M. Kamada
Tokushima University, 2-24 Niikura, Tokushima, Japan
e-mail: kamada@ce.tokushima-u.ac.jp

N. Nakagoshi and J.A. Mabuhay (eds.), *Designing Low Carbon Societies in Landscapes*, Ecological Research Monographs,
DOI 10.1007/978-4-431-54819-5_5, © Springer Japan 2014

environment in urban areas. Furthermore, the children learned about the existence of various ecosystems through playing there and their participation in 60 workshops related to the park for 3 years. They have also actively participated in the development of an accessible environment and have proposed their own ideas for the management of that place.

Keywords Accessible environment • Ecological education • Ecosystems • Natural space • Park design • Urban area

5.1 Introduction

There has been a rapid decrease in the amount of open or natural space, in especially in urban Japan, as a result of the development of housing areas. Preserving these areas as wildlife habitats and spaces where children can play is a very important issue.

A generation ago, children had access to wild lands and used them for exploring, challenging, and exercising the skills needed to master a challenging landscape and unforeseen situations. Focus has been directed on learning effects from the natural environment and its impact on children's development. For example, some Scandinavian studies have described and analyzed how natural environments affect learning qualities in children such as play behavior and motor skills (Fjørtoft 2000a,b, 2001; Fjørtoft and Sageie 2000; Grahn et al. 1997).

These days, children's physical play environments and facilities for play are changing, and the opportunities for free play in stimulating environments seems to be declining. Early studies by Hart (1979), Moore (1986), Moore and Wong (1997), Rivkin (1990, 1995), Titman (1994), and others described the value of complex environments and wild lands for children and how children perceive and experience wild lands as places of their own domain. Physical activity is the number one recommendation for a healthy lifestyle throughout the lifespan, and giving children healthy habits in early years may give a positive payoff in adulthood (Baranowski et al. 1993; Frost Andersen et al. 2005; Fulton et al. 2001; Strong et al. 2005). It is generally accepted that diversity increases activity. The better equipped schoolyards and playgrounds offer a variety of play forms that challenge physical activity. Lindholm (1995) has documented how activities in schoolyards increased with the presence of green structures. Stratton (2000) reports that simple initiatives such as marking the schoolyard with colours have a positive effect on children's physical activity and Zask et al. (2001) found potentials (such as marking hopscotch areas or ball game fields) for increasing physical activity in the schoolyard, particularly for self-organized activities.

In urban areas in Japan in particular, there has been a rapid decrease in the amount of open or natural space in recent years, in particular in urban areas following the development of housing. Preserving these areas as wildlife habitats and spaces where children can play is currently a very important issue: "children's play" is an important experience in learning about the structure of nature, and "environmental education" has been afforded much greater importance in primary and secondary school education in Japan since 2002. Forman (1995) discussed habitat fragmentation and how it occurs naturally as well as being a result of human activity. At this study site, habitat fragmentation has already been caused by the development of housing projects.

If we create a green space such as the park in an urban area, it will serve as a stepping stone for species dispersal (Forman 1995). And even if the site is not large, it can contribute to ecological education in the urban area. Fjørtoft and Sageie (2000) have discussed the natural environment as a playground and learning arena as a way of rediscovering nature's way of teaching or "learning from nature." They also mentioned that landscape diversity was related to different structures in the topography and the vegetation, which were important for children's spontaneous play and activities. It is thus becoming very important to preserve open spaces as biotopes for urban biodiversity these days (Müller et al. 2010; Ito et al. 2010).

On the other hand, public demand for the conservation of cultural landscapes in Japan has increased (e.g., Nakagoshi and Ohta 1992; Kamada and Nakagoshi 1996; Fukamachi et al. 2001). One of the multiple functions of cultural landscapes is their inherent cultural value. The perspectives of cultural properties and heritage have been recognized by the recently established ad hoc committee on cultural landscapes (Nakagoshi 2010). Thus, it is very important to consider the methods of planning the landscape from the point of view of multifunctional use.

Previous studies have focused on children's experience of a place, their particular liking of an unstructured environment that has not yet been developed, and how they interpret a place and space (Hart 1979; Moor 1986; Fjørtoft and Sageie 2000). Consequently, this project started by creating the place for children to play in help restore nature to a small part of Kitakyushu City in the south of Japan. The aim of this project is to create an area for children's play and ecological education that can simultaneously form part of an ecological network in an urban area. Additionally, we discuss how to plan and manage the existing open spaces from a landscape planner's point of view, focusing on the methods used to plan it, the planning process as a whole, and how the schoolchildren participated in this process.

5.2 Planning Site

The planning site is Megurizaka pond, Yomiya City park, Kitakyushu City in Japan. The planning site is surrounded by a residential area. At this study site, habitat fragmentation has already been caused by the development of housing projects. If we create a green space such as an urban forest in an urban area (Fig. 5.1), it will serve as a stepping stone for species dispersal (Forman 1995). There are, however, still a number of streams, ponds and other green spaces remaining within a radius of 5 km of the planning site (Fig. 5.2) and the situation before renewal design

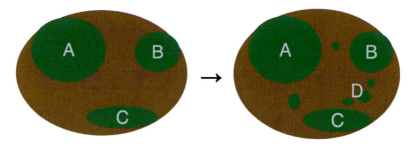

Fig. 5.1 Green space as stepping stones for biotope network in urban area

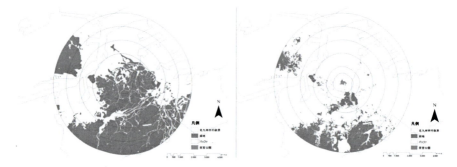

Fig. 5.2 Change of the green areas at 5 km around the planning site in 1922 and 2000. The center of this circle is the planning site

(Fig. 5.3). This area of the city has been developed mainly as a residential area with about 63 % of its original green spaces (forests and grassland) having been lost over the past 70 years (Fig. 5.1). Some ponds and lotus flowers have been planted, and the light of the festival was reflected on the pond in Fig. 5.4.

5.3 Methods for Planning and Design

5.3.1 Process Planning

"Process planning" was used to plan the school garden, given the length of time the process was expected to take. Although we knew in which direction we wanted the project to proceed, it was difficult to predict what kind of flora and fauna would be established there in the future, so we needed to choose a flexible planning method for this project. The architect Arata Isozaki (1970) described three different types of planning process.

5 Landscape Design for Urban Biodiversity and Ecological Education in Japan...

Fig. 5.3 The planning site in the urban park before renewal design. (Photograph taken by K. Ito, February 2008)

Fig. 5.4 The old map around the site in 1932. (Provided by Kitakyushu City government)

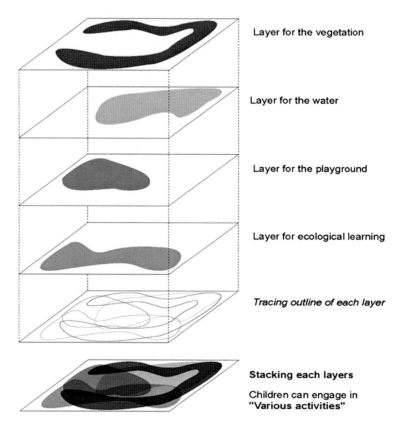

Fig. 5.5 Multifunctional landscape planning (Ito et al. 2003, 2010)

1. "Closed planning," which takes every aspect of the planning process into consideration.
2. "Open planning," which focuses on development for the future.
3. "Process planning," which focuses on the planning process itself and not solely the end form.

"Process planning" was thought to be the best method for planning the school garden when taking into consideration that the space will evolve over time and that its form is likely to change in the future according to the needs of those who use it. Thus, the creation of the biotope involves "process planning."

5.3.2 Multifunctional Landscape Planning

"MFLP" (Ito et al. 2003, 2010) was used to plan the park for space scale planning (Fig. 5.5). In other words, this is a method to think about how to manage the space for various ways. According to this method, the space is divided into a number of layers (layers of vegetation, water, playground, and ecological learning), which overlap each other. However, in contrast to "zoning," MFLP does not divide a space into clear functional areas. The overlapping of layers creates multifunctional areas where, for example, children who are playing by the water can also learn about ecology at the same time. Thus, during the creation of a multifunctional play area, children are able to engage in "various activities" as its different layers are added on top of each other. In addition, they will learn something new about its ecology when they are playing there.

5.3.3 Designing by Use Affordance for Children

For example, the shrubs constituted a mixture of scattered species, which afforded shelter and hiding, as well as social play and construction play (Fjørtoft and Sageie 2000). Very special was the flexible juniper bush, which motivated functional play (getting in and out) and social play (playing house) as well. Some trees were suitable for climbing depending on the branching pattern, the stem diameter, and the flexibility of the tree. The young deciduous trees were easily accessible for climbing. The spruces were more suitable for hiding than for climbing because of their dense branches. The more open areas in the pine and low-herb woodland afforded running, chase and catch, leapfrog, tag, and other games. The shrubs afforded hide-and seek, building dens and shelters, and role playing games such as house-and-home or pirates, and fantasy and function play.

5.3.4 Participation

Children at the school, their teachers, and a number of university students participated in the planning and construction phases of the project and in making improvements to the park (Fig. 5.6). Between September 2008 and August 2011, eight planning workshops were held involving 78 children and 8 teachers from the school and 12 students from the environmental planning course at Kyushu Institute of Technology.

At first, the children were surveyed about the kind of insects and plant life they hoped to find in the park. During workshops, they were asked to make final presentations about their image of the park based on everything that had been talked about in the previous workshops. The children made a number of suggestions for the water environment, in particular regarding the shape of the bridge and the depth of the water (Fig. 5.7). They also came out in favor of planting fruiting trees to attract birds and evergreen and deciduous trees to attract small animals and insects. In this way, they were thus able to gain a basic knowledge of the regional ecosystem and its flora and fauna.

Fig. 5.6 Survey at the pond with children and university student

Fig. 5.7 The children are presenting their ideas to the residents

5 Landscape Design for Urban Biodiversity and Ecological Education in Japan... 81

Fig. 5.8 Model of the park(biotope) by Keitaro ITO Laboratory (1/100). (Photograph taken by K. Ito, June 2009)

Following this, the final drawing and the model were completed by Keitaro ITO's Laboratory (Fig. 5.8) and children's and the university students' cooperation in constructing the landscape element (Fig. 5.9).

5.4 Results and Discussion

"Process planning" (Isozaki 1970) was used in the planning and design phases of this project. This approach does not place emphasis on the finished object but allows changes to be made during the actual process and is thus a very flexible method of design. The children have learned about the existence of various ecosystems when playing in the park and through their participation in the various workshops. Children and teachers at the school, along with a number of local residents, have participated in the planning of the park, and their interest in it continues because they have actively participated in the development of an accessible environment while at the same time being active in proposing ideas for its future management (Fig. 5.10). "Process planning" would thus appear to be well suited for a long-term project such as a city park and school biotope (Ito et al. 2010)

MFLP provides a variety of activities for the children as they are able to learn more about nature when they play in the park (Fig. 5.11). MFLP is thus considered suitable for the planning of a project such as a children's playground, which takes a long time to become established. It was noticeable that children involved

Fig. 5.9 The children and the university students are cooperating and constructing the landscape elements together. (Photograph taken by Y. Shin, February 2011)

Fig. 5.10 Completed site. (Photograph taken by K. Ito, April 2011)

Fig. 5.11 Schoolchildren playing in the park. (Photograph taken by C. Takahashi, June 2010)

themselves in a number of activities in the park. Some children enjoyed running around, jumping from one side of the edge of to the pond, or just sitting there and talking while others were observed trying to catch insects or just looking at the grass and flowers.

Fjørtoft and Sageie (2000) have discussed the concept of affordances, and in this park the children interpreted the affordances and adopted them into functions for play. The children's activities corresponded with Gibson's theory of affordances (Gibson 1979), according to which the composition of the environment as function for use. According to his theory, perception of the environment inevitably leads to some course of action. Affordances, or clues in the environment that indicate possibilities for action, are perceived in a direct, immediate way with no sensory processing: examples include buttons for pushing, knobs for turning, handles for pulling, levers for sliding, etc. (Fjørtoft and Ito 2010).

In this park, an example of this can be seen in the children's idea to plant local plants in the park. As a result, it was suggested that the biotope could become one of a number of habitats for birdlife in this urban area. In a survey conducted in 2011, 12 species of terrestrial insects, 20 species of aquatic insects, and 8 species of birds were observed in the park. It is envisaged that the park will establish itself as one of a network of biotopes in this urban area.

In short, this city park not only provides the children with a place to play in a variety of ways but has also become a habitat for a number of living creatures such as birds, insects, and fish.

5.4.1 Problems and Future Issues

The children have learned about the existence of various ecosystems (Fig. 5.10) by playing in the park and through their participation in the workshops during its planning. Their teachers and a number of local residents have also been active in this process, with the result that their interest in the park remains strong because they have actively participated in the development of an accessible environment and been able to propose ideas for its future management.

Nevertheless, the following problems were encountered during the planning of the park. (1) A great deal of time is needed to plan and manage the project. (2) The cooperative framework in which the park is managed changes every year because the teachers are transferred to other schools every 3–5 years. This change creates added difficulties in attempting to maintain continuity in the planning process each year.

An issue for future consideration is whether it is necessary to produce a manual to provide guidance on the maintenance of the park. Should this be done, there is a fear that the manual could be regarded as the one and only way of maintaining a city park and thus result in a lack of flexibility or diversity in the future.

The city park has gradually changed into a biotope during the past 3 years and the ecosystem contained within it has become more complex every year. It is important that this type of city park can contribute to the ecological network in the city.

A lack of outdoor space to play in, fear of violence in public spaces, the longer working hours of parents, and the artificial nature of most playgrounds have helped create the present-day situation in which young children have gradually lost contact with nature (Herrington and Studtmann 1998). Through this project, we discussed how to design a place that can be an "interface" between nature and people.

It is thus vital that present-day planners and landscape designers consider "landscape" as an "Omniscape" (Ito et al. 2010; Fjørtoft and Ito 2010; Numata 1996; Arakawa and Fujii 1999), in which it is much more important to think of landscape planning as a learnscape, not only embracing the joy of seeing but exciting the five senses as a whole.

Acknowledgments We wish to express our gratitude to all those who gave us the chance to write this paper. We are indebted to all students in Keitaro Ito's laboratory in Kyushu Institute of Technology and the children, teachers, and parents in Tenraiji primary school. This study was supported by Kakenhi, Japan Society for the Promotion of Science (JSPS), Grant-in-Aid for Scientific Research (B) (No. 19300264) in 2007–2010, and Grant-in-Aid for Scientific Research (C) (No.) in 2011–2013.

References

Arakawa S, Fujii H (1999) Seimei-no-kenchiku (Life architecture). Suiseisha, Tokyo

Baranowski T, Thompson WO, DuRant RH, Baranowski J, Puhl J (1993) Observations on physical activity in physical locations: age, gender, ethnicity, and month effects. Res Q Exerc Sport 64 (2):127–133

5 Landscape Design for Urban Biodiversity and Ecological Education in Japan... 85

Fjørtoft I (2000a). Landscape as playscape: learning effects from playing in a natural environment on motor development in children. Doctoral Dissertation. Norwegian University of Sport and Physical Education, Oslo

Fjørtoft I (2000b) Motor fitness in pre-primary school children: the EUROFIT motor fitness test explored on 5–7-year-old children. Pediatr Exerc Sci 12:424–436

Fjørtoft I (2001) The natural environment as a playground for children: the impact of outdoor play activities in pre-primary school children. Early Child Educ J 29(2):111–117

Fjørtoft I, Ito K (2010) How green environments afford play habitats and promote healthy child development. A mutual approach from two different cultures: Norway and Japan, science without borders. Ingunn Trans Int Acad Sci H&E 46–61

Fjørtoft I, Sageie J (2000) The natural environment as a playground for children—landscape description and analyses of a natural playscape. Landsc Urban Plann 48:83–97

Forman RTT (1995) Land mosaics: the ecology of landscapes and regions. Cambridge Press, Cambridge, NY, pp 254–474

Frost Andersen L, Lillegaard ITL, Overby N, Lytle L, Klepp K-I, Johansson L (2005) Overweight and obesity among Norwegian schoolchildren: changes from 1993 to 2000. Scand J Public Health 33(2):99–106

Fukamachi K, Oku H, Nakashizuka T (2001) The change of a Satoyama landscape and its causality in Kamiseya, Kyoto Prefecture, Japan between 1907 and 1995. Landsc Ecol 16:703–717

Fulton JE, Burgeson CR, Perry GR, Sherry B, Galuska DA, Alexander MP et al (2001) Assessment of physical activity and sedentary behavior in preschool-age children: priorities of research. Pediatr Exerc Sci 13:113–126

Gibson J (1979) The ecological approach to visual perception. Houghton Mifflin, Boston

Grahn P, Mårtensson F, Lindblad B, Nilsson P, Ekman A (1997) UTE på DAGIS. Stad & Land nr. 145. Alnarp: MOVIUM./Institutionen för landskapsplanering, Sveriges Lantbruksuniversitet

Hart R (1979) Children's experience of place. Irvington, New York

Herrington S, Studtmann K (1998) The natural environment as a playground for children. Landscape description and analyses of a natural playscape. Landsc Urban Plann 48:83–97

Isozaki A (1970) Kukan e (Toward the space). Bijyutu Shuppan, Tokyo (in Japanese)

Ito K, Masuda K, Haruzono N, Tsuda S, Manabe T, Fujiwara T, Benson J, Roe M (2003) Study on the biotope planning for children's play and environmental education at a primary school—the workshop with process planning methods. Environ Syst 31:431–438 (in Japanese with English summary)

Ito K, Fjørtoft I, Manabe T, Masuda K, Kamada M, Fujuwara K (2010) Landscape design and children's participation in a Japanese primary school—planning process of school biotope for 5 years. Urban biodiversity and design, Conservation Science and Practice Series (eds) N. Muller, P. Werner, J.G. Kelcey, Wiley–Blackwell, Oxford, UK, 441–453

Kamada M, Nakagoshi N (1996) Landscape structure and the disturbance regime at three rural regions in Hiroshima Prefecture, Japan. Landsc Ecol 11:15–25

Lindholm G (1995) Schoolyards: the significance of place properties to outdoor activities in schools. Environ Behav 23(3):259–293

Moore RC (1986) Childhood domain: play and space in child development. Croom Helm, London

Moore R, Wong HH (1997) Natural learning. Creating environments for rediscovering nature's way of learning. The life history of an environmental schoolyard. MIG Communications, Berkley

Müller N, Werner P, Kelcey JG (2010) Urban biodiversity and design, conservation science and practice series. Wiley-Blackwell, Oxford

Nakagoshi N (2010) How to conserve Japanese cultural landscapes: the registration system for cultural landscapes

Nakagoshi N, Ohta Y (1992) Factors affecting the dynamics of vegetation in the landscapes of Shimokamagari Island, southwestern Japan. Landsc Ecol 7:111–119

Numata M (1996) Landscape ecology. Asakura Shoten, Tokyo

Rivkin MS (1990) Outdoor play—what happens here? In: Wortham S, Frost JL (eds) Playgrounds for young children: national survey and perspectives. A project of the American association for leisure and recreation. An association of the American Alliance for Health, Physical education, Recreation and Dance

Rivkin MS (1995) The great outdoors: restoring children's rights to play outside. National Association for the Education of Young Children, Washington, DC

Stratton G (2000) Promoting childrens physical activity in primary school: an intervention study using playground markings. Ergonomics 43:1538–1546

Strong WB, Malina RM, Blimkie CJ, Daniels SR, Dishman RK, Gutin B et al (2005) Evidence based physical activity for school-age youth. J Pediatr 146(6):732–737

Titman W (1994) Special places, special people: the hidden curriculum of school grounds. World Wide Fund for Nature/Learning through Landscapes, Surrey/Winchester

Zask A, van Beurden E, Barnett L, Brooks LO, Dietrich UC (2001) Active school playgrounds–myth or reality? Results of the "move it groove it project". Prev Med 33(5):402–408. doi:10.1006/pmed.2001.0905

Part III
Ecologies in Cultural Landscapes

Chapter 6
Can Satoyama Offer a Realistic Solution for a Low Carbon Society? Public Perception and Challenges Arising

Yuuki Iwata, Takakazu Yumoto, and Yukihiro Morimoto

Abstract This chapter aims to analyze public perception in Japan of rural landscapes, known as Satoyama, and to investigate the possible future role of Satoyama in relationship to the development of a low carbon society in Japan. The data used were from the survey "The Top 100 Japanese Rural Landscapes" conducted by one of the biggest newspaper companies in Japan in 2008 and an additional questionnaire survey conducted in 2010.

The results indicated people's detachment from the productive activities associated with rural landscapes, and their association of it instead with new values particularly related to cultural services such as the beauty of landscapes, and provision of places with a sense of traditional living, where one can be in touch with nature and with a sense of seasons and history.

The main reasons people could not move to rural areas were jobs, money, houses, or lack of human connection to rural areas. For Satoyama to play a realistic role in the development of a low carbon society, these challenges would have to be overcome.

Keywords Cultural service • Ecosystem service • Low carbon society • Public perception • Rural landscapes • Satoyama • Text analysis

---

Y. Iwata (✉)
Agriculture and Food Science Centre, School of Agriculture and Food Science,
University College Dublin, Belfield, Dublin 4, Ireland
e-mail: yuuki.research@gmail.com

T. Yumoto
Department of Ecology and Social Behavior, Section of Ecology and Conservation Primate
Research Institute, Kyoto University, Kyoto, Japan

Y. Morimoto
Bio-Environmental Design, Faculty of Bio-environmental Science,
Kyoto Gakuen University, Kyoto, Japan

N. Nakagoshi and J.A. Mabuhay (eds.), *Designing Low Carbon
Societies in Landscapes*, Ecological Research Monographs,
DOI 10.1007/978-4-431-54819-5_6, © Springer Japan 2014

6.1 Introduction

6.1.1 Rational

Since global warming came to the forefront as an environmental issue of great concern, the establishment of a low carbon society (LCS) has been a pressing task throughout the world. However, urbanized societies with their massive demand for energy and economic growth are struggling to find feasible solutions. Japan is an example of such a society.

At the G8 Summit held in Toyako, Hokkaido in 2008, the participating countries including Japan agreed to reduce CO_2 emissions by 80 % by 2050. To meet such an ambitious target, the Japanese government has developed several visions and action plans for implementation of appropriate countermeasures (Onishi and Kobayashi 2011).

The Japan Low-Carbon Society Project was undertaken under sponsorship by the Ministry of the Environment in 2004, and a report was released with two narrative scenarios: Scenario A ("active, quick-changing, and technology oriented") and Scenario B ("calmer, slower, and nature oriented") (NIES et al. 2008, p. 5).

This research focuses on Scenario B by linking it with the study of "Satoyama," the traditional rural landscape in Japan. The term Satoyama is derived from the Japanese words "Sato (Village)" and "Yama (Mountain)" and refers to the secondary forest attached to agricultural villages, which in the past provided wood, charcoal, and organic fertilizer to every household. From an ecological point of view, "Satoyama" recently came to be regarded as a collective unit of secondary forest and its surrounding elements such as cultivated land, grassland, small rivers, ponds, and reservoirs for irrigation that were once connected with the traditional agricultural system.

The collective body of work on Satoyama has revealed that constant human intervention in the past formerly provided great diversity in habitat for native Japanese species (Washitani 2004). However, these reports also signaled alarm at the shrinkage of rural populations and the loss of traditional agricultural practices, which have led to the danger of loss of biodiversity and traditional landscapes. The role of Satoyama as part of a low carbon society has not been a major strand in the discussion.

The reasons for this exception include the continuous decline of rural populations and the agricultural, fishery, and forestry sectors. Also, there is not enough evidence that the lifestyle in contemporary Satoyama leads to reductions in greenhouse gas emissions.

6.1.2 Aim of the Study

The main aim of this study is to analyze the public perception of contemporary "Satoyama" and to discuss whether it could be one of the potential elements of a low carbon society. It also attempts to outline the potential challenges and difficulties arising from public perception as it stands. To achieve this aim, the following two separate analyses were conducted:

1. Using the data from "The Top 100 Japanese Rural Landscapes" contest conducted by one of the biggest newspapers in Japan in 2008, the words used by the general public who nominated their favorite rural landscapes were extracted and categorized. By comparing the frequency in the appearance of these categories, public perception toward rural landscapes was analyzed.
2. Using the data from a survey conducted by the Research Institute for Human and Nature in 2010, additional information such as the image of the rural landscape, how to relate to the rural landscape, and the reasons preventing people from living in rural areas are investigated with reference to different age groups.

6.2 Background and Requirements of the Study

6.2.1 Movement Toward a Low Carbon Society in Japan and Overseas

In 2008, the G8 summit in Toyako, Hokkaido settled on an agreement to halve global greenhouse gas emissions by 2050, and the following G8 Summit held in Italy in 2009 and a Major Economies Forum on Energy and Climate (MEF) summit meeting brought the G8 members to agreement on a reduction of greenhouse-gas emissions by more than 80 % by 2050, by all developed countries.

In Japan, a new vision, "Towards a Low-Carbon Society Japanese Society," was released by former Prime Minister Fukuda in 2008. In his vision, a long-term goal of reducing carbon emissions by 60–80 % by 2050 was set, with other components such as the establishment of new business in the new energy sector, use of traditional Japanese wisdom, and implementing actions including technology innovations (NIES et al. 2008). In addition, an Action Plan for Achieving a Low-Carbon Society was launched by Japan's Cabinet following the G8 Hokkaido Toyako Summit.

In 2009, the former Prime Minister Yukio Hatoyama announced a mid-term target that Japan should reduce greenhouse gas emissions by 25 % from 1990 levels by 2020. He included forest sinks, carbon credits from other countries, a domestic emission trading system, and green taxation as practical methods. He also emphasized the role of local authorities and other organizations in the creation of a low carbon society (Takemoto 2011).

Table 6.1 Japanese scenarios toward a low-carbon society (*Source:* NIES et al. 2008)

Scenario A	Scenario B
Vivid, technology driven	Slow, natural oriented
Urban/personal	Decentralized/community
Technology breakthrough, centralized production/recycling	Self-sufficient
	Produced locally
Comfortable and convenient	Social and cultural values
2 %/year GDP per capita growth	1 %/year GDP per capita growth
Dependent on import products in agriculture sector	Revival of public interest in agriculture
Shifting production sites to overseas	High-mix low-volume production with local brands
Market deregulation	Adequate regulated rules apply in market

6.2.2 Japan Low-Carbon Society Project

The Japan Low-Carbon Society Project was launched in April 2004 under the sponsorship of the Global Environmental Research Fund (GERF) of the Ministry of the Environment. The project aimed to demonstrate the scientific evidence concerning the potential to reduce CO_2 emissions by 70 % by 2050 while also meeting social and economic demands. The project team named the "Japan Low-Carbon Society Scenario Toward 2050 Team" consisted of about 60 researchers from the National Institute for Environmental Studies (NIES), Kyoto University, and Mizuho Information and Research Institute.

Using a "back-casting approach" in which the target was set first, and a quantitative investigation conducted to set measures required to meet this target, two narrative scenarios were set up: Scenario A ("active, quick-changing, and technology oriented") and Scenario B ("calmer, slower, and nature oriented") (NIES et al. 2008). Based on the two scenarios, the behavior of people, means of transportation, and industrial structure were quantified and calculated into energy-service demand. The appropriate combinations of end-use energy technologies and energy supply technologies were explored, and the amounts of CO_2 emissions were calculated (NIES et al. 2008).

The two narrative scenarios are shown in Table 6.1 and their images are shown in Fig. 6.1.

In Scenario A, the innovations of technologies such as renewable energies, nuclear power, carbon capture and storage (CCS), and other low carbon energy technologies and changes in city structure toward a "pedestrian-friendly city" (NIES et al. 2008, p. 27), are shown to reduce the energy demand and carbon emissions by 70 %.

Alternatively, in Scenario B, people move from the city to seek a slow and self-sufficient lifestyle in rural areas. As a result, decentralization and deurbanization occur. Revitalized rural areas emerge with public and NGO participation.

Fig. 6.1 Images of Scenario A (**a**) and Scenario B (**b**). (From NIES et al. 2008, p. 31)

More people will engage with the agriculture, fisheries, and forestry sectors and more people will have a preference for consuming rural products.

In Scenario B, small-size distributed energy systems, such as wind and biomass, are said to be more suitable, whereas "large-scale centralized energy systems such as nuclear power, carbon capture and storage (CCS), and hydrogen production" are necessary for Scenario A (NIES et al. 2008, p. 12).

Since the Great Tohoku Earthquake Disaster on 11 March 2011, debate on nuclear power has arisen as a result of the accident in the Fukushima Daiichi Nuclear Power Plant. In this context, it is likely that the weighting given to renewable energy will increase going forward.

6.2.3 *Inconvenient Truth for Scenario B*

Regarding a decentralized, deurbanized society, which was central to Scenario B, more barriers remain in the present-day society.

The decline and aging of rural populations has shown no signs of abating since the 1970s. Figure 6.2 shows the change in population from 2005 to 2010 in each prefecture. Tokyo, Kanagawa, Chiba, Okinawa, Shiga, Aichi, Saitama, Osaka, Fukuoka, where mega-cities have developed, are the only prefectures that have experienced population growth over the period. The sharpest decline has been experienced in Akita (5.2 %), Aomori (4.4 %), and Kochi (4.0 %) (Ministry of Agriculture, Forestry and Fisheries 2010).

Figure 6.3 shows the change in the number of households operating commercially in the agricultural sector between 1995 and 2010. This figure shows the change in the ratio of industries between 1955 and 2008. It shows that primary industry declined from 20.1 % in 1955 to 1.6 % in 2008 (Ministry of Health, Labor and Welfare 2010) while secondary industry (mining, construction, and manufacturing) increased between 1995 (36.8 %) and 1970 (46.4 %), and tertiary industry (service, sales, etc.) increased from 42.2 % to 69.6 % over the same period (Ministry of Agriculture, Forestry and fishery 2010).

Figure 6.4 shows the change in the number of agricultural households and the agricultural population. The population of this cohort was about 2.6 million, marking a decline of 22 % since 2005 and 33 % since 2000. The reason for this is the retirement of aging farmers, and the change in status of some farmers from being self-employed to being members of their local agricultural co-operative (Ministry of Agriculture, Forestry and Fisheries 2010). It also shows the change in the ratio of farmers more than 65 years of age. The average age of farmers rose from 61.1 years in 2000 to 65.8 years in 2010. The ratio of those over 75 years old in 2010 was about 30 % (Ministry of Agriculture, Forestry and Fisheries 2010).

6 Can Satoyama Offer a Realistic Solution for a Low Carbon Society...

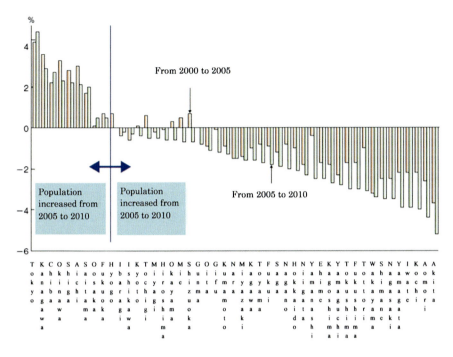

Fig. 6.2 Ratio of population change (from 2000 to 2005, from 2005 to 2010). (From Ministry of Agriculture, Forestry and Fisheries 2010)

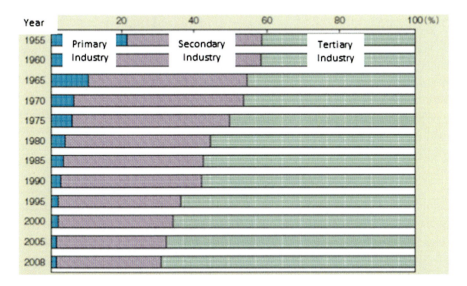

Fig. 6.3 Change in ratio of industrial structure. (From Ministry of Health, Labor and Welfare 2010)

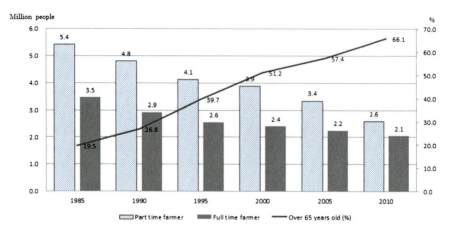

Fig. 6.4 Change in number of agricultural households, full-time farmers, part-time farmers (in millions of people), and ratio of farmers over 65 years old (%) from 1985 to 2010. (From Ministry of Agriculture, Forestry and Fisheries 2010)

6.2.4 Movement of Satoyama Renaissance

Measures to attract urban populations to rural areas have been undertaken in many sectors. The Ministry of Agriculture, Forestry and Fisheries has enhanced the interaction between people in rural areas and urban areas through activities such as green tourism, environmental education, farmers' markets, and short-stay holidays.

In 2009, "the model program for the rehabilitation of rural areas through support of human development (*Noson Kasseika Jinzai Ikusei Shien Model Jigyo*)" was conducted for a year, and 291 people participated. Of this number, 189 people began to live in a rural area. Young people in their twenties made up most of the participants (Ministry of Agriculture, Forestry and Fisheries 2010).

The Ministry of the Environment has been initiating the rehabilitation of "Satochi-Satoyama" areas from an ecological point of view, in which "Satochi-Satoyama" were defined as "conceptual geographic regions located between urban and primitive natural areas where the environment has been formed through the efforts of diverse human activity, and is comprised of secondary forests surrounding villages intermixed with farmland, reservoirs, and grasslands, among other features" (Nature Conservation Bureau, Ministry of the Environment Government of Japan 2004, p. 2). For example, the "Sato-Navi (*The Rural Landscape Navigation*) Project" has been promoted since 2008 and aims to collect good examples of practices conducted in Satochi-Satoyama areas and to share methods and knowledge with people living in different places. From October 2011 to February 2012, several symposiums were held to discuss challenges and potential solutions with local residents, NPOs, local authorities, and experts in ten selected areas.

The valuation of "Satoyama" was conducted under an initiative led by the United Nations University Institute of Advanced Studies (UNU-IAS) known as "Satoyama Satoumi SGA" from 2004 to 2010. This valuation included "Satoumi," which refers to traditional coastal areas in addition to inland areas. Also the "Satoyama Initiative" was introduced internationally as a socioecological production landscape in 2010 when the Convention of Biodiversity COP10 was held in Nagoya (Takeuchi 2011).

However, a breakthrough in the revitalization of rural areas has not been achieved, and the gap between the reality and the situation described in Scenario B remains large.

6.3 Perspectives

6.3.1 *"The Top 100 Japanese Rural Landscapes" and Its Previous Research*

In 2008, Asahi Shimbun Newspaper Company, one of the biggest newspaper companies in Japan, conducted "The Top 100 Japanese Rural Landscapes" contest to select the 100 best Japanese rural landscapes through a system of public nominations. The question asked was "which is your favorite rural landscape, and why?," and people wrote the name of a place and the reasons for its selection in an open answer. The criteria are landscape, biodiversity, and human activities (Forests Culture Association 2009).

Consequently, 3,022 sites were nominated, which were assigned longitude and latitude values (Iwata et al. 2010). Combined with geographic data such as land use (National Land Information Office 1997), mean altitude (National Land Information Office 1981), and height range (National Land Information Office 1981), Iwata et al. (2010) classified the nominated sites into six landscape types using nonhierarchical cluster analysis. Also, keywords that appeared more than 100 times in all the nomination forms were extracted using SPSS Text Analysis for Surveys 3.0 Japanese (SPSS Inc. 2007), and the relationship with the landscape types was examined using cross analysis (χ^2 test; SPSS Inc. 16.0 Japanese 2007).

The results (shown in Fig. 6.5 and Table 6.2) showed that the majority of the nominated sites were categorized as forest type (88 % forest cover in 1 km^2). The most closely correlated keywords were "landscape," "beauty," "traditional culture," and "history." The next most popular rural landscapes were categorized as mixed type (60 % forest, 20 % paddy fields in 1 km^2). The keywords correlated were "biodiversity" and "preservation activities." Paddy field type sites (paddy field >60 % in 1 km^2) were significantly associated with the word "Furusato (home)," and urban and suburban type sites (built-up land, 50 %) were concentrated in the Kanto Region, and they were nominated mainly by local citizens for their nature activities (Iwata et al. 2010).

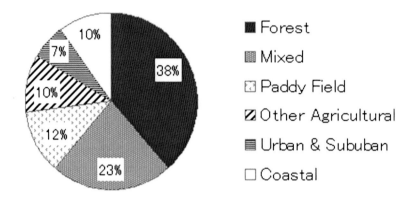

Fig. 6.5 Landscape types of the nominated sites. (From Iwata et al. 2010)

Table 6.2 Landscape types and their characteristics and keywords than appeared frequently in each landscape type

Landscape type	Characteristics	Keywords appeared frequently in each type
Forest type	Forest 88 %	Location & Settings**, Residence**, River**, Mountains*, Water*, Spread*People**, Local**, Local Residence**, Living**, Visit**, Past**, Award for terraced paddies**, Epitome/Original Image (*Genkei*)**, Heart**, Unique*
	High altitude and steep sloped area	
Mixed type	Forest 52 %, paddy field 19 %	Paddy Field*, *Satoyama*, Biodiversity*, Preserve*, Award for terraced paddies*
	Other agricultural field 10 %	
	Mountain ridges, Yatsu	
	Sloped lowland	
Paddy field type	Paddy field 62 %, Built-up land 15 %, Forest 10 %	Spread**, Home (*Furusato*)**, Paddy Field*, Seasons*
	Flat lowland	
Other agricultural type	Other agricultural field 49 %	Agricultural Field**
	Forest 20 %, paddy field 11 %	
	10 % built-up land cover	
	Relatively high altitude and flatland	
Urban and sub-urban type	Built-up land 58 %, forest 9 %	Residence**, Forest**, Greens**, *Satoyama*** Enjoy*
	Other agricultural field 9 %	
	Relatively flat lowland	
Coastal Type	Ocean 80 %, forest 11 %	Sea**, Residence**
	Flat lowland as near coastline	

Source: Iwata et al. (2010)
Correlation with confidence with 1% (marked as **) and 5% (marked as *)

6.3.2 Landscape Services Generated by Keyword Analysis

In this research, further analysis was conducted by extracting the words that appeared more than ten times. This step provided the authors with more detailed insights into the values held by people who nominated their favorite rural landscapes. SPSS Text Analysis for Surveys 3.0 Japanese (SPSS Inc. 2007) was used to extract the words, and they were categorized into groups.

Every noun, verb, and adjective was extracted. Those words that appeared fewer than ten times and those that did not have useful meaning for this study were ignored. Also, individual animal and plant species names were not included, as they were discussed in previous research (Iwata et al. 2010).

The words used in the study were then categorized into 49 subcategories by the authors and were in turn placed in nine larger categories. Table 6.3 shows the categories of words, the number of words in each category, and the frequency of their appearance.

Figure 6.6, a graph of the subcategories, shows that the words categorized as "Village/People's settlement" were most frequently used, indicating that people valued the presence of human life in the rural landscapes rather than natural elements solely. The words categorized as "Individual species name" were used at second frequency, which indicates that specific animals and plants were important elements for the preferable rural landscapes. The words categorized as "Landscape" and "Geographical features" were used at next greatest frequency, which shows that the beauty of landscape and the geographic features that shape rural landscapes were highly valued.

The words with the next largest usage were categorized as "Seasons" and "History," which indicates that the rural landscapes that enable people to feel the passing of the seasons are equally as important as those having a sense of history. This could be a unique tendency for Japan where there are four extremely distinct seasons, and people living in urban areas still long for the feeling of the changing seasons. Those words categorized as "Mountain" and "Water" are also often used, which reflects the fact that the characteristics of typical Japanese rural landscapes are that they are surrounded by mountains and water bodies normally consist of small rivers flowing through mountain valleys.

Words categorized as "Agricultural activity" and "Rice production" were also used more than 500 times, which indicated that many people still value the rural landscape as an agricultural production site.

Figure 6.7, showing the ratio of the large categories from Table 6.3, shows that the most frequently used were words categorized as "Culture/History" and "Nature/Geographical features." The second largest category was "Feeling and Senses."

The authors further attempted to divide the large categories into the Ecosystem Services defined by the United Nations, which include Supporting Services "that are necessary for the production of all other ecosystem services," Provisioning Services that are "products obtained from ecosystems," Regulating Services that are "benefits obtained from the regulation of ecosystem processes," and Cultural

Table 6.3 Categorized words that appeared more than ten times in written appearances

Large category	Subcategory	Individual words used more than ten times in each group ( ) ; number of appearances
Landscape	View (82)	Vista (43), Sweeping View (39)
	Landscape (1,329)	Scenery (547), Landscape (228), Authentic Landscape (131), Terraced Paddy Competition (124), Cultural Landscape (75), Pastoral Landscape (75), Natural Landscape (28), Contrast(25), Rural Scenery(25), Scene (17), Rural Landscape(16), Satoyama Landscape (15), Snowscape (13), Satoyama Scenery (10)
Feelings/Senses	Seasons (1,303)	Seasons (247), Autumn (211), Spring (198), Summer (182), Winter (141), Autumn Leaves (81), Snow (62), Early Summer (41), Spring Shoots (39), The Changing Seasons (17), Each Season (16), A Touch of Spring (14), Layers of Snow (12), Summertime (11), The Eve of Winter (11), Early Spring (10), Winter Time (10)
	Atmosphere (380)	Rusticity (46), Style (44), Atmosphere (42), Distinct (41), Remnant (41), Taste (25), Appearance (24), Uniqueness (21), Individualistic (19), Expression (17), Character of the Locality (14), Secret Places (14), Image (12), Character (10), Climate (10)
	Feeling (275)	Heart (186), Feeling (41), Discovery (24), Moving (14), Spirit (10)
	Senses (595)	Wind (51), Sound (37), Myriad (35), Sky (34), Sunset(33), Murmer (28), Color (24), Light (24), Twitter (23), Fragrance (21), Stars (18), Chirp (17), Yellow (16), Morning (16), Morning Light (15),Night Sky (15), Song (14), Blanket of Cloud (14), Smell (14), Moon (13), The Heavens (12), Starry Sky (12), Sun (12), Memory (12), Fascinate (11), Sensations (11), Innumerable Stars (11), Five Senses (11), Evening Time (11), Carpet (10), Picture (10), Body (10)
	Story (208)	Legend (38), Stage (22), Movie (19), Dream (15), Treasure (15), Memory (13), Evocative (12), Filming Site (12), Story (11), Theme (11), Model (10), *Manyoshu*, ancient song book (10), Treasure (10), Folk Story (10)
	Sense of Belonging (389)	Home (*Furusato*) (118), Culture(84), World (40), Japan (39), Home Town (33), Japanese (23), Nostalgia (16), Origin (14), Japan's Most Famous (11), Proud of (11)
	Life (38)	Life (26), Born (12)
Culture/History	History (1,265)	Past (245), History (182), Since Past (111), Generation (185), Traditional (59), Ancestor (51), Inheritance (45), Folklore (36), Middle Age (35), Ancient (28), Inheritance Site (26), Historic Site (24), Showa Period (24), Cultural Inheritance (22), Old Road (18), Vestige (16), Castle Remains (15), Modern (15), 300 years (14), The Warring State Period (13), Old (12,

		Traditional Culture (12), Thousands Years (12), Historical Culture (12), *Kamakura* Period (12), Heian Period (11), *Jomon* Period (10), Showa 30's (10), Period of Domain State (10)
	Culture (46)	Picture Letter (13), Custom (12), Traditional Art (11), Taste (10)
	Inheritance (63)	Fortune (15), World Heritage (14), Important Cultural Heritage (12), Ancient Tomb (12), Remains (10)
	Religious (185)	Shrine (45), Temple (31), Belief (27), Village Shrine (24),Graveyard (15), Local God (13), Spiritual Site (11), Precinct (11), Invocation (11)
	Place Names (69)	Local Name (31), Name (24), Giving Name (14)
	Village/Settlement (2,523)	Home Village (Sato) (568), Local (501), Settlement (397), Village (135), Mountain Town (89), Village Houses (81), Site of streets (69), Stone Wall (48), Houses (48), Agricultural Village (45), Countryside (35), Small Settlement (31), Mountain Village (31), Homestead Woodland (29), Old Folks House (26), Roof (24), Row of Houses (23), Thatched Houses (21), Thatches (20), Home Village Site (*Satochi*) (19), Marginal Village (29), Boat (21), Inn Town (18), Livelihood Culture (16), Small Village (16), Farming and Mountain Villages (15), Homestead (15), Storehouse (15), Station (15), White Walls (14), Village People (13), Old Home Towns (12), Family of Pedigree (11), Residential Building (11), Waterwheel (11), Town(11), Castle Town (10), Entire Village (10), Human Dwellings (10), Posting Station (10)
Time	Future (157)	Future (37), Succeeding Generations (27), Modern Times (25), After Time (23), Future Generation (12), Descendent (12), Continuation (11), Successor (10)
	Time (89)	Time (41), Year (24), This Year (24)
Tourism/Recreation	Tourism (499)	Hot Spring Site (75), Tourists (53), Symbol (51), Beauty Spot (37), Photograph (29), Tourist Site (25), Highlight (20), Tourism (20), Recreation Trail (19), Mountain Trail (17), Village for Firefly (15), Aesthetic Landscape (15), Upland (15), Museum (13), Viewing Platform (12), Course (11), Exhibition (11), Guide Sign (11), Guest House (10), Spot (10), Hot Spring Facilities (10), Hot Water (10), Forest Road (10)
	Experience/Recreation (412)	Activity (133), Experience (49), Observation (30), Volunteer (26), Tree Planting (24), Biotope (19), Environmental Education (16), Environmental Learning (15), Nature Observation (13), Learning (12), Camp Site (12), Fishing (11), Field Study (11), Agricultural Experience (11), Camp (10), Participant (10), Education (10)
	Event/Contest (242)	100 Sato Contest (38), Village Scenery Contest (38), Festival (33), Important Satochi Satoyama Selected Site (31), Traditional Event (29), 100 Village Scenery (28), Event (28), Select 100 (17)

(continued)

Table 6.3 (continued)

Large category	Subcategory	Individual words used more than ten times in each group ( ); number of appearances
	Research (35)	Examination (24), Research (11)
	Social Circle (357)	Interaction (74), Cooperation (60), Participation (48), Relaxation (41), Together (29), Formation (24), Working Together (21), Coordination (16), Bond (12), Town Planning (12), Regional Planning (10), Regional Revitalization (10)
Resource management/ Preservation	Resource Management (289)	Management (62), Rehabilitation (49), Hands In (45), Open Burning (26), Mowing (18), Cleaning (17), Thinning (14), Work (14), Tree Trimming (13), Maintenance (11), Renovation (10), Burning (10)
	People (505)	People (293), Children (149), Local People (24), Adults (15), Local Farmers (12), Old People (12)
	Preservation (154)	Reservation (55), Preservation (45), Preservation Activity (35), Protected Activity (19)
	Revitalization (81)	Vitalization (39), Creating *Sato* (22), Reconstruction (10), Spread Information (10)
	Organization (264)	Effort (40), Coexistence (30), Organization (22), Establishment (22), Development (20), Project (14), Continuous (14), Inauguration (13), Propellant (13), Excises (13), Securement (12), Contribution (11), Support (10), Accomplishment (10), Planning (10), Sharing (10)
	Exploitation (30)	Exploitation (19), Cultivation (11)
	Resource utilization (105)	Charcoal Burning (33), Resource (24), Charcoal (13), Firewood (12), Extraction (12), Sericulture (11)
Agricultural activities	Application	Application (68)
	Agricultural activity (572)	Agriculture (102), Plowland (82), Cultivation (63), Production (47), Farmland (42), Area of Production (34), Harvesting (34), Cultivation (31), Sales (21), Indigenous product (17), Farming (17), Regular Vocation (14), Agricultural Production (13), Pesticide (12), Ranch (12), Cultivated Field (11), Crop (10), Non-Pesticide (10)
	Rice production (656)	Paddy Field (323), Reservoir (42), Channel (41), Rice Planting (39), Pastoral (36), Rice Cropping (34), Rice Ear (31), Growing Rice (21), Fallow Field (20), Pastoral Region (19), Paths by the Paddies (15), Harvesting Rice (15), Dike by the Paddies (10), Paddy Field Region (10)
	Agricultural field (185)	Agricultural Field (119), Terraced Fields (51), Fruit Farm (15)
Food/Resources	Food in general (67)	Food (18), Cooking Ingredient (16), Food Culture (13), Seaweeds (10), Green Horseradish (10)
	Vegetables (61)	Vegetables (50), Sweet Potato (11)

Category	Subcategory	Items
	Fruits (90)	Apple (25), Orange (17), Japanese Persimmon (16), Sweet Chestnut (12), Fruits Trees (10), Grape (10)
	Grains (141)	Rice (54), Soba Noodle Crop (51), Tea Tree (36), *Koshihikari* (Brand of Rice) (9), Soy (8), Straw (7), Wheat (6), Productiveness of Grain (6)
	Benefit (28)	Benefit (16), Provision (12)
Nature/Geographic features	Virgin Nature (64)	Large Nature (27), Native (19), Untouched (18)
	Geographical features (1,351)	Nature (650), Ground (183), Environment (119), Earth (18), Slope (49), Hill (39), Valley (31), Rock Summit (30), Stone (26), Hillside (22), Canyon (21), Plane Land (21), Slope Site (19), Basin (18), Hillside Land (18), Summit (16), Rock (16), Perpendicular Slope (16), Plateau (13), Bank (13), Valley Site (13)
	Ocean (364)	Ocean (154), Island (128), Coast (38), Coastline (25), Set of Islands(19)
	Mountain (815)	Mountain (344), Foot of Mountain (122), Range of Mountains (112), Between Mountains (78), Mountain Forest (33), Top of Mountain (25), Mountain Ravines (24), Mountain Chain (20), Skirts of Mountain (18), Filed Mountain (10), Between Mountains (10)
	Water (872)	Water (204), River (183), Fresh Stream (81), Stream (64), Pond (50), Waterfall (35), Headstream (34), River (32), Spring (31), Headwaters (30), Waterfront (29), Lake (26), Branch Stream (24), Downstream (19), Freshwater (17), Bog (13)
	Forest (300)	Green (134), Forest (131), Primary Forest (35)
	Woodland (239)	Secondary Woodland (80), Trees (38), Woodland (38), Plantation (33), Bamboo Forest (26), Pine Forest (14), Natural Forest (10)
	Tree (135)	Giant Tree (26), Tree (26), Berry (21), Large Tree (18), Windbreak Forest (16), Broadleaf Tree (16), Old Tree (12)
	Grassland (158)	Grassland (63), Grass (32), Wetland (25), Highland (25), Wilderness (13)
	Parks (93)	Park (35), Planting Ground (29), Row of Cherry Blossom Trees (19), Natural Park (10)
Living organisms	Living Organisms(509)	Animals and Plants (229), Flower (141), Habitat (32), In Clumps (27), Mountain Grasses (21), Vegetation (14), Wild Animals (14), High Mountain Vegetation (11), Grass and Trees (10), Flowers (10)
	Individual species name (1,974)	Tree Species (478), Bacteria Species (28), Grass Species (224), Mammals (346), Birds (135), Amphibians(66), Reptiles (14), Fish (210), Insects (396), Shells (10), Extinctions (81)

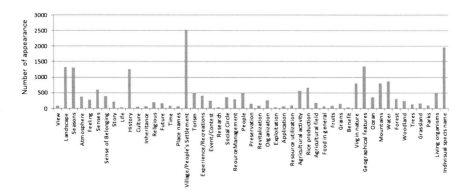

Fig. 6.6 Frequency of words used in nomination sheets shown according to subcategories

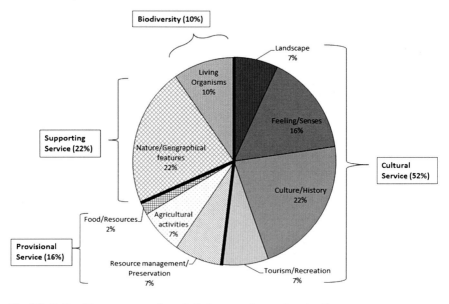

Fig. 6.7 Ratio of larger groups of words that appeared more than ten times

Services that are "nonmaterial benefits people obtain from ecosystems through spiritual enrichment, cognitive development, reflection, recreation, and aesthetic experiences" (Millennium Ecosystem Assessment (MEA), UN 2005).

According to the Ecosystem Service categorization shown in Fig. 6.7, the words categorized as Cultural Services were the most prominent, which accounted for 52 % of the words extracted, followed by Supporting Services including natural factors, which accounted for about 22 %. Provisional Services related to resource management and Agricultural Activities and Biodiversity including individual species names comprised about 16 % and 10 %, respectively.

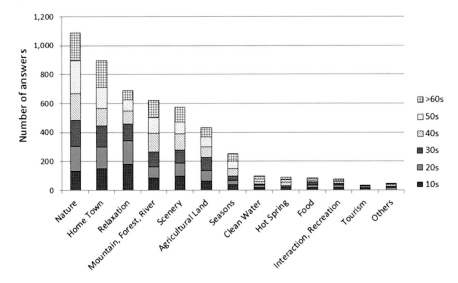

Fig. 6.8 Image of rural landscape in Japan

6.3.3 Questionnaire Survey for People Not Involved in Satoyama Movements

In 2010, an additional survey was conducted to reveal the awareness of people who are not involved in Satoyama movements. A series of questionnaires was sent randomly by Internet to a pool of consumer testers who had agreed to answer surveys in August 2010, and 5,000 answers were obtained: 2,500 from males and 2,500 from females. The spatial distribution of respondents was not biased toward rural or urban areas.

Figure 6.8 shows the results of the survey regarding the image people have of rural landscapes in Japan: it indicated that the majority of people associated "Nature," "Home Town," "Relaxation," "Mountain, Forest and River," "Scenery," and "Agricultural Land" with rural landscapes in Japan. Ratios did not differ greatly according to age group. However, there was a slight significance in the tendency of teenagers and those in their twenties to choose "Relaxation" more than elder populations.

Figure 6.9 shows the answers to the question, "How do you want to relate to rural areas?" The majority of people answered "Want to enjoy nature," followed by "Want to relax," "Want to engage in tourism," and "Want to do outdoor activities." Also, the answer "Want to live in the future" is the fifth biggest answer.

Figure 6.10 shows the answer to the question "What is the reason if you cannot live in rural areas even if you want to?" The majority of people have chosen the answers "Can't leave from current job," "Few shopping options," "Nervous about access to doctors," "There are no interesting jobs," and "Have no relatives in rural areas." The answer "Others" includes "Do not have enough money" and "Have to take care of parents" as additional reasons.

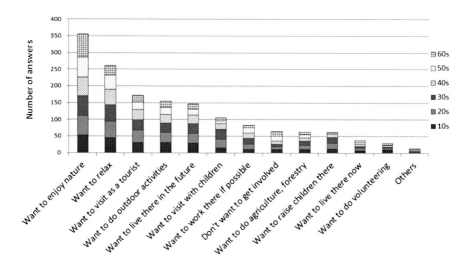

Fig. 6.9 How do you want to relate to rural areas?

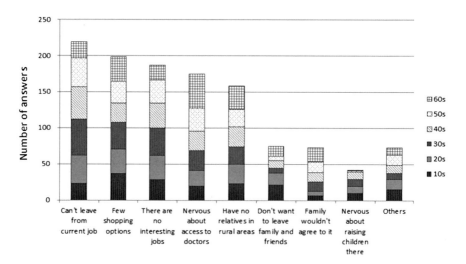

Fig. 6.10 Reasons to prevent living in rural areas

6.4 Discussion and Conclusions

The results from the first dataset revealed that people most valued the Cultural Services provided by rural landscapes, and there was a wide array of such Cultural Services. The majority of people highly valued the presence of human life and the beauty of natural landscapes including mountains and rivers in rural area. People also valued them as places where they could feel the passing of the seasons and a sense of history and traditions. Natural elements including the presence of other animals were also important. It was also revealed that many people still value the rural landscape as an agricultural production site.

The results from the second dataset demonstrated the image of the rural landscape as consisting of nature, mountains, rivers, forest, homeland, and a place for relaxation. The results also showed that 10 % of people felt that they wanted to live in rural areas in the future. In comparison with the answer for the reasons why they cannot live in rural areas, it was revealed that the majority of people were worried about their jobs, money, and practicalities such as access to shopping facilities or doctors.

If the government attempted to follow Scenario B proposed by Japan Scenarios Towards Low-Carbon Society, which aims to shift people's lifestyle toward one "calmer, slower, and more nature oriented" based in rural areas, it is important to understand public attitudes toward rural lifestyle. The results from this study outlined the general tendency and major challenges.

The following areas are the potential options that would bring people to engage more with rural landscapes and could bring gradual revitalization of rural landscapes.

- Beautiful landscapes of nature that provide a sense of history, culture, and traditional living are required for people to visit rural areas.
- A sense of belonging, seasons, events, tourism, a circle of people, Satoyama activities, and social interaction are required for people to get involved in rural areas
- Jobs such as attractive agriculture, the renewable energy sector, or remote access to offices using the Internet, and social interaction are required for people to live there.

These findings show that there are still considerable challenges to be overcome so that "Satoyama" lifestyles could be seen as having a realistic role in the development of a low carbon society.

Acknowledgments This research is partly supported by Environment Research and Technology Development Fund, E-0902 "Ecosystem Services Assessment of Satoyama, Satochi and Satoumi to Identify New Commons for Nature-Harmonious Society" (Project Leader: Masataka Watanabe).

References

Forests Culture Association (2009) Sato no kachi saihakken- Nihon no sato 100 sen ichiran (Re-discovering the value of Sato – list of Japanese top 100 rural landscape). In: Kankyo K (ed) 2009 Seibutsutayosei no Nihon. Asahi Shimbun, Tokyo, pp 120–127 (in Japanese)

Iwata Y, Fukamachi K, Morimoto Y (2010) Public perception of the cultural value of Satoyama landscape types in Japan. Landscape Ecol Eng 7(2):173–184

Millenium Ecosystem Assessment (2005) Ecosystem and human well-being: general synthesis. Island Press, Washington, DC, pp 54–60

Ministry of Agriculture, Forestry and Fisheries (2010) Annual report on food, agriculture and rural communities (Syokuryo Nogyo Noson Hakusyo) (in Japanese)

Ministry of Health, Labor and Welfare (2010) Annual report on assessment of labor economy (Rodo Keizai no Bunseki) (In Japanese)

Nature Conservation Bureau, Ministry of the Environment Government of Japan (2004) Satochi-Satoyama, the old and new, the closest nature. Official pamphlet, Ministry of the Environment Government of Japan, Tokyo (in Japanese)

National Land Information Office (1981) Web-based digital national geographical data. http://nlftp.mlit.go.jp/ksj/

National Land Information Office (1997) Web-based digital national geographical data. http://nlftp.mlit.go.jp/ksj/

National Institute for Environmental Studies (NIES), Kyoto University, Ritsumeikan University, and Mizuho Information and Research Institute ("2050 Japan Low-Carbon Society" scenario team) (2008) Japan scenarios and actions towards low-carbon societies (LCSs), Global Environmental Research Fund (GERF/S-3-1), Japan–UK Joint Research Project "a sustainable low-carbon society (LCS)". http://2050.nies.go.jp/LCS/jpn/japan.html. Accessed Mar 2014

Onishi T, Kobayashi H (2011) Low carbon cities: the future of urban planning. Gakugei Shuppansha, Kyoto

SPSS Inc (2007) SPSS ver. 16.0 for Japanese. SPSS, Chicago

Takemoto K (2011) Measures to address climate change in Japan and overseas. In: Low carbon cities: the future of urban planning. Gakugei Shuppansha, Kyoto, pp 33–58

Takeuchi K (2011) Rebuilding the relationship between people and nature: the Satoyama Initiative. Ecol Res 25(5):891–897

Washitani I (2004) Shizensaisei; Jizokukanou na seitaikei no tameni (Nature rehabilitation; for sustainable ecosystems). Chuokoronshinsya, Tokyo (in Japanese)

Chapter 7
Effects of Sustainable Energy Facilities on Landscape: A Case Study of Slovakia

Katarina Pavlickova, Anna Miklosovicova, and Monika Vyskupova

Abstract Sustainable energy, known as renewable energy, is the provision of energy that meets the needs of the present without compromising the ability of future generations. As sustainable energy sources, those most often regarded are hydroelectricity, solar energy, wind energy, geothermal energy, and biomass energy. The impacts of increased sustainable production and consumption are considerably less than the increased supply and the consumption of conventionally produced energy. However, we have to take into consideration also negative effects of new renewable production energy facilities, mainly on the landscape and its characteristics. Their localization could be considered as a factor that affects future settlements' development: this can be perceived not only as a facility reducing the amount of "green gas emissions" but also as a separate construction in the landscape.

The intent of this contribution is to mention specific aspects of environmental impact assessment of sustainable energy facilities with the emphasis on the role of landscape and landscape ecological evaluation in the Slovak Republic.

Keywords Cumulative effects • Ecological stability • Environmental impact assessment • Landscape • Slovakia • Sustainable energy

7.1 Introduction

Renewable energy implies energy sources that have no undesirable environmental consequences such as combustion of fossil fuels or generation of nuclear energy. Alternative energy sources are renewable and thought to be "free" energy sources

---

K. Pavlickova (✉) • A. Miklosovicova • M. Vyskupova
Department of Landscape Ecology, Faculty of Natural Sciences, Comenius University
in Bratislava, Mlynska dolina B-2, 842 15 Bratislava, The Slovak Republic
e-mail: pavlickova60@gmail.com; miklosovicova.anna@gmail.com;
monikavyskupova@gmail.com

N. Nakagoshi and J.A. Mabuhay (eds.), *Designing Low Carbon Societies in Landscapes*, Ecological Research Monographs,
DOI 10.1007/978-4-431-54819-5_7, © Springer Japan 2014

because they all have lower carbon emissions compared to conventional energy sources: these include biomass, wind, solar (thermal, photovoltaic, and concentrated), geothermal, hydroelectric, and tidal energy sources (AES 2011).

After the Kyoto Protocol entered into force on February 16, 2005, investment in rational uses of energy, savings, and efficiency became the main premise to support the development of new energy. If energy consumption decreases, renewable sources could cover a significant part of the demand of energy, in particular, electricity; if consumption remains uselessly high because inefficient and less energy consuming (acting also on final uses), renewable energy would become a reality, a feasible method even in these sectors. With investments being equal (today all are in the sector of generation from fossil sources), if there was parallel research on how to reduce consumption and wastes considerably (at least 35 %) and on power plants from renewable sources, there would be also a reduction of gas emissions, without any negative influence on development (Iacomelli 2005).

Two slightly differentiated concepts can be identified in the discussion on energy transition. At first, all kinds of energy based on renewable resources found their way into the scientific debate and public discussion. With rising concern for a socially fair, environmentally friendly, and economically feasible future, the focus has shifted to include sustainable energy sources (van Etteger and Stremke 2007).

The growth of renewable sources of energy also stimulates employment in Europe, the creation of new technologies, and improved trade balance (EC 2011). The renewable sources are also considered as a basis for building new energy facilities (new energy power stations), which can be called "sustainable energy facilities" because of their global and regional impacts on the environment. But on the local level also, all these facilities can have their adverse impacts on the environment and especially influencing the landscape. For instance, the emplacement of high wind power stations into the country influences the scenery, and solar parks require spacious open ground.

The capacity for sustainable energy production is affected by geographic location and climate as well as geology and is therefore limited. This perception is based on ecological understanding. But according to Kozova and Pauditsova (2010), landscape suitability must be assessed not only in ecological terms but also in the terms of social and cultural carrying capacity. Environmental impact assessments from simple methods such as checklists to complex predictive models can also evaluate landscape suitability. This idea became a supporting topic for our authentic research. Its methods and knowledge are stated in this chapter, "Case Study from Slovakia" (sect. 7.5).

7.2 The Concept of Sustainable Energy Facilities Within the Slovak Republic

The Slovak Republic, as a member state of the European Union, has to follow European strategies, plans, and programs. In the frame of the Energy Sector, the "Green Paper: A European Strategy for Sustainable, Competitive and Secure Energy" (COM 2006) can be considered as a leading document. According to this document, renewable energy is already the third electricity generation source worldwide (after coal and gas) and has the potential to grow still further, with all the environmental and economic advantages that would follow. For renewable energy to fulfill its potential the policy framework needs to be supportive and in particular to stimulate increasing competitiveness of such energy sources while fully respecting the competition rules. Although some sources of low carbon indigenous energy are already viable, others, such as offshore wind, waves, and tidal energy need positive encouragement to be realized. The full potential of renewable energy will only be realized through a long-term commitment to develop and install renewable energy facilities.

How is the Slovak Republic prepared to use sustainable energy sources? In July 2004, the Government approved the document "Progress Report on the Development of Renewable Energy Sources, including the Identification of National Indicative Targets for the Use of Renewable Energy Sources." According to this document, based on Slovak natural and economic conditions, it is realistic to produce electricity from renewable energy sources at approximately 5.9 TWh in 2010. Then, in 2006, the "Energy Policy of the Slovak Republic" was approved by resolution of the Government of the Slovak Republic No. 29 from January 11, 2006. Under this policy, the obtainable ratio of all the renewable energy sources had to share in the overall electricity production at 19 % (5.9 TWh) in 2010, 24 % in 2020, and 27 % in 2030. Biomass is considered to be the most promising renewable source for heat and electricity production.

The energy policy was the starting point for the development of electro-energy, the thermal power industry, the gas industry, extraction, processing, and transit of oil, coal extraction, and the use of renewable energy sources. It defined three objectives:

1. Safeguarding, as effectively as possible, a secure and reliable supply of all forms of energy in desired volume and quality;
2. Decreasing the share of gross domestic energy consumption in gross domestic product—reducing the energy intensity;
3. Ensuring the volume of energy generation that would cover demand on a cost-effective principle.

To achieve the objectives of energy policy, 11 fundamental priorities were set in "Energy Policy of the Slovak Republic" (Ministry of Environment of the Slovak Republic 2006). To achieve point 8, "Increasing the share of renewable energy sources in the generation of electricity and heat with the aim of creating adequate

Table 7.1 Conservative scenario for the use of renewable energy sources (RES) in the Slovak Republic

Type of RES (total joules, TJ)	2010	2015	2020	2025	2030
Biomass	31,000	48,000	66,000	85,000	120,000
Solar energy	300	1,000	6,000	14,000	20,000
Geothermal energy	200	1,000	3,000	4,500	7,000
Hydroenergy	18,000	20,000	22,000	23,000	24,000
Wind energy	300	x	x	x	x
Energy waste	200	x	x	x	x
Total amount	50,000	73,000	100,000	130,000	175,000
Share of RES in total energy consumption (%)	6.4	9.0	12.0	16.0	21.0

x, no increase is expected; 1,000 TJ = 278 GWh (in electricity)
Source: Ministry of Environment of the Slovak Republic (2008)

Table 7.2 Optimistic scenario for the use of renewable energy sources (RES) in the Slovak Republic

Year	2010	2015	2020	2025	2030
Type of RES (TJ)					
Biomass	31,000	50,000	74,000	90,000	120,000
Solar energy	300	3,000	12,000	22,000	37,000
Geothermal energy	200	2,000	7,000	10,000	14,000
Hydroenergy	18,000	20,000	22,000	23,000	24,000
Wind energy	300	x	x	x	x
Energy waste	200	x	x	x	x
Total amount	50,000	78,000	120,000	150,000	200,000
Share of RES in total energy consumption (%)	6.4	9.5	14.0	18.0	24.0

x, no increase is expected; 1,000 TJ = 278 GWh (in electricity)
Source: Ministry of Environment of the Slovak Republic 2008

additional sources in order to cover domestic demand," many documents were established. As the most important, we mention here the "Strategy of Higher Usage of Renewable Sources of Energy" (Ministry of Environment of the Slovak Republic 2007) and the "Energy Security Strategy" (Ministry of Environment of the Slovak Republic 2008).

The future of using renewable energy sources in the Slovak Republic is illustrated in Tables 7.1 and 7.2, which represent the so-called "conservative" and "optimistic" approaches. To achieve both the proposed numbers the Slovak Republic had to build and still has to build new power stations based on renewable energy sources.

The evaluation of expected impacts on the environment and sustainable development, resulting from the new developed power stations, is governed in compliance with the relevant legislation. The environment is also affected by the development of related networks and systems. Therefore, the development and placement of new system should be located especially in areas where a sufficient system and network is already present. The development of new facilities and the modernization of existing energy facilities should be realized only under the rule of law, the implementation of recommendations and comments from the environment impact assessment process: Act No. 24/2006 Coll. on environmental impact assessment, Ministry of the Environment, and according to the decision of authorities.

7.3 Specific Feature of Environmental Impact Assessment in the Slovak Republic: The Role of Landscape

7.3.1 Environmental Impact Assessment/Strategic Impact Assessment of Sustainable Energy Facilities

Many projects concerning energetic constructions were carried out in the Slovak Republic. To maintain the typical landscape of our country and to conserve or improve the actual conditions of the environment there, it was necessary to engage some tools that can help to avoid these problems. As many others countries have done, Slovakia has adopted the structured approaches of Environmental Impact Assessment (EIA) and Strategic Environmental Assessment (SEA). These processes, as the base elements of sustainable development, have been helpful in creating energetic politics and strategies and in finding appropriate locations for many different projects, including renewable facilities.

Formally, the EIA and SEA are tools for obtaining and evaluating environmental information before its use in decision making in the development process. This information consists basically of predictions and the evaluation of social, economic, health, and environmental impacts and advice as to how best manage these changes if one alternative is selected and implemented (Abaza et al. 2004). An EIA focuses on proposed physical developments (constructions, facilities, and activities), and an SEA focuses on proposed documents such as new laws, policies, strategies, and plans.

The institution of impact assessment was first established in the Slovak Republic by Act No. 127/1994 Coll., which was effective from September 1, 1994. This law covered not only EIA, but also in a rather simple form, SEA. To harmonize the Slovak legislation with that of the EU, the law was later modified by the Act No. 391/2000 Coll. On February 1, 2006 the law was replaced by the new Act No. 24/2006 Coll. This law covers all requirements from the relevant EU directives and related international agreements. This act has been many times revised. In this current law, the processes of EIA, SEA, and transboundary assessment are equally represented. The individual steps of processes, the structure of documentation, and public participation are all specified.

	Wind energy	Solar energy	Biomass energy	Water energy	Geothermal energy
Act No. 127/1994 Coll.	0	0	0	2	2
Act No. 391/2000 Coll.	9	0	0	16	4
Act No. 24/2006 Coll.	65	12	3	16	9

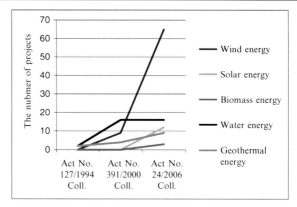

Fig. 7.1 Proposed projects of sustainable energy facilities assessed under environmental impact assessment (EIA) in the Slovak Republic in years 1994–2011

The main part of the EIA and SEA is the impact assessment. The actual law in Slovakia characterizes it as a comprehensive finding, description, and assessment of presumed impacts of the strategic documents and proposed activities on the environment, including comparison with the existing state of the environment in the given locality and in the area of presumed impacts. The assessment includes the preparation of environmental impact statements, consultations, and the final statement (Act No. 24/2006 Coll. in wording of later issued provisions).

Together, more than 5,900 constructions, facilities, and activities and more than 700 strategic documents were assessed during the force of the EIA and SEA legislation from 1994 (based on www.enviroportal.sk). Among them were four assessed documents related to energetics, specifically, the "Draft of Conception on Utilization the Hydroenergetic Potential of the Water Flows," the "Energy Security Strategy," the "Strategy of Final Part of Nuclear Energetics," and the "Strategy of Higher Usage of Renewable Sources of Energy." Also, the conceptions of the development of heat energetics in cities and towns have been evaluated. In focusing on renewable sources of energy under EIA, altogether 138 proposed projects of renewable facilities have been assessed (see Fig. 7.1).

The majority of these intentions were proposed to build up the wind power stations situated in wind parks, which were located not only in wide open spaces but also in built-up areas. Other projects suggested some wind parks in the same areas or nearby areas. Nevertheless, they usually have been assessed individually. A challenge for impact assessment is therefore to distinguish the limits of the study area and to determine all the connected projects to judge effects as well as possible.

7.3.2 Specific Feature of Environmental Impact Assessment: The Role of Landscape

The specific feature of EIA in the Slovak Republic is that great emphasis is put on the landscape, which coincides with the environment (see the definition of the landscape, following). The quality of the landscape is considered as equivalent to the quality of the environment. Interpretation of landscape in the assessment is fully compatible with the concept of sustainability, which is the basis of impact assessment. The process is based on three dimensions in terms of sustainable development: the environmental dimension, the social dimension, and the economic dimension.

Generally, the relationship of humans to the landscape is complicated. Mankind was born in the natural landscape and its biological and spiritual existence depends on it. The human population is an inseparable part of the landscape. Positioning of people in the landscape is determined by their physiological dependence on it (air, water, food, and all other landscape elements necessary for humans). Landscape also provides shelter to humans and is their home. It is an existential relationship. On the other hand, the landscape is the direct (primary production) or indirect (secondary production and the following production levels) object of human work. This relationship to the landscape is given by humans, creators of material things. The first relationship is of primary importance for humans and corresponds with the concept of sustainability. Failure to consider this connection as a criterion for decision making about the use of natural resources and the landscape has led the present society to the global environmental crisis. This is the reason why this specific approach, which relies on the concept of sustainability, is in the Slovak Republic a determining one in valuation of the landscape for any purpose (Drdos 2005).

As the landscape is the living environment of humans, the environmental impact assessment also considers humans (affected population) and their activities along with the landscape. Also, without knowledge regarding humans and their activities it is impossible to identify correctly anthropogenic phenomena and changes in the landscape.

The effect (impact) influences (Drdos 2005) the natural landscape components (natural effects); landscape (as geosystem and ecosystem), its structure and use (geosystemic effects); protected territories and elements of the territorial system of ecological stability (ecozoological effects); image of the landscape (visual effects); population (social effects); economy and its branches (economic effects); and material and immaterial components of culture (cultural effects). Such classification of impacts fully reflects the analytical procedures applied to the landscape-ecological studies.

7.4 The Most Important Impacts of Sustainable Energy Facilities

The most important impacts of sustainable energy facilities are based on the general knowledge of the authors, who utilized their practical experiences from the Slovak Republic.

7.4.1 Hydroelectric Power Plants

7.4.1.1 Negative Impacts

Hydroelectric water plants are relatively expensive and are associated with significant negative environmental and social impacts. These impacts significantly depend on the type of hydroelectric power plant: accumulation, derivative, flow, pumped, or combined type. The main negative impacts, listed below, are mainly associated with the following significant effects:

– Changes of water flow and water quality in a river modify living conditions for aquatic organisms, especially fish;
– Dam lakes separate fish populations living in the lower and upper parts of a water course and block migration routes;
– Construction of a dam causes changes in local climate and groundwater levels with high territorial radius;
– Changes of water flow may result in changes of sediment transfer; sedimentation in a reservoir can lead to erosion in the lower part of the water course;
– Construction of a water dam causes increased transfer of mud and sediment, thereby reducing water quality in the lower part of the water course;
– Construction of the water dam changes visual effects.

7.4.1.2 Positive Impacts

In addition to the fundamental socioeconomic impacts typical for energy constructions, it is possible to consider contribution to summer tourism, potential recreational space near water increasing water sports and fishing, as positive impacts of hydroelectric power plants and therefore resulting in significant increase of tourism in the chosen area. Flood protection and possible use of more environmentally friendly shipping can be also considered as a significant positive impact.

7.4.2 Wind Power Plants

7.4.2.1 Negative Impacts

It is necessary to point out the assumption that small turbines do not affect the surrounding environment. Larger turbines (with a tower height of 110–130 m) are considered as a problem with such parameters as noise, visual impact, collisions with birds and bats, and interference with the electromagnetic field. From the energy aspect it is a source dependent on weather conditions, power limited, and unstable.

7.4.2.2 Positive Impacts

Cheap power supply: Wind parks do not require a larger area and in terms of land use are shown to be most economical.

7.4.3 Solar Power Plants

7.4.3.1 Negative Impacts

Environmental impacts are linked to land occupation, change of landscape structure, and landscape character. Another disadvantage is the strong dependency of solar power plants for effectiveness on climatic conditions, especially the intensity of solar radiation and the number of sunny days per year.

7.4.3.2 Positive Impacts

The use of solar energy could be the cleanest way of energy use, and in contrast with other renewable sources, its impacts on the surrounding environment are lower. Direct use to produce heat is also another advantage of solar power plants.

7.4.4 Biomass Power Plants

7.4.4.1 Negative Impacts

Electricity production from biomass is similar to the thermal power plants with fossil fuel combustion (coal, gas). Therefore, correspondingly similar negative effects can be observed in biomass power plants in relationship to the emergent flue gas and solid and gaseous pollutants, as in the case of thermal power plants. The difference, in comparison with thermal power plants, is in significantly lower values of carbon dioxide (CO_2). According to other information, the greatest danger represents escape of PM10 and PM2.5, because filters cannot capture them and these particles are inhaled by humans. Impacts on the landscape usually have a visual character.

7.4.4.2 Positive Impacts

The technology of direct combustion of biomass is the most common way of its energy exploitation; methods of biomass modification to biofuels such as pyrolysis, gasification, aerobic putrefaction, and fermentation are also possible. Modern combustion facilities are able to burn almost any treated or untreated organic material. Given the impacts on the environment, burning of wood waste, agricultural production waste, and municipal waste especially has significance.

7.4.5 Geothermal Power Plants

7.4.5.1 Negative Impacts

There is a risk of possible release of toxic compounds from the bore (e.g., boric acid) or radioactive radon directly from the thermal water or steam. There are also other frequently identified impacts, especially on the landscape character, and common impacts on the landscape are usually visual.

7.4.5.2 Positive Impacts

A geothermal power plant is characterized by high performance in a permanent work regime and does not produce any technological waste or pollutants.

7.5 Case Study in Slovakia

In any EIA system, a definitive record of EIA reports undertaken should be maintained and made public (Wood 1995). We have chosen four of them as a basis for our research: Wind Park Svabovce I with two wind power stations, Wind Park Svabovce II with five wind power stations, Wind Park Horka pri Poprade with four wind power stations, and Wind Park Straze pod Tatrami with five wind power stations (Fig. 7.2).

7 Effects of Sustainable Energy Facilities on Landscape: A Case Study... 119

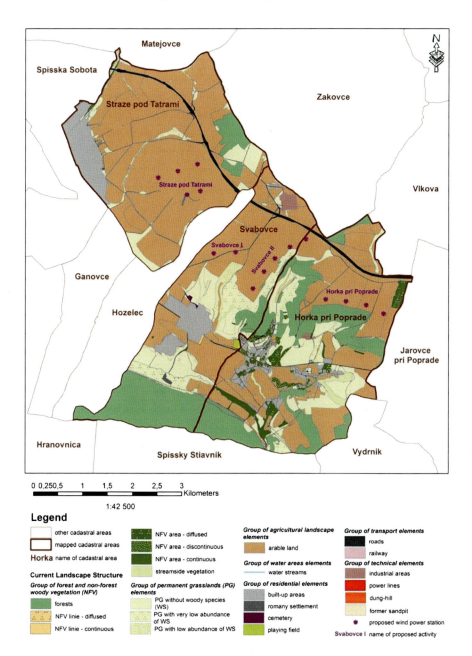

Fig. 7.2 Current landscape structure of all cadastral areas and the placement of proposed wind parks

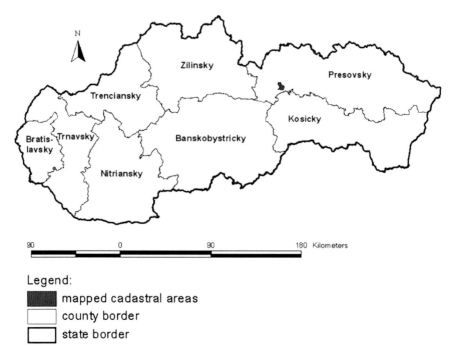

Fig. 7.3 Location of mapped cadastral areas within the Slovak Republic

All these parks are situated (Fig. 7.3) in the east part of the Slovak Republic (Presov region, Poprad district), covering an area of 3,061.33 ha. According to the geomorphological division of Slovakia (Mazur and Luknis 1980), cadastral areas belong to the unit Podtatranska kotlina Basin. These cadastral areas fall under the moderately cool subregion: basins with high altitude are cool. Average temperatures vary between 4 ° and −6 °C in January and from 14 ° to 17 °C in July. The total rainfall varies from 600 to 900 mm/year (Fasko and Stastny 2002). According to the phytogeographical division of Slovakia (Futak 1980), all cadastral areas belong to the subregion of Pannonian flora (*Pannonicum*), ward of West Carpathian flora (*Carpaticum occidentale*), and district Inner Carpathian Basin (*Intercarpaticum*). According to the zoogeographical division of Slovakia, the terrestrial biocycle, Svabovce, Horka pri Poprade, and Straze pod Tatrami cadastral areas belong to the West Carpathian District of Carpathian Mountain subprovince and Central European mountain province (Cepelak 1980). According to the zoogeographical division of Slovakia, the limnic biocycle (Hensel and Krno 2002), these cadastral areas belong to the Poprad area of Atlantic province.

Among the most common impacts on landscape evaluated are impacts on landscape structure, protected areas, landscape image (visual impact), and ecological stability. The proposed wind turbine plants were evaluated with a matrix method, using assessment of impact significance (in positive or negative sense) as follows:

- Without impacts (the proposed activity will not affect components of natural environment, population, and landscape in any way): value, 0;
- Insignificant impact (mainly impact with character of risk, coincidence, or with a negligible influence or contribution): value, 1;
- Small significant impact (impact with low influence from quantitative point of view, local impact, with low reception): value, 2;
- Significant impact (the impact on the wider environment, reception is high): value, 3;
- Very significant impact (reception is high to very high): value, 4.

Localization of proposed wind parks was incorporated to the map of current landscape structure by using the transposition methodology there through was reached illustration of their localization in cadastral areas. Impacts were different especially in their significance in which they are affecting landscape character. Actually, they were evaluated separately with the following results: Horka pri Poprade, 1; Svabovce I, 3; Svabovce II, 2; and Straze pod Tatrami, -1. But you should realize that all four wind parks are proposed in three neighboring cadastral areas, and this is why a joint map of landscape structure and proposed wind park localization was created for all three cadastral areas. We came to the logical conclusion that this impact is multiplied so that it is necessary to evaluate cumulative effects. The impact of all wind parks on the landscape character is insignificant, but the cumulative impact has a final value of 3, a significant impact.

In the conditions of the Slovak Republic, in Environmental Impact Assessment a specific evaluation methodology aimed at ecological landscape stability is used. By evaluation of impacts on ecological landscape stability, we came to the conclusion that it is necessary to evaluate it as one unit. Because this methodology is not commonly known, we describe it next.

7.5.1 Evaluation of Landscape Stability

To evaluate ecological landscape stability as the most frequent indicator in the assessment of environmental landscape quality, several methodological approaches were developed that are mostly based on defining the coefficient of ecological stability, the basic definition and mathematical expression of which were introduced by Michal (1982) in his work. The equation for calculation of this coefficient of ecological stability has undergone several revisions and modifications for different types of work.

The most frequently used term for characterization of landscape ecological quality is the term ecological stability. Ecologists have proposed several incompatible definitions of ecosystems and landscape ecological stability. The most available definitions for this chapter are found in articles by Michal (1982, 1992, 1994), Mician and Zatkalik (1986), Miklos (1992), Forman and Godron (1993), Voloscuk (2001), and Rehackova and Pauditsova (2007).

According to Michal (1982), the stability of any kind of system is not in its unchanging state, but in its ability to retain its own dynamic equilibrium. Ecological stability is the ability of ecological system to persist even under a disruptive influence and reproduce their essential features as well as in terms of outside distortions. This ability is expressed by minimal change in the case of destructive impact or spontaneous return to its initial state, respectively, the development of the original trajectory after eventual change. Landscape ecological stability preservation is the most general and comprehensive condition for preservation of the gene pool, biological diversity, equilibrium, flexibility, natural ecosystem behavior, and the natural productive ability of the landscape (Izakovicova et al. 2008).

7.5.2 The Coefficients of Ecological Stability

7.5.2.1 Coefficient of Ecological Stability 1

This coefficient of ecological stability 1 (Low 1984) is based on assignment of ecological significance coefficient from A to E to each secondary structure element. The highest value of ecological significance coefficient achieves areas of forests and water areas, which are hereby the most stable landscape elements, and the formula for the coefficient (CES) is as follows:

$$CES_1 = \frac{1.5A + B + 0.5C}{0.2D + 0.8E}$$

where A is areas with ecological stability (ES) degree 5 (forest, water areas); B is areas with ES degree 4 (bank overgrown, greenways); C is areas with ecological stability degree 3 (meadow, pastureland); D is areas with ecological stability degree 2 (arable land); and E represents areas with ecological stability degree 1 (built-up areas).

Attributes of the coefficient are interpreted thus:

$CES_1 < 0.1$	Degraded landscape
$CES_1 < 1$	Disrupted landscape
$CES_1 = 1$	Balanced landscape
CES_1 1–10	Landscape with dominating natural elements
$CES_1 > 10$	Natural or almost natural landscape

7.5.2.2 Coefficient of Ecological Stability 2

According to Rehackova and Pauditsova (2007) the CES_2 calculation considers the total area of particular landscape structure elements and the degree of their

7 Effects of Sustainable Energy Facilities on Landscape: A Case Study... 123

ecological stability (the ecological stability degree varies between 0 and 5). These degrees are assigning in accordance to Low (1984). In comparison with Low's work, there are amplified other landscape structure elements with their ecological stability degree. For the assessment of ecological stability coefficient is proposed the use of following formula, which considers acreage deal of landscape structure elements and the degree of their ecological stability in the assessment area:

$$CES_2 = \sum_{i=1}^{n} \frac{P_i \times S_i}{p}$$

where P_i is the area of secondary landscape structure (SLS) elements; S_i is the ecological stability degree of land-use elements; p is the total area; and n is the number of SLS elements in territory.

Attributes of the coefficient CES_2 are interpreted thus:

1.00–1.49	Landscape with very low ecological stability
1.50–2.49	Landscape with low ecological stability
2.00–3.49	Landscape with medium ecological stability
3.00–4.49	Landscape with high ecological stability
4.00–5.00	Landscape with very high ecological stability

7.5.3 Landscape Ecological Stability in Horka pri Poprade, Svabovce, and Straze pod Tatrami Cadastral Areas

In the Slovak Republic is EIA documentation executed on the basis of landscape protection and landscape ecological stability preservation. Landscape ecological stability coefficients in cadastral areas were evaluated according two mentioned methodologies. At first, they were evaluated particularly and thereafter together as one unit. Particular results are as follows:

Horka pri Poprade:	$CES_1 = 0.76$	$CES_2 = 2.32$
Svabovce:	$CES_1 = 0.51$	$CES_2 = 1.68$
Straze pod Tatrami:	$CES_1 = 0.08$	$CES_2 = 1.17$

When evaluating landscape ecological stability, the choric aspect should be considered also; this means horizontal relationships between particular cadastre areas, because landscape evaluated only from the topical aspect represents landscape as a discontinuous system. Thereafter, particular cadastral areas with different degrees of ecological stability are evaluated as separate segments regardless of their horizontal relationships with adjacent cadastral areas. Especially, horizontal relationships affect in a high degree the value of landscape ecological stability.

Table 7.3 Representation of individual land structure elements and values of ecological stability coefficients in Horka pri Poprade, Svabovce, and Straze pod Tatrami Cadastral Areas

Component of land structure	Area (ha)	Area (%)	CES_1	CES_2
Built-up areas	87.21	2.85	E	0
Permanent grassland	478.52	15.63	D	2
Gardens areas	112.01	3.66	D	2
Arable land (large-area)	1,664.58	54.37	E	1
Industrial area	11.98	0.39	E	0
Nonforest vegetation (area)	74.71	2.44	B	2
Nonforest vegetation (line)	7.6	0.25	B	2
Forests	512.69	16.75	B	3
Roads	69.09	2.26	E	0
Railway	3.54	0.12	E	0
Nonforest vegetation by the river	21.2	0.69	B	3
Agricultural dung-pit	0.33	0.01	E	0
Former sandpit	1.6	0.03	E	0
Streams	11.44	0.37	A	4
Playing field	2.7	0.07	E	0
Romany settlement	2.1	0.07	E	0
Cemetery	1.38	0.04	E	1
Total	3,061.33	100.00	0.4	1.52

CES_1, degree of ecological stability; CES_2, degree of ecological stability

So, finally, the cadastre areas were evaluated together as one unit (Table 7.3): the first coefficient of ecological stability, which regards the total area of secondary landscape structure elements and also their ecological stability degree, reaches a value of 0.4, which means that this cadastral area is disrupted landscape type. When specific structure is deliberate, the coefficient of ecological stability reaches 1.52, which means that this cadastral area is classified as landscape with low ecological stability.

According to the degree of human extended intensity to the landscape and the ability of landscape to regenerate to the previous landscape condition (Ruzicka et al. 1978), this cadastral area represents a disrupted landscape, which rises from antirational landscape use and impairing of natural resources; hence, human economic activities are negatively affecting its natural conditions. Disturbance of biological equilibrium is evident, so there are impending changes in secondary landscape structure, but its regeneration may still be possible by natural-biological or technical means.

7.6 Cumulative Effects of Project Assessment

Many environmental problems faced today result from the accumulation of multiple small, often indirect effects rather than a few large obvious ones. Examples include loss of tranquility, heathland, and wetland, changes in landscapes, depletion of fish stocks, and global warming. These effects are very hard to manage on a project-by-project basis through EIA. The EIA comes too late, is too detailed, and is too focused on the short term. As such, despite the difficulties of doing so, SEA should make a special effort to consider cumulative, indirect, and long-term impacts (Therivel 2004).

For accumulation of effects from several projects, cumulative effects are used. According to Canter (1996), they are impacts on the environment that result from the incremental impact of the action when added to other past, present, and reasonably foreseeable future actions regardless of what agency or person undertakes such other actions. Cumulative impacts can result from individually minor, but collectively significant, actions taking place over a period of time.

Cumulative effects may result from combined impacts from many, varied sources or repeated impacts from a single source. Cumulative impacts may be (Treweek 1999) additive (incremental), aggregated (synergistic), and associated (connected).

To evaluate cumulative impacts it is necessary to create a specific conceptual framework on which basis impacts will be identified and thereafter evaluated. Clark (1994) focused attention on the potential of cumulative effects assessment to serve as a predictive tool for gauging the sustainability of proposed development projects. According to this work, we propose the following steps for evaluating these effects:

1. Establish the geographic scope for the analysis and the time frame for the analysis
2. Describe the affected environment(this part includes evaluation of landscape character and ecological stability)
3. Identify other actions affecting the environment
4. Determine the environmental consequences of cumulative effect
5. Determine the magnitude and significance of cumulative effects

7.7 Conclusions

Our society and industry relies on large amounts of energy and the world is becoming increasingly dependent on fossil fuels (oil, gas, coal, etc.). The industrialized nations of Western Europe and North America, China, and India depend almost entirely on these fuels, and the developing nations are also increasing their use. It is understood that there is a direct link between the way we produce energy and damage caused by pollution. Finding cleaner and alternative ways of producing energy are now looked upon as being very important for the future of our planet.

On the basis of dozens of evaluated sustainable energy facilities (activities) assessed in the past 10 years as the specific aspects were determined: choice of locality—"brown fields" versus "green areas," increased emphasis on cumulative impacts evaluation, clear suggestions are strengthening the proposed activities including their visualization and visibility evaluation.

To focus attention on the possibility of cumulative impacts is a way to call attention to disrupting the landscape where humans live. It is important to realize all the projects in harmony with landscape quality and respect its limits. To achieve this objective, it is necessary to accumulate knowledge about the environment from the beginning of the EIA process. A very appropriate way is to present its stability through the coefficients as we have shown in the case study in the Slovak Republic.

Acknowledgments We express our gratitude to the Slovak VEGA grant system for supporting our project 1/0544/11 as a basis for our research.

References

Abaza H, Bisset R, Sadler B (2004) Environmental impact assessment and strategic environmental assessment: towards an integrated approach. UNEP, Geneva

Act of National Council of the Slovak Republic No. 127/1994 Coll. on environmental impact assessment

Act of National Council of the Slovak Republic No. 24/2006 Coll. on environmental impact assessment in later wording

Act of National Council of the Slovak Republic No. 391/2000 Coll. amending and supplementing

AES (2011) Alternate Energy Systems, Inc., Georgia (http://altenergy.com/Technology/Tech.htm)

Canter LW (1996) Environmental impact assessment. McGraw-Hill, Singapore

Cepelak J (1980) Zoological regions 1:1 000 000. In: Atlas of the Slovak Socialist Republic. Slovak Academy of Science, Bratislava (in Slovak)

Clark R (1994) Cumulative effects analysis: Final Draft. President's Council on environmental quality, Canada

COM (2006) Green paper: a European strategy for sustainable, competitive and secure energy. Commission of the European Communities, Brussel

Drdos J (2005) Environmental impact assessment. In: Drdos J, Michaeli E, Hrnciarova T (eds) Geoecology and environmentalistics: environmental planning in regional development. FHPV PU, Presov (in Slovak)

EC (2011) Energy. http://www.ec.europa.eu/energy/renewables/index_en.htm

Fasko P, Stastny P (2002) Mean annual precipitations totals. In: Bratislava ME SR (ed) Landscape atlas of the SR. Slovak Environmental Agency, Banska Bystrica (in Slovak)

Forman RTT, Godron M (1993) Landscape ecology. Academia, Prague (in Czech)

Futak J (1980) Phytogeographical classification; MapVII/14. In: Atlas of the Slovak Socialist Republic. Slovak Academy of Science, Bratislava (in Slovak)

Hensel K, Krno I (2002) Zoogeographic classification: limnic biocycle. In: ME SR (eds) Slovak environmental agency: landscape atlas of the SR. Bratislava, Banska Bystrica (in Slovak)

Iacomelli A (ed) (2005) Renewable energies for Central Asia countries: economic, environmental and social impacts. Springer, Dordrecht

Izakovicova Z, Moyzesova M, Bezak P, Dobrovodska M, Grotkovska L, Hrnciarova T, Kenderessy P, Krnacova Z, Majercak J, Miklosovicova Z, Moyses M, Pavlickova K, Petrovic F, Spulerova J, Stefunkova D, Valkovcova Z (2008) Agricultural landscape evaluation in transitive economics. Slovak Academy of Sciences, Institute of Landscape Ecology, Bratislava (in Slovak)

Kozova M, Pauditsova E (2010) Development, current state and trends of further improvement of landscape planning (comparative analysis of different approaches). In: Barancokova M,

Krajci J, Kollar J (eds) Landscape ecology: methods, applications and interdisciplinary approach. Institute of Landscape Ecology, Slovak Academy of Sciences, Bratislava

Low J (1984) Principles for determination and design of territorial systems of ecological stability. AgroprojektBrno (in Czech)

Mazur E, Luknis M (1980) Geomorphological units 1:1 000 000. In: Atlas of the Slovak Socialist Republic. Slovak Academy of Science, Bratislava (in Slovak)

Mician L, Zatkalik F (1986) Landscape theory and environment maintenance. Comenius University in Bratislava, Faculty of Natural Sciences, Bratislava (in Slovak)

Michal I (1982) Principles of territory landscape evaluation. Architecture Urbanism 16(2):65–87 (in Czech)

Michal I (ed) (1992) Restoration of forests ecological stability. Academia, Prague (in Czech)

Michal I (1994) Ecological stability. Veronica, Brno (in Czech)

Miklos L (1992) Ecologization of spatial organization, utilization and protection of landscape. Slovak Technical Library, Bratislava (in Slovak)

Ministry of Environment of the Slovak Republic (2006) Energy policy of the Slovak Republic. Ministry of the Economy of the Slovak Republic, Bratislava (in Slovak)

Ministry of Environment of the Slovak Republic (2007) Strategy of higher usage of renewable sources of energy. Ministry of the Economy of the Slovak Republic, Bratislava (in Slovak)

Ministry of Environment of the Slovak Republic (2008) Energy security strategy. Bratislava: Ministry of the Economy of the Slovak Republic, Bratislava (in Slovak)

Rehackova T, Pauditsova E (2007) Methodology of landscape ecological stability coefficient establishment. Acta Envir Univ Com (Bratislava) 15(1):26–38 (in Slovak)

Ruzicka M, Ruzickova H, Zigrai F (1978) Landscape components, elements and structure in biological planning. Quaestiones Geobiologicae (Problems of biology of landscape) (Bratislava) 23:69–77 (in Slovak)

Therivel R (2004) Strategic environmental assessment in action. Earthscan, London

Treweek J (1999) Ecological impact assessment. Blackwell, Oxford

van Etteger R, Stremke S (eds) (2007) ReEnergize South Limburg: designing sustainable energy landscapes. Regional Atelier of Wageningen University and Research, The Netherlands

Voloscuk I (2001) Theoretical and practical problems of forest ecosystems ecological stability. Scientific Studies 1/2001/A, Technical University, Zvolen. In: Drdos J (2004) Geoecology and Environmentalistics; Landscape ecology/geoecology, its environmental commitment and tasks. University of Presov, Presov(in Slovak)

Wood C (1995) Environmental impact assessment: a comparative review. Longman Scientific & Technical, Michigan www.enviroportal.sk

Chapter 8
Low Carbon Society Through *Pekarangan*, Traditional Agroforestry Practices in Java, Indonesia

Hadi Susilo Arifin, Regan Leonardus Kaswanto, and Nobukazu Nakagoshi

Abstract *Pekarangan*, as a traditional homestead garden, an optimal and sustainable land-use type of agroforestry system in the tropical region of Indonesia, has been researched since 1996. As greenery open space, which is located in the surroundings of a house or residential building, it has spread from rural to urban areas, from the upper to the downstream reaches of watersheds. The area of *pekarangan* varies with the owners and depends on the socioeconomic level, profession, and their distance from the city. However, sustainable and abundant bioresources are expected to be available. Through local wisdom and local knowledge of the community, *pekarangan* have been practiced as agro-forestry, agro-silvo-pastura, and agro-silvo-fishery systems. Agricultural biodiversity and sustainable material circulation are maintained in *pekarangan*. *Pekarangan* is potential land for ecosystem services, such as carbon sequestration, water resource management, agrobiodiversity conservation, and landscape beautification. Multistory levels of vegetation structures and species richness of *pekarangan* not only can be proposed to mitigate global warming and global climate change impacts, but also can be promoted as supporting agricultural land for food security at the household level. The number of species in a *pekarangan* varies according to local physical circumstances, ecological characteristics of the plants, kinds of animal species, and socioeconomic and cultural factors. Results showed that the size of the open space area of *pekarangan* has decreased, and the number of species has also become less, during the 10-year period of research. If *pekarangan* systems and other smallholder tree-based systems were to expand in currently degraded and

---

H.S. Arifin (✉) • R.L. Kaswanto
Landscape Management Division, Landscape Architecture Department,
Faculty of Agriculture, Bogor Agricultural University (IPB), Kampus IPB,
Dramaga-Bogor 16680, Indonesia
e-mail: hsarifin@ipb.ac.id; dedhsa@yahoo.com

N. Nakagoshi
Graduate School for International Development and Cooperation, Hiroshima University,
Higashi-Hiroshima 739-8529, Japan

N. Nakagoshi and J.A. Mabuhay (eds.), *Designing Low Carbon Societies in Landscapes*, Ecological Research Monographs, DOI 10.1007/978-4-431-54819-5_8, © Springer Japan 2014

underutilized lands, such as Imperata grasslands, the C sequestration potential would be about 80 Mg C ha^{-1}. On the other hand, *pekarangan* as an agroforestry system contributes significantly to a region's carbon budget while simultaneously enhancing the livelihoods of the rural community.

Keywords Agrobiodiversity • Agro-silvo-fishery • Agro-silvo-pastura • Species richness • Watershed

8.1 Introduction

The global crisis has been affecting Indonesia in all aspects, such as a social crisis, political crisis, and economic crisis, as well as the environmental and ecological crisis. Those impacts have already touched most of Indonesian communities from the rural to the urban areas. To increase the ecological-social-cultural-economic welfare of the rural community in Indonesia, urgent action is needed to develop environmental conservation through traditional or complex agroforestry practices; thus, community welfare can be gained by eco-village implementation, which is balanced among the ecological, socioeconomic, and spiritual values of the community (Arifin and Arifin 2010). In the micro-level of landscape, *pekarangan*, a piece of land surrounding the house, is potential land for ex situ agrobiodiversity conservation through agroforestry, agro-silvo-fishery, and agro-silvo-pastura system practices.

As greenery open space, *pekarangan* has permanent vegetation. Therefore, ecologically the *pekarangan* is supposed to sequestrate carbon dioxide (CO_2) from the air to be stocked in tree leaves, branches, trunks, roots, and soils. *Pekarangan* has a role not only in carbon (C) sequestration, but also in water resources management, agrobiodiversity conservation, and landscape beautification as part of the scheme of the payment for environmental services (PES) (Kaswanto and Nakagoshi 2012). The PES scheme is being proposed and tested in different contexts as a way to involve the local people in conservation practice (Nurhariyanto et al. 2010). Furthermore, the low carbon society (LCS) can be achieved through *pekarangan*, the traditional agroforestry practices in Java, Indonesia

The *Pekarangan* area was studied mostly in Java island because of the 5,132,000 ha of *pekarangan* in Indonesia, 1,736,000 ha are on Java (Prosterman and Mitchell 2002) (citing 2000 Statistical Yearbook of Indonesia, Table 5.1.1.). As in the distribution of croplands, the distribution of *pekarangan* is very unequal. Thus, for Indonesia as a whole, 40.28 % of households have less than 100 m^2 of *pekarangan*, 25.24 % have 100–200 m^2, 11.72 % have 200–300 m^2, and 22.76 % have 300 m^2 or more (Arifin 1998). Table 8.1 shows the distribution for the four provinces of Java. *Pekarangans* areas spread from rural, to suburban, to urban areas. The LCS could be achieved through *pekarangan*; so long as housing development is constructed by the horizontal building system, it is assumed the more built-up housing, the larger the numbers and area of *pekarangan*.

8 Low Carbon Society Through *Pekarangan* 131

Table 8.1 Size distribution of *pekarangan* land in agricultural provinces of Java (percentages of households that have *pekarangan* in the size groups shown)

Provinces in Java	<100 m^2 (%)	100–200 m^2 (%)	200–300 m^2 (%)	>300 m^2 (%)
West Java-Banten	52.29	25.00	8.77	8.95
Central Java	27.50	27.57	13.20	31.73
East Java	34.52	25.83	13.33	26.31
D.I. Yogyakarta	33.51	17.48	14.61	34.40

Source: Arifin (1998) [Appendix Table 2 (citing 1995 Housing and Settlement Statistics, Indonesian Statistics Center Bureau 1996)]

8.2 Objectives of Research

The multiyear research on *pekarangan* has the objective to reconstruct and to revitalize traditional Indonesian agroforestry to achieve sustainable bioresources management systems on Java. Furthermore, this research calculates and assesses C sequestration, water resources management, agrobiodiversity conservation, and landscape beautification from *pekarangan*.

8.3 Methods

This study has been ongoing since 1996 in some watersheds of West Java Province for a period of 10 years. This study was divided into four stages with the activity targets in each stage as follows:

1. *Stage I (1996–2000)*: Survey on traditional *pekarangan* bioresources in rural areas.
2. *Stage II (2000–2003)*: Analysis interrelationships among components in *pekarangan* bioresource management system and evaluation.
3. *Stage III (2003–2005)*: Reconstruction of a *pekarangan* bioresources management system.
4. *Stage IV (2005–2007)*: Adaptation of the new biomanagement system and proposal of the reconstruction of the *pekarangan* bioresources management system.

Simultaneously, these *pekarangan* studies have been extended on Java Island under joint research with the Rural Development Institute (2006–2007), *Hibah Penelitian Tim Pascasarjana* Directorate General of Higher Education (DGHE) of Indonesia (2006–2008), *Hibah Kompetensi* DGHE of Indonesia (2008–2010), and joint research with the Global Environmental Leaders (GEL)s Education Program for Designing a Low Carbon Society (LCS) of Hiroshima University, Japan (2009–2013).

In this chapter, those results were demonstrated to show the conditions and the significant roles of *pekarangan* in Java, Indonesia. Several settlements of hamlets or villages within administrative boundaries were chosen as the study sites of a microscale research unit. Selection of the study sites in each small-scale catchment area was based on several considerations, as follows:

1. Elevation gradient: 200–500 m, 500–1,000 m, and >1,000 m above sea level (a.s.l.)
2. These study sites are located in the linear slope.

8.4 Results and Discussion

8.4.1 Traditional Agroforestry of Pekarangan

Pekarangan is the traditional and privately owned home garden, and an integrated system with an intimate relationship among human, plants, and animals. It is well known that this garden has multiple functions, such as conservation of genetic resources, soil, and water, crop production, and sociocultural relationships in the rural area. It is thought that *pekarangan* is an optimal and sustainable land use with high productivity in tropical regions (Arifin 1998). Arifin and Arifin (2010) stated that *pekarangan* is a kind of traditional agroforestry practice that is found in rural and agricultural landscapes beside *kebun campuran* (mixed gardens) and *kebun talun* (forest gardens). The design and structure depends on local and ecological knowledge of the surrounding communities. The survey showed that the western part of Indonesia practices agro-silvo-fishery, as there are many water resources, and in contrast, the eastern part of Indonesia practices agro-silvo-pastura because of lack of water (Arifin et al. 2008a).

Pekarangan fulfills an ecological function in that its multilayered vegetation structure resembles that of natural forests and offers habitats and niches for a diverse community of wild plants and animals (Albuquerque et al. 2005; Karyono 1990). This study has confirmed the performance of *pekarangan* at the smallest scale. Those provisions are the contribution of *pekarangan* for nutrition intake, income, wealth assets, family status, access to credit, control of production, and product marketing.

Some research, particularly *pekarangan* biodiversity based on urbanized vegetation structures, was conducted in the landscape ecological unit of Ciliwung and Cisokan Watershed, which covers the Bogor-Puncak-Cianjur (BOPUNJUR) region (Arifin 2004; Arifin et al. 2001). Species richness was elucidated for *pekarangan* starting from the upper stream reaches to the downstream portion of the watersheds. Landscape structure in the traditional agroforestry of the *pekarangan* system has horizontal and vertical diversity (Arifin et al. 1998). Based on plant function, horizontal diversity has been classified into eight groups: ornamental plants, fruit plants, vegetable crops, starchy crops, medicinal plants, spices crops, industrial plants, and others (Arifin 1998). It is found that the size of *pekarangan* and percentage of plant canopy coverage are larger from the upper stream reaches to downstream. However, the highest averages of individual numbers per *pekarangan*

8 Low Carbon Society Through *Pekarangan* 133

Table 8.2 Number of species and individual numbers per *pekarangan* in Cianjur Watershed

Research area	Average *Pekarangan* size (m^2)	Average plant canopy area^a (m^2)	Average number of species	Total number of species	Average number of individuals	Total number of individuals	Shannon–Wiener diversity index
Upper stream	188.2	167.0	26.7	90	280.0	1,680	1.17
Middle stream	218.7	629.0	40.4	166	491.5	4,915	1.31
Downstream	562.0	1,733.2	44.0	116	346.2	1,731	1.24

^{a}Only trees and shrubs with dbh > 2.5 cm were measured *Pekarangan* size, size of the open space area
Source: Arifin et al. (2001); Arifin (2004)

and species diversity index were found in the middle streams of Cianjur watershed (Table 8.2). This area is a transition zone between the lowland and mountainous areas (Arifin et al. 2001).

Based on plant function, the lower parts of watershed have a smaller ornamental plant ratio (Table 8.3). Fruit plants were found in the downstream predominantly (30.4 %), followed by others (17.1 %), such as fuel wood species, wood for handicrafts material, and wood for building materials (Arifin 2004).

In the BOPUNJUR region, changes of *pekarangan* plant diversity were studied along an urban–rural continuum as well as along an elevation gradient. The vegetation structure and composition of 115 *pekarangans* in six villages were investigated to determine the urbanization effects (Arifin et al. 1998). The six villages differed in urbanization level: one is a rural village, three are characterized as intermediately urbanized, and two are urban villages. In each *pekarangan*, both ornamental and crop plants were inventoried. *Pekarangan* sizes ranged from 30 m^2 to 4,000 m^2; the average size was 270 m^2. In total, 440 plant species were grown in the 115 *pekarangans*; about half the species were ornamentals. The number of species in a *pekarangan* varies according to local physical circumstances, ecological characteristics of plants, kinds of animal species, and socioeconomic and cultural factors. Plant species numbers varied largely among the 115 *pekarangans* studied. Average species number per *pekarangan* were not markedly different between the rural, the intermediate, and the urban *pekarangans* (Arifin 1998 and Arifin et al. 1998). However, the average number of nonornamental plant species per *pekarangan* was markedly higher in rural than in urban *pekarangan*. The proportion of ornamental plants from total species increased with a higher level of urbanization (40 % in rural to 70 % in urban). *Pekarangan* size decreased continuously from rural to urban areas. In many densely populated tropical regions, *pekarangans* appear to be the last forest-like islands surrounded by increasingly extended, uniform staple crop fields. In these areas, *pekarangans* with their multi-layered vegetation structure serve as an important habitat for wild flora and fauna. *Pekarangans* fulfill not only important ecological but also many social and cultural functions (Kehlenbeck et al. 2007).

Table 8.3 Ratio of species number by *pekarangan* plant function in Cianjur Watershed

Plant function	Species number (%)		
	Upper stream	Middle stream	Downstream
Ornamental plant	47.5	48.9	24.4
Nonornamental plant			
Fruit plant	16.9	20.8	30.4
Vegetable crop	11.9	12.2	8.3
Spice crop	3.1	4.5	4.6
Medicinal plant	3.1	1.7	4.1
Starchy crops	8.8	5.5	3.7
Industrial plant	3.1	1.5	7.4
Others	5.6	5.1	17.1
Total	100.0	100.0	100.0

Source: Arifin (2004)

Furthermore, a homestead plot survey on Java (Arifin et al. 2008b) was conducted in 144 *pekarangan* samples from three provinces: West, Central, and East Java provinces. The *pekarangan* samples covered two watershed units per province. *Pekarangan* size was divided into two groups: smaller than 120 m^2 (small *pekarangan*) and between 120 and 400 m^2 (moderate-size *pekarangan*). The total species number is 196 (Table 8.4), consisting of ornamental plants (103 species), fruit plants (29 species), vegetable crops (21 species), medicinal plants (13 species), spice crops (9 species), industrial plants (9 species), other plants (7 species), and starchy crops (5 species).

8.4.2 The Dynamics of Pekarangan

Vegetation structure dynamics in *pekarangan* was analyzed between years 1996 and 2006 (Mayanti et al. 2007). The sample sites were located in BOPUNJUR, West Java Province. The samples were taken at the selected sites with different levels of urbanization, that is, the least urbanized sites, less urbanized sites, and urbanized sites. In 2006, there are 362 plants species in *pekarangans*. The result showed that between 1996 and 2006, the size of open space areas of *pekarangan* decreased, and the number of spesies also became less. However, the number of individual was increased because some plants, especially shrubs and ground covers can reproduce by themselves vegetatively. Some factors that influenced the changes of vegetation structure at *pekarangan* are (1) small open space area, (2) land fragmentation, (3) different owner, (4) changes in function of some part of the *pekarangan*, (5) plant popularity trend, and (6) economic condition changes.

Regarding vegetation stratification, it was observed that the first stratum of vegetation such as grasses and herbs was predominant in each level of urbanization, both in 1996 and 2006. In the intervening 10 years, the availability of tree strata was

8 Low Carbon Society Through *Pekarangan* 135

Table 8.4 Number of species by plant function in 144 *pekarangan* samples on Java Island

Category	No.	Latin name	Family name	English name
I Starchy crops				
	1	*Ipomoea batatas*	Convolvulaceae	Sweet potato
	2	*Manihot esculenta*	Euphorbiaceae	Cassava
	3	*Oryza sativa*	Poaceae	Asian rice
	4	*Solanum tuberosum*	Solanales	Potato
	5	*Zea mays*	Poaceae	Maize
II Fruit plants				
	1	*Ananas comosus*	Bromeliaceae	Pineapple
	2	*Annona muricata*	Annonaceae	Soursop
	3	*Annona squamosa*	Annonaceae	Sugar-apple
	4	*Artocarpus altilis*	Moraceae	Breadfruit
	5	*Artocarpus heterophyllus*	Moraceae	Jack fruit
	6	*Averrhoa carambola*	Oxalidaceae	Starfruit
	7	*Carica papaya*	Caricaceae	Papaya
	8	*Citrullus lanatus*	Cucurbitaceae	Watermelon
	9	*Citrus sinensis*	Rutaceae	Orange
	10	*Cucumis melo*	Cucurbitaceae	Melon
	11	*Dimocarpus longan*	Sapindaceae	Longan
	12	*Durio zibethinus*	Malvaceae	Durian
	13	*Fragaria xananassa*	Rosaceae	Strawberry
	14	*Garcinia mangostana*	Clusiaceae	Mangosteen
	15	*Lansium domesticum*	Meliaceae	Dookoo
	16	*Malus domestica*	Rosaceae	Apple
	17	*Mangifera indica*	Anacardiacea	Mango
	18	*Manilkara zapota*	Sapotaceae	Sapodilla
	19	*Musa paradisiaca*	Musaceae	Banana
	20	*Nephellium lappaceum*	Sapindaceae	Rambutan
	21	*Passiflora edulis*	Passifloracea	Passionfruit
	22	*Persea americana*	Lauraceae	Avocado
	23	*Phoenix dactylifera*	Arecaceae	Date palm
	24	*Punica granatum*	Lythraceae	Pomegranate
	25	*Salacca zalacca*	Arecaceae	Snake fruit
	26	*Sandoricum koetjape*	Meliaceae	Santol or Sandorica
	27	*Spondias dulcis*	Anacardiaceae	Golden apple
	28	*Syzygium samarangense*	Myrtaceae	Wax apples
	29	*Vitis vinifera*	Vitaceae	Grape
III Vegetable				
	1	*Allium fistulosum*	Alliaceae	Spring onion
	2	*Amaranthus* spp.	Amaranthaceae	Amaranth
	3	*Apium graveolens*	Apiaceae	Celery
	4	*Archidendron pauciflorum*	Fabaceae	Jengkol
	5	*Brassica oleracea*	Brassicaceae	Cabbage
	6	*Brassica rapa*	Brassicaceae	Chinese cabbage
	7	*Citrus aurantifolia*	Rutaceae	Key lime
	8	*Cucumis sativus*	Cucurbitaceae	Cucumber
	9	*Daucus carota*	Apiaceae	Carrot

(continued)

Table 8.4 (continued)

Category	No.	Latin name	Family name	English name
	10	*Gnetum gnemon*	Gnetaceae	Melinjo
	11	*Ipomoea aquatica*	Convolvulaceae	Water spinach
	12	*Luffa acutangula*	Cucurbitaceae	Silk squash
	13	*Momordica charantia*	Cucurbitaceae	Bitter melon
	14	*Parkia speciosa*	Fabaceae	Stink bean
	15	*Phaseolus lunatus*	Fabaceae	Lima bean
	16	*Sauropus androgynus*	Phyllanthaceae	Sweet leaf
	17	*Sechium edule*	Cucurbitaceae	Chayote
	18	*Solanum lycopersicum*	Solanaceae	Tomato
	19	*Solanum melongena*	Solanaceae	Eggplant
	20	*Solanum nigrum*	Solanaceae	Black nightshade
	21	*Vigna unguiculata*	Fabaceae	Yardlong bean
IV Spice crops				
	1	*Alpinia galanga*	Zingiberaceae	Blue ginger
	2	*Capsicum annuum*	Solanaceae	Chili
	3	*Curcuma longa*	Zingiberaceae	Turmeric
	4	*Cymbopogon citratus*	Poaceae	Lemon grass
	5	*Etlingera elatior*	Zingiberaceae	Torch ginger
	6	*Myristica fragrans*	Myristicaceae	Nutmeg
	7	*Pandanus amaryllifolius*	Pandanaceae	Pandan
	8	*Syzygium polyanthum*	Myrtaceae	Bay leaf
	9	*Zingiber zerumbet*	Zingiberaceae	Shampoo ginger
V Medicinal plants				
	1	*Andrographis paniculata*	Acanthaceae	Creat
	2	*Blumea balsamifera*	Asteraceae	Sambong
	3	*Chloranthus erectus*	Chloranthaceae	Cryphaea
	4	*Hydrocotyle sibthorpioides*	Apiaceae	Lawn pennywort
	5	*Melastoma polyanthum*	Melastomataceae	Grass jelly
	6	*Morinda citrifolia*	Rubiaceae	Great morinda
	7	*Orthosiphon aristatus*	Lamiaceae	Cat's whiskers
	8	*Phaleria papuana*	Thymelaeaceae	God's crown
	9	*Piper betle*	Piperaceae	Betel
	10	*Pluchea indica*	Asteraceae	Marsh fleabane
	11	*Sonchus arvensis*	Asteraceae	Swine thistle
	12	*Tinospora crispa*	Menispermaceae	Guduchi
	13	*Zingiber officinale*	Zingiberaceae	Ginger
VI Industrial plants				
	1	*Camellia sinensis*	Theaceae	Tea
	2	*Ceiba pentandra*	Malvaceae	Kapok
	3	*Cocos nucifera*	Arecaceae	Coconut
	4	*Coffea arabica*	Rubiaceae	Coffee
	5	*Hevea brasiliensis*	Euphorbiaceae	Rubber tree
	6	*Paraserianthes falcataria*	Fabaceae	Albizia
	7	*Saccharum officinarum*	Poaceae	Sugar cane
	8	*Syzygium aromaticum*	Myrtaceae	Clove
	9	*Theobroma cacao*	Malvaceae	Cacao

(continued)

8 Low Carbon Society Through *Pekarangan*

Table 8.4 (continued)

Category	No.	Latin name	Family name	English name
VII Ornamental plants				
	1	*Acalypha macrophylla*	Euphorbiaceae	Copperleaves
	2	*Adenium obesum*	Apocynaceae	Desert-rose
	3	*Adiatum* spp.	Pteridaceae	Maidenhair ferns
	4	*Agave* spp.	Agavaceae	Agave
	5	*Aglaonema* spp.	Araceae	Aglaonema
	6	*Aloe vera*	Asphodelaceae	Aloe vera
	7	*Alternanthera amoena*	Amaranthaceae	Alternanthera
	8	*Anthurium scherzeranum*	Araceae	Flamingo plant
	9	*Araucaria heterophylla*	Araucariaceae	Norfolk island pine
	10	*Axonopus compressus*	Poaceae	Lawn grass
	11	*Bauhinia purpurea*	Fabaceae	Hongkong orchid tree
	12	*Begonia* spp.	Begoniaceae	Begonia
	13	*Bougenvillea* spp.	Nyctaginaceae	Paper flower
	14	*Caesalpinia pulcherrima*	Fabaceae	Peacock flower
	15	*Caladium* spp.	Araceae	Caladium
	16	*Calathea makoyana*	Marantaceae	Peacock plant
	17	*Cananga odorata*	Annonaceae	Cananga tree
	18	*Canna edulis*	Cannaceae	Canna
	19	*Carex morrowii*	Cyperacea	Japanese sedge
	20	*Catharanthus roseus*	Apocynaceae	Vinca
	21	*Chlorophytum comosum*	Agavaceae	Spider plant
	22	*Chrysalidocarpus lutescens*	Arecaceae	Golden cane palm
	23	*Chrysanthemum* spp.	Asteraceae	Chrysanths
	24	*Clerodendron paniculatum*	Clerodendron	Pagoda flower
	25	*Codiaeum variegatum*	Euphorbiaceae	Garden croton
	26	*Coleus blumei*	Lamiaceae	Coleus
	27	*Cordyline fruticosa*	Asparagaceae	Red palm lily
	28	*Cordyline terminalis*	Asparagaceae	Green palm lily
	29	*Crinum* spp.	Amaryllidaceae	Crinum
	30	*Cuphea hyssopifolia*	Lythraceae	Mexican heather
	31	*Cupressus papuana*	Cupressaceae	Italian cypress
	32	*Cycas rumphii*	Cycadaceae	Queen sago
	33	*Datura mollis*	Solanaceae	Trumpet flower
	34	*Delonix regia*	Fabaceae	Flamboyant
	35	*Dieffenbachia seguine*	Araceae	Dumb cane
	37	*Dracaena angustifolia*	Dracaenaceae	Dracaena
	38	*Dracaena sanderiana*	Ruscaceae	Ribbon dracaena
	39	*Duranta* spp.	Verbenaceae	Golden dewdrop
	40	*Epiphyllum oxypetalum*	Cactaceae	Night queen
	41	*Epipremnum aureum*	Araceae	Silver vine
	42	*Episcea cupreata*	Gesneriaceae	Flame violet
	43	*Eugenia uniflora*	Myrtaceae	Surinam cherry
	44	*Euphorbia milii*	Euphorbiaceae	Christ plant

(continued)

Table 8.4 (continued)

Category	No.	Latin name	Family name	English name
	45	*Euphorbia pulcherrima*	Euphorbiaceae	Poinsettia
	46	*Ficus benjamina*	Moraceae	Ficus tree
	47	*Ficus elastica*	Moraceae	Rubber plant
	48	*Ficus lyrata*	Moraceae	Fiddle-leaf fig
	49	*Gardenia augusta*	Rubiaceae	Gardenia
	50	*Gerbera* spp.	Asteraceae	Daisy
	51	*Gomphrena* spp.	Amaranthaceae	Globe amaranth
	52	*Helianthus annuus*	Asteraceae	Sun flower
	53	*Heliconia* spp.	Heliconiaceae	Heliconia
	54	*Hemigraphis alternata*	Acanthaceae	Metal leaf
	55	*Hibiscus rosa-sinensis*	Malvaceae	Shoe flower
	56	*Hydrangea* spp.	Hydrangeaceae	Hortensia
	57	*Hyophorbe lagenicaulis*	Arecaceae	Bottle palm
	58	*Impatiens balsamina*	Balsaminaceae	Garden balsam
	59	*Impatiens walleriana*	Balsaminaceae	Balsam/busy lizzy
	60	*Ixora javanica*	Rubiaceae	Javanese ixora
	61	*Jasminum multiflorum*	Oleaceae	Indian jasmine
	62	*Jasminum sambac*	Oleaceae	Jasmine
	63	*Kalanchoe pinnata*	Crassulaceae	Miracle leaf
	64	*Lilium* spp.	Liliaceae	Lily
	65	*Livistona* spp.	Arecaceae	Fan palms
	66	*Maihuenia* spp.	Cactaceae	Cactus
	67	*Manihot esculenta* "variegata"	Euphorbiaceae	Variegated tapioca
	68	*Maranta leuconeura*	Marantaceae	Maranta
	69	*Michelia alba*	Magnoliaceae	White champaca
	70	*Michelia champaca*	Magnoliaceae	Champaca
	71	*Mirabilis jalapa*	Nyctaginaceae	Four o'clock flower
	72	*Neoregelia* spp.	Bromeliaceae	Bromelia
	73	*Nerium oleander*	Apocynaceae	Oleander
	74	*Nothopanax scutellarium*	Araliaceae	Saucer-leaf
	75	*Nymphaea* spp.	Nymphaeaceae	Water lily
	76	*Pachystachys lutea*	Acanthaceae	Golden shrimp plant
	77	*Pedilanthus tithymaloides*	Euphorbiaceae	Christmas candle
	78	*Phalaenopsis amabilis*	Orchidaceae	Moon orchid
	79	*Philodendron* spp.	Araceae	Philodendron
	80	*Pilea cadierei*	Urticaceae	Aluminium plant
	81	*Pinus merkusii*	Pinaceae	Pine tree
	82	*Platycerium bifurcatum*	Polypodiaceae	Elkhorn fern
	83	*Plumeria alba*	Apocynaceae	Caterpillar tree
	84	*Polianthes tuberosa*	Agavaceae	Tuberose
	85	*Portulaca* spp.	Portulacaceae	Moss roses
	86	*Rhapis excels*	Arecaceae	Bamboo palm
	87	*Rhoeo discolor*	Asteraceae	Oyster plant
	88	*Ricinus communis*	Euphorbiaceae	Castor oil plant
	89	*Rosa* spp.	Rosaceae	Rose

(continued)

8 Low Carbon Society Through *Pekarangan* 139

Table 8.4 (continued)

Category	No.	Latin name	Family name	English name
	90	*Sansevieria trifasciata*	Agavaceae	Snake plant
	91	*Saraca asoca*	Fabaceae	West Indian jasmine
	92	*Schefflera arboricola*	Araliaceae	Dwarf umbrella tree
	93	*Scindapsus* spp.	Araceae	Scindapsus
	94	*Spondias pinnata*	Anacardiaceae	Common hog plum
	95	*Stachytarpheta mutabilis*	Acanthaceae	Keji beling
	96	*Stenochlaena palustris*	Blechnaceae	Epiphytic fern
	97	*Syngonium podophyllum*	Araceae	Syngonium
	98	*Tagetes erecta*	Asteraceae	African marigold
	99	*Thuja occidentalis*	Cupressaceae	Graveyard cypress
	100	*Wedelia biflora*	Asteraceae	Beach sunflower
	101	*Yucca guatemalensis*	Agavaceae	Spineless yucca
	102	*Zepiranthes* spp.	Amaryllidaceae	Fairy lily or Rain lily
	103	*Zinnia* spp.	Asteraceae	Zinnia
VIII Others				
	1	*Albizia saman*	Fabaceae	Saman tree
	2	*Bambusa* spp.	Poaceae	Bamboo
	3	*Canarium ovatum*	Burseraceae	Cesnut
	4	*Maesopsis eminii*	Rhamnaceae	Umbrella tree
	5	*Pterocarpus indicus*	Fabaceae	Narra
	6	*Swietenia mahogany*	Meliaceae	Mahogany
	7	*Tectona grandis*	Lamiaceae	Teak wood

reduced in *pekarangan* (Fig. 8.1). Most of these factors were thought to be correllated with the impact of urbanization. This study also proposed some actions to utilize *pekarangan* effectively through a Participatory Rural Appraisal (PRA) approach, such as in plant selection, recycling systems, and revitalizing mixed agroforestry practices.

8.4.3 *Low Carbon in* Pekarangan

Pekarangan, a traditional biodiversity–low carbon system in Indonesia to establish green procurement, promote greening, and set green guidelines, species diversity, or biodiversity plays an important role in sustaining the ecosystem at present and in the future (Arifin and Nakagoshi 2011). *Pekarangan* is a common smallholder agroforestry system in Indonesia and throughout the tropics, from the rural to the urban areas (Arifin 1998). These species-rich, tree-based systems produce non-wood and wood products for both home use and market sale. High biodiversity is an intrinsic property of the home gardens (Kumar 2006), which presumably favors greater net primary productivity (NPP) and higher C sequestration potential than monospecific production systems. Projections by Roshetko et al. (2002) revealed that, depending

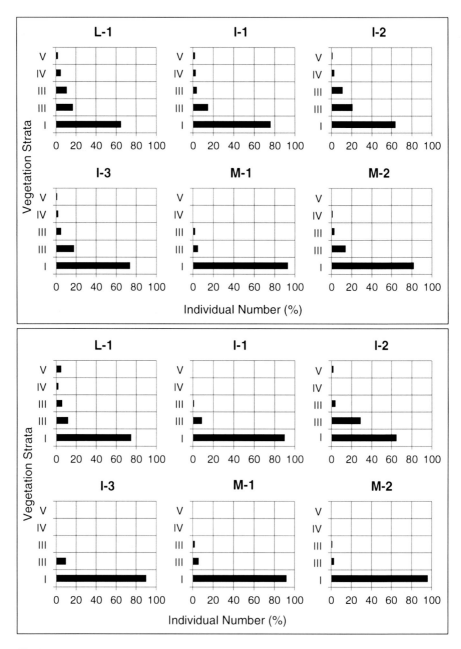

Fig. 8.1 Percentage of individual plant number in each stratum by urbanization level in 1996 (*above*) and in 2006 (*bottom*). L-I, the least urbanized area; I-1, I-2, I-3, intermediate urbanized areas; M-1, M-2, the most urbanized area); 1st strata, 0–1 m; 2nd strata, 1–2 m; 3rd strata, 2–5 m; 4th strata, 5–10 m; 5th strata, >10 m

on management options, the time-averaged above-ground C stocks of *pekarangan* systems could vary from 30 to 123 Mg C ha^{-1}. These projected time-averaged above-ground C stocks of *pekarangans* are substantially higher than those of Imperata–cassava systems (2.2 Mg C ha^{-1}), which is an extensive vegetation type in the Lampung study area. *Pekarangan* research showed these systems simultaneously offer potential for C storage because of their high biomass. Although small size limits the amount of C stored by individual smallholder agroforestry systems, on a per area basis these systems can store as much C as some secondary forests. In aggregate, smallholder *pekarangan* agroforestry systems can contribute significantly to a regional C budget while simultaneously enhancing smallholder livelihoods. A field study in other areas outside Java Island, that is, Lampung, Indonesia indicates that *pekarangans* with an average age of 13 years store 35.3 Mg C ha^{-1} in their above-ground biomass, which is on par with the C stocks reported for similar-aged secondary forests in the same area (Roshetko et al. 2002).

Some experimental evidence also suggests that plant diversity and composition influence the enhancement of biomass and C acquisition in ecosystems subjected to elevated atmospheric CO_2 concentrations (Kumar 2006). Reich et al. (2001) reported that biomass accumulation was greater in species-rich than in species-poor experimental populations under conditions of CO_2 and N fertilization. By extension, home gardens, which are inherently species rich, may trap progressively greater quantities of atmospheric CO_2 under rising levels of this gas.

If *pekarangan* systems and other smallholder tree-based systems were to expand in currently degraded and underutilized lands, such as Imperata grasslands, the C sequestration potential would be about 80 Mg C ha^{-1}, with considerable variation depending on species composition and management practices. Clearly, opportunity exists to induce management that leads to higher C stocks at the systems level. However, incentive mechanisms are needed that assure smallholders will benefit from selecting management practices that favor higher C stocks (Arifin and Nakagoshi 2011).

8.5 Summary

Published articles and a dissertation (Arifin 1998) such as those on *pekarangan* defined that Indonesian home gardens are generally regarded as a complex, species-rich agroforestry system, a diverse mixture of perennial and annual plant species arranged in a multilayered vertical structure, often in combination with raising livestock managed in a sustainable manner over decades or even centuries. A wide spectrum of multiple-use products can be generated with relatively low labor, cash, or other external inputs. In many densely populated tropical regions, *pekarangan* appear to be the last forest-like islands surrounded by increasingly extended, uniform staple crop fields. Some research sponsored by the Core University Research Program JSPS Japan/DGHE Indonesia, and STORMA Germany (1998–2007), concluded that with their multilayered vegetation structure, *pekarangan* serves as an important habitat for included wild flora and fauna.

Pekarangan fulfils not only important ecological but also many social and cultural functions (Arifin et al. 1998; Arifin et al. 2001). However, the major purposes of *pekarangan* are subsistence production and income generation, particularly in rural areas (Kehlenbeck et al. 2007). At forest margins, high production levels in *pekarangan* might help to reduce deforestation. Furthermore, *pekarangan* should be considered as a model for sustainable agroforestry systems, integrating both economic and ecological benefits.

Acknowledgments The authors express our gratitude to the Directorate General of Higher Education (DGHE/DIKTI), Republic of Indonesia for its support of our research through a competency grant (*Hibah Kompetensi*) for 2008–2010 and Graduate Team Research Grant (*Hibah Penelitian Tim Pascasarjana* - HPTP) for 2006–2008; and thanks to Rural Development Institute—Seattle US for *pekarangan* research on Java during 2006–2008. Finally, thanks to the Global Environmental Leaders (GELs) Education Program for Designing a Low Carbon Society (LCS) of Hiroshima University, Japan for research collaboration for 2009–2013.

References

Albuquerque UP, Andrade LHC, Ceballero J (2005) Structure and floristic of homegardens in northeastern Brazil. J Arid Environ 62:491–506

Arifin HS (1998) Study on the vegetation structure of *Pekarangan* and its changes in West Java, Indonesia. Doctoral dissertation for the Graduate School of Natural Science and Natural Science, Okayama University, Japan

Arifin HS (2004) An overview of a landscape ecology study on a sustainable bio-resources management system in Jakarta–Bogor–Puncak–Cianjur (JABOPUNJUR). Proceedings of the international seminar: towards rural and urban sustainable communities: restructuring human–nature interaction, Bandung, Indonesia, 6–7 Jan 2004, p 13

Arifin HS, Arifin NHS (2010) Local wisdom and ecovillage oriented agroforestry development for enhancing creative economy. Seminar of the managing of environment: learning from the past, reaching for the future. Workshop on the international world conference WISDOM 2010, University of Gadjah Mada, Yogyakarta

Arifin HS, Munandar A, Arifin NHS, Takeuchi K, Sakamoto K (2008a) Integrated rural and agricultural landscape management on the watersheds of Bogor – Puncak – Cianjur, Indonesia. In: Proceedings of the 4th seminar toward harmonization between development and environmental conservation in biological production, Tokyo, Japan

Arifin HS, Munandar A, Mugnisjah WQ, Budiarti T, Arifin NHS, Pramukanto Q (2008b) Final Report: Homestead plot survey on Java. Department of Landscape Architecture IPB and Rural Development Institute, Bogor

Arifin HS, Nakagoshi N (2011) Landscape ecology and urban biodiversity in tropical Indonesian cities. Landsc Ecol Eng 7:33–43

Arifin HS, Sakamoto K, Chiba K (1998) Effects of the urbanization on the vegetation structure of home gardens in West Java, Indonesia. Jpn J Tropic Agric 42:94–102

Arifin HS, Sakamoto K, Takeuchi K (2001) Study of rural landscape structure based on its different bio-climatic conditions in middle part of Citarum Watershed, Cianjur District, West Java, Indonesia. In: Proceedings of the first seminar: toward harmonization between development and environmental conservation in biological production, Tokyo University, Japan, pp 99–108

Karyono (1990) Home gardens in Java. Their structure and function. In: Landauer K, Brazil M (eds) Tropical home gardens. The United Nations University, Tokyo, Japan, pp 138–146

8 Low Carbon Society Through *Pekarangan*

Kaswanto, Nakagoshi N (2012) Revitalizing *Pekarangan* home gardens, a small agroforestry landscape for low carbon society. Hikobia 16:161–171

Kehlenbeck K, Arifin HS, Maass B (2007) Plant diversity in homegardens in a socio-economic agro-ecological context. In: Tscharntke T, Leuschner C, Zeller M, Guhardja E, Bidin A (eds) Stability of tropical rainforest margins. Springer, Berlin Heidelberg, pp 295–317

Kumar BM (2006) Carbon sequestration potential of tropical homegardens. In: Kumar BM, Nair PKR (eds) Tropical homegardens: a time-tested example of sustainable agroforestry. Springer, Dordrecht, pp 185–204

Mayanti, R, Arifin NHS, Arifin HS (2007) Study on the vegetation structure dynamics of *Pekarangan* in West Java (Case studies: Cibakung, Sirnagalih-Pagentongan, Babakan Sukaningal, Tegal Gundil Old Settlement, Tegal Gundil New Settlement, and Baranangsiang Indah). Proceedings of the 13th national seminar of PERSADA, Bogor, 9 August 2007

Nurhariyanto, Nugroho P, Jihad, Joshi L, Martini E (2010) Quick biodiversity survey (QBS) guideline: for rapid agro-biodiversity appraisal (RABA) (TUL-SEA project flyer). World Agroforestry Centre—ICRAF, SEA Regional Office, Bogor, p 4

Prosterman R, Mitchell R (2002) Concept for land reform on Java. Rural Development Institute, Seattle, WA

Reich PB, Knops J, Tilman D, Craine J, Ellsworth D, Tjoelker M, Lee T, Wedin D, Naem S, Bahauddin D, Hendrey G, Jose S, Wrage K, Goth J, Bengston W (2001) Plant diversity enhances ecosystem responses to elevated CO_2 and nitrogen deposition. Nature (Lond) 410:809–812

Roshetko JM, Delaneya M, Hairiah K, Purnomosidhi P (2002) Carbon stocks in Indonesian homegarden systems: can smallholder systems be targeted for increased Carbon storage? Am J Altern Agric 17:138–148

Chapter 9
Challenges and Goal of the Sustainable Island: Case Study in UNESCO Shinan Dadohae Biosphere Reserve, Korea

Sun-Kee Hong, Heon-Jong Lee, Bong-Ryong Kang, Jae-Eun Kim, Kyoung-Ah Lee, Kyoung-Wan Kim, and Dae-Hoon Jang

Abstract The Republic of Korea has more than 3,400 large and small islands. About 60 % of these islands are located in the southwestern Jeollanam-do Province, which also includes a huge tidal flat wetland. Because of high biodiversity in the tidal flat ecosystem and a healthy oceanic ecosystem, this area was designated as Dadohae Haesang National Park in 1981. Shinan Dadohae, including Heuksan Do-Hong Do (-Do corresponds to Island) and Bigeum Do-Docho Do, are well known for their island vegetation, migratory birds, and biodiversity. Jeung-Do, famous for its tidal flat ecosystem and biodiversity, was designated a Provincial Park of Jeollanam-do. The excellence of ecosystem, landscape, and cultural attributes gave significant reasons to designate these areas as the 3rd UNESCO Biosphere Reserve in the Republic of Korea in 2009. Since this designation, research has been carried out to develop a management plan for sustainable development based on a balance of human and natural systems in biosphere reserve areas. In the management plan, several special strategies related to global climate change and a low carbon society were adopted, such as to monitor changing socioeconomic standards as well as to monitor changing ecosystems of island and coastal environments. Because education on sustainable use of energy and resources is also an important issue in the island system for accomplishing a low carbon society, this was also included. The most important issue in the management plan, however, is related to the environmental adaptation process of human society on islands, given that these areas are limited resource areas.

S.-K. Hong (✉) • J.-E. Kim • K.-A. Lee • K.-W. Kim • D.-H. Jang
Institution for Marine and Island Cultures, Mokpo National University, Jeonnam 530-841, Republic of Korea
e-mail: landskhong@gmail.com

H.-J. Lee
Department of Archeology, Mokpo National University, Jeonnam 534-729, Republic of Korea

B.-R. Kang
Department of History, Mokpo National University, Jeonnam 534-729, Republic of Korea

N. Nakagoshi and J.A. Mabuhay (eds.), *Designing Low Carbon Societies in Landscapes*, Ecological Research Monographs, DOI 10.1007/978-4-431-54819-5_9, © Springer Japan 2014

Keywords Comprehensive management system • Korea • Shinan Dadohae Bioshpere Reserve • Sustainable Island

9.1 Introduction

Designated in 2009, the UNESCO Shinan Dadohae Biosphere Reserve (SDBR) located in the southwestern area of the Korean peninsula in East Asia consists of about 1,000 islands. It also encompasses the majority of the areas that fall under the jurisdiction of Dadohae Marine National Park located in Shinan-gun, an area which was established with an eye toward preserving nature and contributing to the sustainable development of human life. The UNESCO Shinan Dadohae Biosphere Reserve (SDBR) is divided into several sub-areas. Its total area, including its transitional zone of 39,746 ha, has been estimated at 57,312 ha. The area is home to many islands, as well as the tidal flats that surround these islands. In this regard, the difference between the ebb and flow of the tide and the presence of complex waterways has created unique geographic and biological diversity within the area under conservation. In addition, during the process of adjusting to the comprehensive ecosystem composed of the sea, tidal flats, land, and forests, local people have formed a unique island culture that has been rooted in this geographic and biological diversity. The SDBR is at once a warm temperate broad-leaved forest zone and a warm temperate evergreen broad-leaved forest zone. There is growing concern that the rise in sea levels and changes in vegetation and fish species occasioned by recent climate variability may impact these islands in a complicated manner. As such, there is an urgent need to conserve this complex ecosystem, which has long been maintained in a unique manner, and to establish a human lifestyle that facilitates the conservation of the ecosystem. A biosphere reserve area can be regarded as an area in which the preservation of elements of biodiversity and cultural diversity, such as outstanding landscapes, varied ecosystems, and indigenous biological species, has been deemed important at the global level (UNESCO MAB 2008). In this regard, the protection of biodiversity and the ecosystems deemed to be of high value for conservation is predicated on the establishment of a mechanism through which to study the change in the ecosystems within the core zone (UNESCO 2008). The ecosystem management plan should include the basic guidelines for a long-term monitoring system that can be used to predict and analyze changes in the island ecosystem, marine environment, and the residents "life system," including vegetation in the BR (biosphere reserve) area, in accordance with climate change. In addition, measures to implement ecosystem services such as the installment of a management center to manage monitoring should also be included in the BR management plan.

9.2 Characteristics of the SDBR by Zone

The core zone includes the Dadohae Marine National Park, an inland area designated as a natural monument, and the tidal flat area, for which steps have been taken to have it designated as a wetland protection area. The total area of core zone has been estimated at about 3,242 ha (Table 9.1). The buffer zone includes 2,384 ha of inland area, including the Dadohae Marine National Park and Mud-Flat Provincial Park, 2,242 ha of tidal flats, and 9,698 ha of sea area. Its total area encompasses 14,324 ha. The transition zone includes the marine area designated as the Dadohae Marine National Park as well as some farmland and residential areas in Heuksanmyeon, Bigeum-myeon, Docho-myeon, and Jeungdo-myeon. Its total area is 39,746 ha. The area of the core zone is relatively smaller than that of the buffer and transition zone. However, as the majority of the tidal flats that preserve natural resources and purify the pollution caused by man are surrounded by islands and the sea, they run a lower risk of being damaged by potential industries that can be set up in the area.

The biosphere reserve is divided into the core zone, which has been legally designated as such because it meets the prevailing conditions in terms of conservation objectives, the buffer zone adjacent to the core zone that requires long-term protection, and the transition zone within which sustainable development and resources management can be carried out (UNESCO MAB 2008).

The core zone includes the Dadohae Marine National Park and three natural monuments designated and protected in accordance with the *National Parks Act* and *Cultural Properties Protection Act*. Laws related to the conservation of this core zone include the *Coast Management Act*, *Natural Environment Conservation Act*, *Marine Pollution Prevention Act*, *Water Quality Conservation Act*, and *Framework Act on Environmental Policy*. To this end, in accordance with Article 28 of the Natural Parks Act, *Special Protected* Areas within Korean National Parks are to be designated based on the following process. Areas such as habitats of endangered species that will have to be identified as protected areas in the future should be added based on the *Rest-year Sabbatical System*, which has been implemented

Table 9.1 Size of the Biosphere Reserve by functional area (unit, ha)

		Heuksando	Hongdo	Bigeum Docho	Jeungdo	Overall
Core zone	Inland	732	409	120	15	1,276
	Tidal flat	–	–	937	499	1,436
	Sea	–	–	530	–	530
Buffer zone	Inland	685	246	1,080	373	2,384
	Tidal flat	–	–	1,474	768	2,242
	Sea	70	–	9,440	188	9,698
Transition zone	Inland	1,222	28	8,429	1,157	10,836
	Tidal flat	–	–	–	–	–
	Sea	3,660	1,125	22,950	1,175	28,910
Totals		6,369	1,808	44,960	4,175	57,312

since 1999. Thereafter, the candidate areas should be reclassified and organized based on the purposes of protection. Finally, the *Special Protected Areas* in Korean National Parks can be selected. Special Protected Areas in Korean National Parks represent a mechanism designed to limit unwanted human behavior through such means as the placing of restrictions on access to main resources areas, such as wild animal habitats, wild plant areas, wetlands, and valleys, whose protection from artificial and natural damage is deemed of high value. In this regard, the Jeungdo area of Shinan-gun, a tidal flat area that sits within the core zone, was determined to have a high conservation value. This area was designated as Mud-Flat Provincial Park in June 2008. Meanwhile, steps have been taken, in cooperation with the related administrative agencies, to bring about the designation of the tidal flats near Bigeum-myeon and Docho-myeon as a Wetland Protected Area.

The buffer zone is an area within which activities that are deemed not to hinder the objectives of conservation are permitted. It either lies adjacent to the core zone or surrounds the latter. All in all, the buffer zone consists of 14,324 ha, or 9,698 ha of sea area, 2,384 ha of forested and farming lands, and 2,242 ha of tidal flats. The sea area has been designated as the Dadohae Marine National Park. Although residents have engaged in the collection of Korean medicinal herbs in the forested land, various agricultural products such as spinach, onion, garlic, and rice have been cultivated in the farmland. Meanwhile, local denizens have engaged in fishing without gear for octopus, crab, and shellfish in the tidal flats. These human activities that have made use of indigenous knowledge have played an important role in the preservation and conservation of the ecosystem and helped to maintain harmony with nature. These methods of resources appropriation will in the future greatly contribute to the establishment of an indigenous knowledge transmission system through such means as the development of eco-tourism programs and the management of fishing experience villages.

The transition zone refers to an adjacent area where sustainable resources management practices are encouraged and developed (Lee et al. 2010). Located outside the buffer zone, the transition zone encompasses residential areas and privately owned forests. For the most part, the area consists of spaces in which the everyday activities of residents unfold. The residents in the transition zone have earned their living based on indigenous knowledge for the most part related to the sea. Economic activities within this zone are mainly based on primary industries such as agriculture, fishing, and the collection of marine products. Accommodations and restaurants have also developed as secondary economic activities geared toward visitors who flock to the area in search of a clean natural environment. In particular, the tidal flats within the transition zone function as a mechanism that conserves the natural resources and purifies pollution caused by man. The majority of the tidal flats are surrounded by islands and seas (Koh 2001). The possibility of any of the potential economic activities damaging these tidal flats is very low. For instance, the saltpans not only help bring about the regeneration of the tidal flats, but also serve as nutrient salts within this natural ecosystem known as tidal flats. In addition, the saltpans are also a source of nutrition for shallow fish species that live along the West Coast. These tidal flats and saltpans can be regarded as a treasure trove of biodiversity, an ecological corridor that connects the mainland to the sea, and as a source of livelihood for the people who collect marine products in the area (Hong and Kim 2007). The excellent economic attributes of tidal flats and limited risk of additional environmental destruction makes this area one in which a sustainable economy can be brought about (Hong et al. 2010). The transition zone is an area in which sustainable development is guaranteed under the Coast Management Act and the Fishing Villages and Fishery Harbors Act. These regulations will help prevent development that leads to the destruction of nature, as well as environmental destruction caused by an explosive increase in the number of tourists. In addition to the conservation of natural resources, the designation of the area as a biosphere reserve area can help identify the regional economy, which has been relatively backward, as a successful case of activation through sustainable development.

9.3 The Functional Characteristics of the SDBR

In keeping with the Statutory Framework of the World Network of Biosphere Reserves and the Seville Strategy for Biosphere Reserves, the UNESCO Shinan Dadohae Biosphere Reserve (SDBR) can be divided into three functional characteristics, namely, conservation, development, and logistic support. Each of these functional characteristics is now discussed in turn.

9.3.1 Data Collection

A comprehensive ecosystem is made up of various topographies and biological diversity. The core zone of the SDBR includes two islands and two natural monuments that have been designated as the Dadohae Marine National Park. This particular zone is composed of a terrestrial ecosystem, a marine ecosystem, and a tidal flat ecosystem that connects the two previous ecosystems in a comprehensive manner. The area is home to a significant number of animal and plant species that have adjusted to various topographies ranging from abyssal to intertidal and coastal terrestrial ecosystems and also including plains, forests, and mountains.

- The plant species and vegetation found in this area that encompasses everything from a subtropical to a temperate zone can be regarded as the core element of these island and marine ecosystems: geographic isolation results in the territorial ecosystem within the core zone boasting scores of subtropical plants, coniferous trees, and rare plants. As this area exhibits the characteristics of warm-humid oceanic climate, it is home to both temperate deciduous broad-leaved forest and evergreen broad-leaved forest ecosystems, and as such can be regarded as a very important eco-region from an academic standpoint. This area, which belongs to

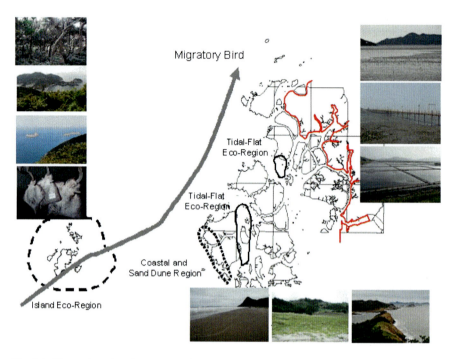

Fig 9.1 The major migration routes of migratory birds within the Shinan Dadohae Biosphere Reserve (SDBR)

the coast and island axis, represents one of the most important national ecological networks on the Korean peninsula (Yim and Kira 1975, Choung and Hong 2006).

- Numerous kinds of migratory birds as well as animals and plants inhabit the tidal flats (Fig. 9.1): these include various types of marine organisms, such as zoobenthic invertebrate animals and algae. The presence of a highly diverse range of biota that includes halophytes, sand dune plants, and animals in the tidal flats, as well as sand dune insects, helps to form an abundant food web that contributes to heightening the stability and soundness of the ecosystem within the tidelands (Koh 1997, Koh and Shin 1988). Thirteen kinds of birds designated as natural monuments, including the black-faced spoonbill, *Egretta eulophote*, and Eurasian oystercatcher, have been observed in this tideland area. In addition, 337 kinds of rare migratory birds are known to inhabit or pass through this area. This list includes endangered bird species such as the golden eagle, osprey, and honey buzzard.
- Habitats and migratory homes for animals and plants can be found from the coastal neritic zone to the abyssal zone. A rare type of marine algae, genus *Haliptilon*, the indigenous species of which is believed to be extinct, inhabits this marine area. In addition, 24 kinds of algae, 117 kinds of invertebrate animals,

and 233 kinds of fish also inhabit the area. In his "*Jasan eobo*" (in Korean), a work that can be regarded as an encyclopedia of marine biology, the seventeenth-century Joseon scholar Jeong Yakjeon (1758–1816), who was an adherent of the Sil-hak (Practical Learning) School, a branch of Confucianism known for its pragmatism, described the area as being home to various kinds of fish and marine products. This situation, he concluded, was the result of the diversity and abundance of marine organisms found in the area. Animal and plant selection of their habitats and migratory homes has been carried out in accordance with the prevailing marine conditions, which are characterized by the presence of tidal flats as well as neritic and abyssal zones and food conditions. As part of efforts to enhance the efficiency with which resources are obtained and develop indigenous natural knowledge and technology that will not alter the balance of resources, local residents have also adjusted to these ecological characteristics.

- The area is also home to an ecological transition zone that connects the sea to the forests. The area's landscape is continuously connected by the environmental gradient of the ecotone that links together the sea-tidal flats-beaches-sand dunes-vegetation (Hong 2007, Hong and Kim 2007). The presence of various bioorganisms in this area can be regarded as the result of continuous adaptation to the multispectrum. This multispectrum has heightened the conservation value of the area's landscape diversity and biodiversity.

9.3.2 Development

Based on the diversity of the landscape and biological diversity, the determination was made that environmental conservation and sustainable development could be implemented in conjunction with a total of 54,070 ha. In this regard, economic development can be achieved in this area by actively developing culture in accordance with nature and developing in a creative manner. In other words, this area can creatively inherit the values and wise usage practices that local people have traditionally displayed vis-à-vis the landscapes within these terrestrial and marine ecosystems as well as natural products, and develop local industries and tourism in a manner that seeks to harmonize the lifestyle and aesthetic awareness of contemporary people with the nature and culture found in this area.

- Local fish species and salt, both of which benefit from the nutritional circulation within the tidal flats, represent important natural products. The fishermen in this area determine the value of fish species that they have obtained based on the extent to which the tidal flats have impacted their nutritional level and flavor. For instance, the influence of the tidal flats results in the fish and shellfish and algae that inhabit them, as well as those which reside in the sea adjacent to the tidal flats and in the clefts between rocks, possessing gustatory elements that can clearly be distinguished from the natural products found in other areas.

Salt is another gift that emanates from the tidal flats. Whether created by the sun in conjunction with the abundant minerals found in the tidal flats or by man, salt was long perceived as a symbol of wealth. The fact that the quality of the salt produced in this area has not only been reevaluated in Korea but also gradually accepted abroad can be construed as a sign that the development of a local industry linking the tidal flats with salt would help to maximize the value of natural products and activate the local economy. In addition, that the local practice of pickling and fermenting fish species indigenous to this area has now been accepted at the national level as a unique cultural attribute means that there exhibits a strong possibility for the formation of a new industry that combines local fish and salt to emerge as the main local industry (Ministry of Oceanic and Marine Products 2002).

- The establishment of a sustainable fishing industry and of a production and distribution system for local products can also be envisioned. The collection of marine products in tidal flats and coastal areas has traditionally been controlled by the village organization called '*jubi*.' The main functions of this organization have been to ensure equal access to resources and ensure the presence of conditions conducive to the reproduction of resources. Currently, a cooperative body called the *eochongye* has played the role of the '*jubi*'. In addition to the collection of marine products, the *eochongye* has also been in charge of coastal fishing and aquaculture. The further entrenchment of a consumption-based economy and society has resulted in the functions of these organizations declining, weakening, and being distorted. Under these circumstances, the UNESCO Shinan Dadohae Biosphere Reserve (SDBR) must play an important role in reviving these organizations and functions so as to alleviate poverty among the lower class. In addition, the exploitive fishing activities that lead to resources depletion must be countered by guaranteeing the reproduction of resources, and fishing methods and distribution structures through which to stabilize the income of residents must be established.

- The relevant technology and information systems developed up to this point by specialized organizations and administrative agencies can be utilized to develop a sustainable fishing industry. Local leaders also possess knowledge and experience, as well as their own awareness, of the resources problem. It is now vital that the designation of the UNESCO Shinan Dadohae Biosphere Reserve (SDBR) be utilized as an opportunity to secure the area's status as a prominent place, and that a sustainable fishing industry be connected to the obtainment of economic goods emanating from these values. Furthermore, it is also necessary to transform existing fishery organizations into something akin to the Korean co-ops and producer-consumer cooperative organizations that have greatly influenced the market economy and other sectors.

- It is also important to link the fishing industry to the food culture and to distinguish it. At the heart of the Honam area, which has been recognized as having the most advanced food culture in Korea, are foods that directly and indirectly include marine products (Ministry of Oceanic and Marine Products 2002). In addition to using these traditional foods, a nationwide slow food

9 Challenges and Goal of the Sustainable Island

system can be established by extending the slow city activities taking place within the SDBR to the rest of the country. At this point, the focus can be placed on the traditions established up to the present. Moreover, a series of food culture coordination methods that revolve around the serving of marine products prepared by local residents as part of the local food culture for everyday and ritual purposes to visitors and guests can be established. Furthermore, by developing environment-friendly storage and distribution technologies, a nationwide marine products supply system can be forged and a distribution and consumption system linking together fishing villages and urban areas established. As such, the SDBR can represent an opportunity to eventually lead to the establishment of a nationwide 'slow food' system.

- In addition to marine products, active use can also be made of the agricultural products and herbs grown amidst the area's unique environment. For example, although abundant sunlight and soil conditions found on the islands have allowed spinach production to develop; the isolated nature of these islands, which results in minimizing the risk of a virus spreading, is also conducive to the development of underground stem and root crops. Moreover, the evergreen broad-leaved forests found in the area are ideal for the production of herbs. In this regard, such products should be cultivated as part of an environment-friendly production and collection system that is rooted in adaptation to the local environment. As such, the value of this area, as well as of the goods produced therein, can be enhanced by directly introducing a slow food system or through the promotion of body-local area-food and body-local area-herbs linkages.

- The developmental potential of the SDBR in large part revolves around the 1,000 or so islands located within its confines. The landscape of these islands, which float in the sea during high tide, is uniquely altered during low tide when they suddenly find themselves surrounded by deep and widespread tidal flats and tidal waterways. Every day, the landscape made up of the islands scattered in the SDBR is altered as low tide gives way to high and vice versa. Fishing activities, rituals, and leisure are carried out in accordance with the lunar calendar that is itself based upon tidal currents. These islands boast forests called *Woosil* that surround the villages and, in their capacity as a belt, protect them from unwanted influences. These forests are composed of black pine and evergreen broad-leaved trees (Hong 2011).

- Another resource that can help foster the economy of the SDBR is its own unique landscape. The uniqueness of the ecosystem can attract tourism. Furthermore, it is possible to develop unique and original programs such as silver retreats that focus on tidal flat and sand geology, nature and cultural resorts that combine traditional culture with the natural calendar, tracking that makes use of geological diversity and waterways, cruises and adventures through tidal waterways, observation of animals and plants in migratory places, sustainable fishing, and marine product collection experiences. Under such conditions, the following plans should be drawn up to use the UNESCO Shinan Dadohae Biosphere

Reserve (SDBR) as a means to achieve strengthened conservation and sustainable development.

First, this plan should be established in a manner that allows humans to access the landscape and resources in this area, should have cultural implications, and be able to foster the formation of an interdependent conservation and development structure linking together man and nature.

Second, there is a need to establish a system that can utilize space and resources in a manner that reflects the increasingly nature-friendly desires of contemporary people. This system should encompass the area's complex ecosystems that range from sea to forests; residents' knowledge and cultural activities prepared in accordance with these ecosystems (Maffi 2001); and plans to interpret customs in a modern manner. For example, in keeping with the cultural expectations of those who visit this island area, a visitor center can be established as a resort retreat or for cultural enjoyment purposes. This center is designed to, on the outside, connect complex ecosystems ranging from the sea to forests, and on the inside, based on the use of the garden concept that serves as one of the main elements of traditional culture, introduce the borrowed scenery method as a garden technique through which to establish a garden design that symbolizes the aesthetics of island ecosystems. It is thus necessary to inherit the stream of traditional culture and island nature, and create spaces for resorts and cultural enjoyment that emphasize ecological spaces and functions. Another aesthetic factor that can be used in conjunction with such goals is the tulips and other bulbs and tubers whose growth has been made possible by the environmental conditions affecting the islands in this area.

Third, the comprehensive ecosystems within the SDBR and the roads in these ecosystems should constitute the main elements employed in conjunction with ecotourism and cultural tourism. In this regard, the marine routes can be linked to forest roads, the watersheds on the mainland, roads of *Woosil* (island village forests), and tidal waterways. The tidal waterways and marine routes, as well as the wide-open marine routes, can be used as elements for ecological and cultural tourism. Shaped like a fishing net and featuring various currents, the tidal waterways and marine routes are closely related to the indigenous knowledge of fishermen. Further, the wide-open marine routes connect this area in which the 'Shinan ship' was sunk during the fourteenth century to the marine routes that were plied by this trading ship in Korea, China, and Japan.

- The development of the SDBR must be carried out in a manner that contrasts with the existing method of establishing tourism facilities that lead to the destruction of ecosystems and reckless development. This aim can be achieved by creating, based on the wisdom amassed in accordance with the adaptive culture that has taken root within this ecosystem, nature-friendly spaces that can be used for residential, tourism, or resort purposes, and by creating wealth for local residents through the heightening of the value of this area. The aforementioned measures are in keeping with these objectives.
- Low carbon urban functions should be ensured by developing and using renewable energy (Fig. 9.2). Solar, wind, tidal, and biogas energy have been identified

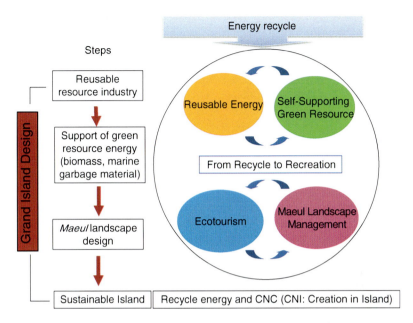

Fig 9.2 Scheme of Grand Island Design of SDBR to realize a low carbon island

as very important energy sources with which to cope with climate variability. Alternative energy sources have been emphasized as a means of not only resolving the environmental crisis, but also of mitigating the increase in the price of fossil fuels such as oil. It is expected that the value of these alternative energies as economic resources will also increase. The difficulty associated with providing electricity to isolated islands within the SDBR has raised the necessity for these islands to produce their own electricity. These conditions have led to the development of solar photovoltaic power generation facilities in Jeungdo that make use of the island's abundant sunlight. In this regard, the designation of the SDBR should serve as an opportunity to further develop the tradition of renewable energy. In addition, potential energy shortages resulting from the geographic environment of islands should be mitigated through the transformation of marine waste into energy.

9.3.3 Logistic Support

Various monitoring and investigation projects have been implemented by domestic and international research institutes within the UNESCO Shinan Dadohae Biosphere Reserve (SDBR). For example, elements of the central and local governments, namely, the Ministry of Environment, Ministry of Land, Transport and Maritime Affairs, Cultural Heritage Administration, Jeonnam Province, and

Shinan-gun, have conducted various projects such as the monitoring of tideland ecosystems, monitoring of specific islands, investigation of uninhabited islands, investigation of the national natural environment, investigation of humanities and social environment, and the investigation of historical cultural indictors. The tidal flat in Jeundo-myeon was designated as the Mud-Flat Provincial Park in June 2008. The tidal flats in the Bugeum-myeon and Docho-myeon areas are now in the process of being designated as wetland protected areas. Currently, various programs are being developed and implemented that allow students and citizens to experience and learn about the unique ecosystems, landscapes, historical relics, and culture of the Dadohae area. To this end, most of this activity has centered on university institutes and civil environmental groups. Especially, the NGOs in the area have established national networks through which to exchange regional monitoring data and operate a conservation strategy monitoring system for the tidal flats that constitute a key marine resource within the biosphere reserve area.

As such, attempts to bring about a sustainable society have included various monitoring efforts and studies designed to educate and convey the importance of marine ecosystems and the indigenous knowledge system of local residents at the industry–university collaboration level. The interest of overseas research institutes in this particular area has also increased. Members of the Bird Identification Association of Japan have regularly visited Hongdo and Heuksando twice a year as part of a joint study with the National Park Migratory Birds Center to investigate the migration route of migratory birds by attaching rings to these birds. Such continuous monitoring results can help to better understand changes in natural populations and changes in habitat environments. They also greatly contribute, based on migratory birds, to international networks between countries.

9.4 The Functions of the SDBR and the Basic Management Plan for Each Zone

The UNESCO Shinan Dadohae Biosphere Reserve (SDBR) boasts an ecological landscape and biodiversity of an outstanding quality (Lee et al. 2010). The wise uses of the various elements that make up the natural environment have allowed this area to achieve a history of harmonious development. The fact that the protection and conservation of nature has been heavily geared toward the survival and development of man has created an atmosphere under which reckless development has been accompanied by damage and destruction to natural ecosystems. The destruction of natural ecosystems whose biodiversity cannot be guaranteed will be further accelerated by the presence of a circular feedback system. The destruction of ecosystems will eventually determine the very survival of mankind. As such, the goals for the management of the SDBR should be prepared in accordance with the theme 'conservation equals sustainable development.'

A biosphere reserve area can be regarded as a sphere in which we can actualize values and ideals related not only to global problems such as global warming, climate change, biodiversity, and food shortages, but also to national issues such as environmental conservation, balanced national development, and low carbon green growth (Table 9.2). To this end, this study established, in accordance with the 'conservation of biodiversity and cultural diversity' (see Maffi 2001), the development of various programs that make use of indigenous knowledge as the management goals of the SDBR, and, based on the three functions of biosphere reserve areas, established the following fundamental objectives.

Given the spatial attributes of the coastal sea, near sea, and distant sea, the management plan should be divided into short-, mid-, and long-term strategies (Table 9.3). Over the short term, there is a need to implement an SDBR management plan that divides the overall area into the coastal sea (Jeungdo), near sea (Bigeumdo), and distant sea (Hongdo in Heuksan-myeon). Over the mid- to long term, attention should be increased to encompass the coastal sea (Dochodo), distant sea (Daeheuksan in Heuksan-myeon) and distant sea (Heuksando and Hongdo), while simultaneously implementing discussions on the expansion of the biosphere reserve areas. Bigeum-myeon is regarded as playing a key role within the SDBR in terms of the ecological culture of islands in that it can connect the distant and coastal seas. Moreover, as the small- and medium-sized saltpan industry that includes fishing without gear has been concentrated here, this area is also well placed to play a leading role in fostering the participation of small-scale local merchants and businesses and activating the community. Furthermore, Bigeum-myeon, which belongs to the Dadohae Marine National Park, boasts strong biodiversity and an excellent ecological landscape.

The presence of traditional villages on local islands makes it necessary to develop a program that revolves around the establishment of a UNESCO information center in Bigeum-myeon. Although Hongdo has long been perceived as a symbol of marine tourism in Korea, it also falls within the core zone where warm temperate evergreen forests can be found. Hongdo is the perfect place to monitor the conservation of biodiversity and ecosystems and ecosystem services, both of which can be considered as the core plan for UNESCO biosphere reserve areas. Jeungdo's unique characteristics were damaged when the Jeungdo Bridge connecting the island to the mainland was opened in March 2010. In addition, Jeungdo's designation as a Slow City may cause confusion in terms of the plans for the management of the area as UNESCO biosphere reserve. To this end, it becomes necessary to entrench UNESCO social culture monitoring and ESD in conjunction with local residents in this area. After having determined whether the short-term plan can be successful, the opinions of people in areas surrounding Shinan Dadohae regarding the extension of the biosphere reserve area should be collected, with detailed discussions and collaboration with the UNESCO Korea MAB undertaken. Thereafter, discussions pertaining to the establishment of mid- and long-term plans that are based on the collected opinions should be commenced regarding the extension of the biosphere reserve area.

Table 9.2 Basic goals of Shinan Dadohae Biosphere Reserve (SDBR) management according to three functions of Biosphere Reserve of Madrid Action Plan

Three functions of the SDBR	Basic goals of SDBR management		
	Ecosystem service	Sustainable systems	Activation of indigenous industry
Conservation	Landscape ecological planning-vegetation, geology, land use, ecosystem mapping, conservation and development planning	Specification of the functions of each zone-division into core, buffer, and transition zones and identification of the functions and roles of each zone	Primary industry (indigenous industries) Examination of resources (by year, type, and region) Distribution of the main fish species, shellfish, algae, and aquaculture Traditional shellfish utilization methods Spatial distribution, production of saltpans
	Eco-monitoring	Social culture monitoring	Monitoring of economic situation (development of statistical data and monitoring by local residents)
Development	Measures to use the main ecosystems within each zone	Categorization and land-use measures for each zone	Support programs for indigenous industries run by local residents
Logistic support	ESD-education of specialists responsible for conducting special programs related to UNESCO, NGO education and activity-support programs, formation of a local specialists community	Reorganization of management and steering committee, establishment of activation measures Organization of management body, PR strategy and planning that includes the creation of signs identifying the area as a UNESCO-designated biosphere reserve, guidebooks and pamphlets; creation of emblems; and measures for utilization	Standardization: BR standardization, industrial standardization, HACCP
	International issues related to marine pollution and the establishment of international networks		
	Application for the extension of biosphere reserve area and registration as a World Mixed Cultural and Natural Heritage		

9 Challenges and Goal of the Sustainable Island 159

Table 9.3 Basic management plan by zoning of SDBR

Three functions of the SDBR	The basic plan for the management of each zone of the SDBR		
	Core zone	Buffer zone	Transition zone
Conservation	Basic research on the general biota and cultural resources within the core zone	Basic research on the general biota and cultural resources within the buffer zone	Basic research on the general biota and cultural resources within the transition zone
	Planning and implementation of long-term-based eco-monitoring	Social and cultural monitoring	Monitoring of economic situation (development of statistical data and monitoring by residents)
	Establish guidelines regarding detailed limitations for development within each zone		
Development	The division of land uses by zone and suggestion of land-use measures		Indigenous industries run by local residents
			Support programs
Logistic support	Extension of the research base for specialized researchers	Establishment of a support system to activate eco-cultural tourism	Exchanges between residents through ESD
	Expansion of exchanges between local residents, NGOs, and specialized researchers	Support for ESD conservation of biological and cultural diversity programs	Education of specialists, NGO education, and activity support programs
			Implementation and expansion of integrated support programs for local indigenous industries

9.5 Conclusion: Direction of the Implementation of the Management Plan

The majority of the areas that were initially designated as biosphere reserve areas were large mountainous regions such as Seolak Mountain and Jeju Island: these regions were characterized by a high level of biodiversity because their geographic situation led to their ecosystems remaining largely untarnished by man. The Seville Strategy for Biosphere Reserves recommended that so long as damage to the ecological environment could be avoided, various development strategies could be established (UNESCO MAB 2008). For instance, by expanding the zones, for example, the buffer and transition zones, the proclivity for biological diversity-centered isolated biosphere reserves could be overcome and sustainable use of such areas could be ensured. The SDBR boasts the minimum-sized core zone needed to qualify for the conservation of biospecies and ecosystems. The core zone is currently protected under existing institutions established by the national and provincial parks. The buffer and transition zones are regarded as experimental spaces in the biosphere reserve area within which the economic system and

Table 9.4 Roles of the main actors involved in the management of the SDBR

Main actors in SDBR management	Roles and contents
State, local autonomies	Establishment and implementation of management policy, assuring of the budget
Academic research institutes	Think tanks, field study/data accumulation, measures for optimum utilization
Local resident community/NGOs	Actualization of the policy measures and research findings developed by the government and research institutes
Advising organizations	Provision of advice regarding measures related to the management of development, suggestion of measures to facilitate the use of international networks

indigenous culture that has long prevailed can be augmented while promoting the harmonious development of nature and man, for instance, agriculture, fishing, and sea salt. As such, a sustainable economic system is expected to develop in the buffer and transition zones.

The actual management of the biosphere reserve area is predicated on the main actors involved in its management playing their role in an organic manner. Although a management committee has been established in Jeonnam Province, it remains necessary to supplement the functions of this committee. To this end, the main actors involved in the conservation and management of UNESCO biosphere reserves should be the 'residents.' Domestic and international case studies have shown that the participation of residents has been instrumental in determining the success of the management of UNESCO biosphere reserve areas. Therefore, it becomes very important to establish the structure and functions of a management committee that can be used to conserve the ecosystem and traditional culture within the UNESCO SDBR; to establish a sustainable economic system; and to encourage community activities designed to improve the residents' quality of life. The Korean National Commission for UNESCO should cooperate with research institutes to ensure that local communities voluntarily participate in the conservation and management of UNESCO biosphere reserve areas.

Jeonnam Province established the 'Rules Regarding the Management of Biosphere Reserve Areas in Jeonnam Province' (March 19, 2010). The organization and objectives of this management committee were spelled out as follows: "A Management Committee for Biosphere Reserve Areas in Jeonnam Province shall be established to ensure the effective management of biosphere reserve areas in Jeonnam Province and review important management measures." The management committee is responsible for reviewing important matters related to biosphere reserve areas, and in particular makes decisions pertaining to the three following matters: first, matters related to designation and changes within the biosphere reserve area; second, matters related to the establishment and execution of management and implementation plans for biosphere reserve areas; and third, matters related to the improvement of the biosphere reserve area-related collaboration between government agencies and other organizations (Table 9.4).

9 Challenges and Goal of the Sustainable Island

Table 9.5 Functional classification of the comprehensive management system (CMS) for the SDBR

Conservation	Investigation and accumulation of basic materials through integrated monitoring
	Development and use of Islands Geographic Information Systems (IGIS)
	Integrated management measures designed to conserve biological diversity
Development	Establishment of a residential community based on the participation and organization of local residents
	Search for measures to develop and activate indigenous industries run by local residents as a scenery industry
	Measures for the commercialization through products management
	Attachment and distribution of eco-labeling
	Development of eco-tourism and sustainable development education programs using indigenous knowledge
	Training of local eco-cultural tourism guides
	Development of education-oriented tourism programs
	Educational program: programs for local residents/regular people
	Development of various cultural contents using local cultural prototypes
Logistic support	Establishment of a Visitor Center and related management body
	Establishment of an integrated monitoring center
	Preparation of a landscape ecological planning map
	Measure for SDBR PR (CI development, installment of signs and boards, etc.)
	Development of renewable energy
	Identification of the marine pollution problem as an international issue
	Establishment of international networks
	Application for expansion of the biosphere reserve area and designation as a World Mixed Cultural and Natural Heritage

According to the 'Rules Regarding the Management of Biosphere Reserve Areas in Jeonnam Province,' the management committee should be composed of 25 members (3 official positions and 22 appointed positions) selected for 2-year terms. A look at the composition of the management committee reveals that the number of members hailing from local and state governments outnumbers those whose background is related to academic research institutes and NGOs, thus meaning that the ability to implement and evaluate the actual management plans is lessened. Furthermore, the fact that no local representatives have participated in the management committee can also be pointed out as a problem. To this end, it is necessary to reorganize the composition of the management committee in a manner that facilitates the actual management of biosphere reserve areas (Table 9.5). Moreover, it is essential that a comprehensive management system (CMS) that makes possible the management of biosphere reserve areas in an integrated manner be established. The items that fall under the individual functions of SDBR management include the following.

The SDBR management plan must include the forging of organic relationships between various programs, such as those related to local communities, the activation of indigenous industries, the development of cultural contents, IGIS systems, landscape planning and management, and integrated monitoring program (Table 9.5). This goal can be achieved by entrenching a comprehensive management system (CMS).

The comprehensive management system (CMS) for the SDBR has been focused on the actual operation and management of individual programs. To this end, the comprehensive management system (CMS) for the SDBR is designed in a manner that revolves around the functions of conservation and development. Meanwhile, the logistic support function, which is strongly related to policy making, is designed to foster collaboration outside the comprehensive management system (CMS).

Acknowledgments This work was funded by Jeonnam Green Environment Center, Jeollanam-do. This work was also supported by the National Research Foundation of Korea Grant by the Korean Government (MEST)"(NRF-2009-361-A00007). Our thanks to UNESCO MAB Korea and Jeollanam-do Province, Republic of Korea for their valuable information. This paper was presented in 8th IALE World Congress at Beijing, China (18–23 August 2011).

References

Choung HL, Hong SK (2006) Distribution patterns, floristic differentiation and succession process of *Pinus densiflora* forest in South Korea: a perspective from nation-wide scale. Phytocoenologia 36(2):213–229
Hong SK (2011) Eco-cultural diversity in island and coastal landscapes: conservation and development. In: Hong SK, Wu J, Kim JE, Nakagoshi N (eds) Landscape ecology in Asian cultures. Springer, Tokyo, pp 11–28
Hong SK (2007) Linking man and nature landscape systems: landscaping Blue-Green network. In: Hong SK, Nakagoshi N, Fu B, Morimoto Y (eds) Landscape ecological applications in man-influenced areas: linking man and nature systems. Springer, Dordrecht, pp 505–523
Hong SK, Kim JE (2007) Circulation and network of forest-stream-coastal ecosystems. Island Cult 30:267–286
Hong SK, Koh CH, Harris RR, Kim JE, Lee JS, Ihm BS (2010) Land use in Korean tidal wetlands: impacts and management strategies. Environ Manag 45:1014–1026. doi:10.1007/s00267-006-0164-3
Koh CH (2001) Tidal of Korea: environment, biology and man. Seoul National University Press, Seoul
Koh CH (1997) Korean megatidal environments and tidal power projects: Korean tidal flats-biology, ecology and land use by reclamation and other feasibilities. Houille Banche 3:66–78
Koh CH, Shin HC (1988) Environmental characteristics and distribution of macrobenthos in a mud flat of the west coast of Korea (Yellow Sea). Netherlands J Sea Res 22:279–290
Lee HJ, Cho KM, Hong SK, Kim JE, Kim KW, Lee KA, Moon KO (2010) Management plan of UNESCO Shinan Dadohae Biosphere Reserve (SDBR), Republic of Korea: integrative perspective on ecosystem and human resources. J Ecol Field Biol 33:95–103
Maffi L (2001) On biocultural diversity linking language, knowledge, and the environment. Smithsonian Institution Press, Washington, DC
Ministry of Oceanic and Marine Products (2002) Korean Maritime Culture 2, the West Sea Area (in Korean), vol 1. Institute of Islands Culture, Mokpo National University, Korea
UNESCO (2008) Links between biological and cultural diversity: concepts, methods and experiences. Report of an International Workshop. UNESCO, Paris
UNESCO MAB (2008) Madrid action plan for Biosphere Reserves (2008–2013). UNESCO, Paris
Yim YJ, Kira T (1975) Distribution of forest vegetation and climate in the Korean Peninsula. 1. Distribution of some indices of thermal climate. Jpn J Ecol 25:77–88

Chapter 10
The Neglect of Traditional Ecological Knowledge on Wild Elephant-Related Problems in Xishuangbanna, SW China

Zhao-lu Wu, Qing-cheng He, and Qiu-jun Wu

Abstract Elephants have roamed into areas of human settlement and destroyed crops, damaged houses and infrastructures, and occasionally injured or killed people, resulting in elephant-related problems in southwestern China. Based on data collected in the field during July 2009 to March 2011 as well as in the literature, we discuss the effects of neglecting traditional ecological knowledge (TEK) on wild Asian elephant-related problems. We had three findings. First, the environment-friendly TEK, including the simple ecological views of forest priority, biocultural rural landscape, and elephant culture, played an important role in conservation of forests and all the wildlife in the forests. Second, indigenous people were forced to abandon traditional beliefs and cultures, which were considered as blind worship in the late 1950s. The neglect of TEK increased human–elephant conflicts. After the armed "capturing elephant" approved by government during 1972–1973, illegal elephant poaching occurred yearly. Elephants lost habitats and began to cause problems: 140 people were injured or killed as a result of elephant attacks during 1991–2008. Third, the prevailing selfish consciousness encouraged people to move into Xishuangbanna to cultivate tropical crops including rubber and to build a highway network, worsening the elephant-related problems. The traditional culture of adoring the wild elephant became weak; 95.9 % of the 412 interviewees did not like to coexist with wild elephants. In conclusion, it could be understood clearly that humans had approached the elephants' habitat and provoked the elephants' aggressive behavior. It is necessary to reconfigure moderate and environment-friendly TEK to mitigate elephant-related problems.

Z.-l. Wu • Q.-j. Wu
Institute of Ecology and Geobotany, Yunnan University, Kunming 650091, China
e-mail: zlwu@ynu.edu.cn; zlwu1958@qq.com

Q.-c. He (✉)
Editorial Department of Journal, Yunnan Agricultural University, Kunming 650201, China
e-mail: he_qingcheng@163.com

N. Nakagoshi and J.A. Mabuhay (eds.), *Designing Low Carbon Societies in Landscapes*, Ecological Research Monographs,
DOI 10.1007/978-4-431-54819-5_10, © Springer Japan 2014

Keywords Cultural landscape • Human–elephant conflict (HEC) • Rubber cultivation • Sacred site • Selfish consciousness • Traditional ecological knowledge (TEK) • Wild elephant-related problems • Xishuangbanna

10.1 Introduction

It was believed for many years that there were no wild elephants in China (Chen et al. 2006), until a Scientific Investigation Team from the Chinese Academy of Sciences found wild elephants living in forests of Southern Yunnan Province during 1957–1958. Wild elephants "inhabited in primary forests and never attacked humans" (Shou et al. 1959), revealing the harmonious relationship between humans and elephants at that time. By the end of the twentieth century, however, although there was only 200 to 250 wild elephants in China, elephant-related problems happened frequently, causing serious human–elephant conflict, ecological problems, and social troubles (Chen et al. 2006). Elephant-related problems, such as destroying crops, damaging houses and infrastructures, disturbing daily life and agricultural production, and attacking people and livestock (Hoare and Du Toit 1999; O'Connell-Rodwell et al. 2000; Sukumar 2006), occurred when elephants appeared. From 1991 to 2008, 140 people were injured or killed as a result of elephant attacks; the number was even increased year by year (Wu 2008). For example: on July 7, 2010, in Mengyang sub-reserve, a passerby was injured by two elephants; on Nov. 3, 2010, in Mengman Town, Mengla County, a 67-year-old man was trampled to death by elephants when he was working in his cultivated field; on Feb.13, 2011, in Yunxian Town, Puer Prefecture, four elephants fled from Xishuangbanna, destroyed pigpens and kitchens, and threatened villagers; on Aug.16, 2011, in Mengyang sub-reserve, five motor cars, four cars, and one truck were damaged by a young elephant that roamed in the highway. According to the Bureau of Xishuangbanna Nature Reserve, elephant-related problems caused directly pecuniary loss, more than 4.3 million Yuan RMB, in 136 villages of Xishuangbanna in 2010.

Usually, based on anthropocentricism, human–wildlife conflicts have been appraised according to material and mental damages or threats to humans caused by wildlife and were called "wildlife troubles" (He and Wu 2010). If humans withdraw from the conflict, however, the named wildlife troubles would just be the ecological adaption of wildlife to the human-dominated environments. In Xishuangbanna, humans and human activities approached the elephants' habitat unceasingly (Wu et al. 2007), making encounters between humans and elephants likely. Elephant-related problems, therefore, are unavoidable (He et al. 2011).

Different approaches from technical aspects to cultural aspects had be applied to the mitigation of human–wildlife conflict. Currently, much research is focused on human population growth, habitat fragmentation (Lang et al. 2008), change of elephants' ecological behaviors (Feng et al. 2010), diet and water availability (Loarie et al. 2009), as the technical approach(Nyhus and Tilson 2004; WWF 2008).

We discuss here another approach, the cultural approach. We hypothesized that elephant-related problems occurred because of the occupation of elephant habitats by humans, and that this occupation was the result of the neglect of local traditional ecological knowledge (TEK). We then analyzed the roles of TEK in protecting wild elephants and mitigating elephant-related problems.

10.2 Natural and Social Environment

On the Asia map, one can find a large slope that spans southward from the Southeast Tibetan Plateau and extends down to the Indo-China Peninsula. Midway on this slope, between 99°58′-101°50′ E and 21°09′-22°36′ N is the Xishuangbanna region, a Dai Ethnic Nationality Autonomous Prefecture of Yunnan Province, China. Historical records showed that the ancestors of the Dai people resided in this region more than 2,000 years ago. In 1180, the chieftain of a Dai tribe conquered other tribes and founded a local Kingdom, which conquered 12 local districts. In the Dai language, "Xishuang" means 12 and "Banna" means districts; hence, "Xishuangbanna" implies the 12 districts that existed historically.

Climatically, Xishuangbanna is best described as having a transition monsoon climate of the tropics and subtropics. The monsoons from the southwest carry large amounts of warm and moist air masses into Xishuangbanna, but with no risk of typhoons. The area is very mountainous, reaching 2,429 meters above sea level (m a.s.l.) to the north, and sloping southward to a low point of 477 m a.s.l. The general topography consists of basins or valleys alternating with hills or mountains. A tropical climate prevails in the basins and low valleys, with a subtropical climate in the mountains and higher hills. Annually, summers are characterized by high temperatures and humidity, whereas winters offer little rain and much heavy fog, and then gentle winds bode spring. The region is free from frost all year round. However, cold currents occasionally intrude in higher elevations, rendering a cool and dry climate in relationship to the tropics of Southeast Asia.

Xishuangbanna is a kingdom of fauna and flora biodiversity. The main vegetation types are tropical rainforests, tropical monsoon forests, and subtropical evergreen broad-leaved forests. According to incomplete statistics, approximately 4,600 species of higher plants have been discovered, of which 80 % belong to tropical and subtropical floral species. There are 343 species of higher plants that have been considered rare species. There is also an extraordinary abundance of fauna in Xishuangbanna, and 758 species of vertebrates have been identified. Of the 108 mammals, 36 species have been listed in the Chinese Fauna Red Book (Xu et al. 1987).

There are 13 ethnic groups with a larger population recognized in Xishuangbanna. Different ethnic groups occupy different environments. Dai, Han, and Hui people resided in the basins and valleys and cultivated paddy rice in addition to earning income from handicrafts. Hani, Lahu, Jino, and Blang people inhabited the upland areas and cultivated upland rice and tea, and collected or grew other cash crops. The social environment has been changed since a large number of immigrants moved in to develop tropical agriculture in the 1960s. The human population increased from 199,300 in 1949 to 905,000 in 2010.

10.3 Method

10.3.1 Data Collection

Literature analysis was used to collect historical data about the TEK of different ethnic groups in Xishuangbanna and to learn the roles of such TEK in the perception, protection, and utilization of wild elephants in the past half century.

Investigation research carried out in villages, where elephant-related problems happened, to collect and to analyze data of local communities' view on the wild elephant-related problems, elephant conservation, and the social and economic development of the villages. A participatory rural appraisal (PRA) was applied in investigation research. PRA is a method of "hearing communities, learning and perceiving their needs and development with them" (Liu 2005). It was widely used to analyze and estimate the current local situation and development planning through informal interviews with inhabitants (Yu et al. 2009), as well as the change of species and landscape in rural areas (Wu 1997a). A questionnaire was used to collect data in detail. It contained two parts: (1) the interviewees' view on the environment, resources, and development of the villages they lived in; and (2) the interviewees' knowledge of wild elephant conservation and problems, including attitudes toward conservation and who benefited from it, their views on the causes of elephant-related problems, and ways to mitigate elephant-related problems.

Investigations were carried out in areas where elephant-related problems happened, from July 2009 to March 2011, with the assistance of the reserve staff who had been working in local communities for several years and well understood the situation of the studied communities. We received 423 questionnaires from 22 villages, of which 412 (97.40 %) were valid. The generality of interviewees were (1) age: from 18 to 80 years old with an average age of 39; (2) gender: 350 men and 62 women; (3) eight ethnic groups: 118 were Jino, 84 Hani, 74 Yi, 51 Dai, 50 Blang, 21 Han, 13 Yao, and 1 Lahu; and (4) school education level: 2 graduated from colleges, 15 from senior high schools, 175 from junior high schools, 212 from elementary schools, and 8 were illiterate.

10.3.2 Data Analysis

To analyze roles of TEK in local communities attitudes toward wild elephants, we used interviewee age to signify the change in TEK; then, we considered factors

10 The Neglect of Traditional Ecological Knowledge on Wild Elephant-Related... 167

Table 10.1 Categories of factors influencing local communities' attitudes toward wild elephants

Category	Levels			
	1	2	3	4
Methods to mitigate elephant-related problem	Planting food resources	Building obstacles	Expelling	Unknown
Coexistence with elephants	Indifferent	Impossible	Opposition	Unknown

Table 10.2 Cultivated land area and costs of elephant-related problems in 16 villages (2010)

Village	Cultivated land area (hm^2)	Elephant-related problems cost (¥)	Village	Cultivated land area (hm^2)	Elephant-related problems cost (¥)
Xiamancha	248.20	250,000	Manwaxinzhai	427.27	72,000
Cicaitang	124.00	55,000	Nanlang	424.67	30,000
Shangguokou	129.07	3,000	Nanbeng	120.00	15,000
Xiaopingzhang	175.47	24,000	Mandan	207.93	7,000
Xinlongshan	215.33	46,000	Manlang	144.00	6,000
Xintianba	370.73	220,000	Shangzhongliang	203.80	50,000
Xinshan	413.20	100,000	Hetu	442.00	87,000
Tiaobahe	343.00	82,000	Nanping	203.73	40,000

Note: hm^2, square hectometer (hectare)

including community knowledge regarding methods to mitigate elephant-related problems and coexistence with elephants. The influencing factors were grouped into four levels, as listed in Table 10.1. To analyze effects of agricultural expansion on elephant-related problems, we analyzed the relationship between cultivated land area and costs of elephant-related problems (Table 10.2). All statistical analysis used SPSS Statistics 17.0 with bivariate correlation analysis.

10.4 Results and Analysis

10.4.1 Unique Traditional Ecological Knowledge (TEK) Benefiting Conservation of Wild Elephants and Their Habitats

According to the literature, wild Asian elephants were distributed historically northward to both sides of the Yangtze River in China (Elvin 2004). In the middle of the nineteenth century, however, the northern boundary of their distribution moved southward to the south and southwest of Yunnan Province, including Lincang, Puer, and Xishuangbanna (Wen 1995). The social and natural environments in Xishuangbanna were suitable for the survival of wild elephants in the early

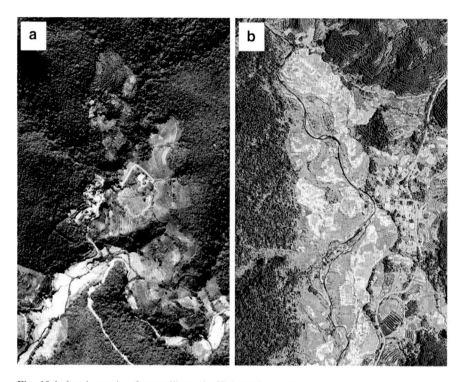

Fig. 10.1 Land mosaic of two villages in Xishuangbanna based on remote sensing image in December 2010 (see Fig. 10.3 for geographic positions of village). (**a**) A Blang village, Nanmuhe, was moved from the mountain down to a narrow valley in 1973, but the new village was still surrounded by natural forests; wild elephants sometimes visited the cultivated fields. (**b**) A Dai village, Dalongha, was set up in the 1900s in a small basin with a stream running through it. Rubber plantations not only replaced natural forests (on the right side) but also reached into holy hill (the triangular forest patch close to houses on the north); wild elephants visited the rubber plantations and cultivated fields almost yearly

twentieth century. The Dai people lived in flat basins and other ethnic groups such as Blang, Jino, and Hani resided in the mountains; surrounding the residential areas in basins and mountains were vast tropical and subtropical forests that were home to wild elephants and other wildlife (Fig. 10.1).

Such an environmental pattern was maintained by the unique TEK, a special knowledge system, created by the indigenous people, on forests, land use, and harmony of humans and nature, which may be described as follows.

(1) The forest priority bioculture. Entrenched in the Dai philosophy are concepts about the intimate relationships between humanity and environment. They believe that humanity is part of a unified whole that consists of humans, various gods, natural components (such as hills, forests, animals, plants, and water), and artificial components (such as buildings, paths and roads, planted forests, and cultivated lands). Of all these components, the Gods are aweful and rule over all

other components in the whole, the forests are most important, and the artificial components are imitations of nature. Forests provide water running in rivers, and the water then irrigates cultivated land from which humans obtain their food. Hence, forests tend to occupy the most important position in the hierarchy: forests → water → cultivated fields → grains → humanity. Such a human–nature relationship, stated in a Dai proverb as "No water if without forest; no field if without water; no grain if without field; no people if without grain," was handed down for generations. Humans, therefore, should restrict themselves when they take subsistence materials from their surroundings (Dao 1996).

Because of the prolonged Dai rule, this perspective of nature had a great impact on other indigenous people. The supreme feudal lord owned all lands of Xishuangbanna; chieftains at different levels managed villages authorized to them and took land rents from villagers who resided in these villages. There were district boundaries between villages as well as cultivated fields of households. Villagers planted crops, collected wild plants, and hunted or fished within their territories (Ma and Miao 1989). This territorial system made possible the maintenance of local rainforests. As a result, indigenous people respected and consciously protected forests as well as the animals and plants in the forests, benefiting the survival of wildlife including elephants.

(2) The unique bioculture landscape. Rural landscapes in the basin and mountains of Xishuangbanna were characterized by special local biocultures. People who inhabited the upland areas set up villages on the tops of hills where sunshine, water sources, and dense forests were ample. This tradition has been kept until today. Most of the people have moved down to low valleys to develop paddy rice agriculture but their new villages are still built on the ridges of hills (Wu et al. 1997). The Dai people who resided in basins and valleys set up their villages next to streams with clean water and hills covered with forests; the flourishing trees in the villages provide shade during the hot summer (Dao 1996). One can find that a typical Dai village landscape was composed of houses, vegetable and fruit gardens, the Buddhist temple, a holy hill, graveyard forest, fuel-wood forest, bamboo forest, farmlands, rivers, and natural forests (Fig. 10.2). From such a complex mosaic, the Dai people obtained most of what they required daily.

The holy hill was a strictly protected natural forest located near the village. All the animals and plants inhabiting the holy hill were considered to be either companions of the Gods or creatures of Gods' gardens. Elders, especially the heads of clans or villages, taught youths by personal example as well as verbal instructions, making the codes of Gods to pass down from generations to generations (Wu 1997b). As a result, up to the 1950s, more than 400 sites of holy hill forests with a total area of 50,000 km^2 were still well protected in Xishuangbanna (Gao 1998).

Indigenous people cultivate fuel-wood (*Cassia siamea*) to meet energy demands. They harvested the canopy of trees, leaving the stumps to coppice abundantly during the year. Our study indicated that the indigenous people benefited greatly from their fuel-wood trees for ease in collecting wood for fuel,

Fig. 10.2 The unique bioculture landscape of a typical Dai village in Xishuangbanna, indicating the mosaic of houses *(1)*, vegetable and fruit garden *(2)*, Buddhist temple *(3)*, holy hill *(4)*, graveyard forest *(5)*, fuel-wood forest *(6)*, bamboo forest *(7)*, farmlands *(8)*, rivers *(9)*, and natural forests *(10)*

beautifying the settlement environment and providing shade in the hot summer, protecting banks of rivers and streams, and serving as visible ethnic symbols.

(3) The elephant culture. Elephants were regarded as Gods. Although large in stature, elephants were so modest, gentle, and friendly that they never attacked human and small animals actively. The Dai people adored and loved elephants, forming an elephant culture with peace, friendliness, and properties of diligence. The elephant culture infiltrated into myth, legend, sculpture, murals, and the arts (Yan 1990).

Therefore, indigenous people in Xishuangbanna not only had an environmentally friendly TEK but also perfected being gentle and friendly in practice. They enjoyed the peaceful life harmonizing with nature, which benefited the conservation of forests and wildlife, including the wild elephants in the forests.

10.4.2 The Neglect of Traditional Ecological Knowledge (TEK) Increasing Human–Elephant Conflicts

In the late 1950s, agricultural technique and medical aids were introduced to improve living conditions. Indigenous people were encouraged to acquire scientific knowledge and to use scientific techniques in their daily life and production. Some people inhabiting the uplands were transferred to lower valleys and foothill belts.

Traditional belief and cultures including TEK as just described were considered to be superstitions (which meant blind worship based on the prevailing culture at that time). Indigenous people were forced to abandon such "superstitions." The coexistence between humans and elephants was broken.

The news of finding wild elephants encouraged some zoos in China to capture wild elephants in Xishuangbanna. During 1972–1973, an armed "capturing elephant" team organized and approved by government, with the assistance of local people, caught wild elephants in forests of Mengyang sub-reserve. They caught a young elephant at the cost of shooting five elephants to death and injuring four other elephants. In March 2011, we interviewed adult villagers who participated in the "capturing elephant" event. They said that the capturing was only known to several local government cadres and villagers, being afraid of protests from indigenous people. They recollected many elephant corpses remained in the forests because of inexact doses of anesthetic and a shooting rampage by a soldier in self-defense. Later, a "three-leg elephant," who fled from the poaching, roamed in the wilderness and attacked people who came close to it (Fig. 10.3).

People involved in the capturing got rewards from the team instead of punishment from the Gods, which told the indigenous people that the majesty of their Gods was so limited. The traditional beliefs and superstitions in the indigenous people's minds collapsed. Some local people began to hunt wild elephants. By 1985, 15 wild elephants had been killed by local people (Chen et al. 2006) and nine people were injured by the "three-leg elephant" (Chen 2005).

Human–elephant conflicts occurred. Humans became big and powerful in the people's spiritual world and daily life. International illegal elephant poaching took place in Xishuangbanna. During 1991–1995, several criminal groups from China and Laos sneaked into the forests and killed 30 elephants (Chen et al. 2006).

By 2010, the traditional culture of adoring the wild elephant became weak but elephant-related problems became a crisis. Of the 412 interviewees, 19.7 % hated wild elephants and 10.7 % liked them. Many (66.0 %) of these people indicated they did not like wild elephants because of the serious elephant-related problems that happened annually. Most of them (95.9 %) did not like to coexist with wild elephants (Table 10.3).

10.4.3 Prevailing Selfish Consciousness Worsening Wild Elephant-Related Problems

Accompanying the neglect of TEK was the rapid spread of exotic selfish consciousness, resulting in bloody elephant poaching and savage destruction.

The "capturing elephant" in Mengyang sub-reserve during 1972–1973 indicated nothing but human selfish consciousness. The capturing team did not learn traditional capturing experiences or consider the feelings of the indigenous people; they just went into the forests with guns and took what they wanted. The young elephant

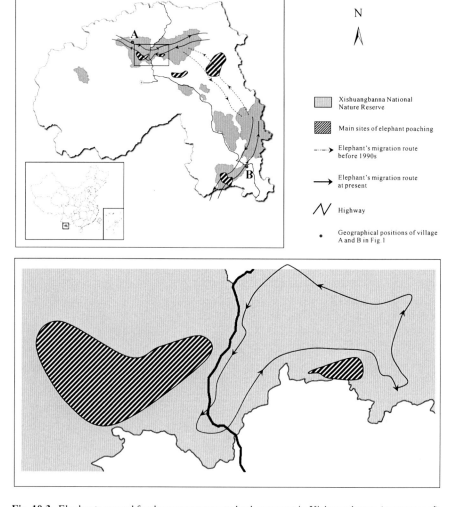

Fig. 10.3 Elephants moved freely among separated sub-reserves in Xishuangbanna (*upper panel*). After the 1990s, however, the large populations of elephants were divided into small isolated populations by habitat degeneration, elephant poaching, and other human disturbances. The highway, passing through or stretching across sub-reserves, especially disrupted elephant movement, making elephant-related problems more serious. In the Mengyang sub-reserve, for example, obstructed by the highway, elephants moved at random on the eastern side of the highway, creating more damage (*bottom*)

caught at that time was sent to Shanghai Zoo as a exhibition animal. The whole procedure of "capturing elephant" was recorded and edited as a scientific and education film that was shown all over China for many years after 1973. The film told the public about the untamed animals and tropical forests and rich biological

10 The Neglect of Traditional Ecological Knowledge on Wild Elephant-Related... 173

Table 10.3 Perception and attitude of interviewees toward wild elephants

Items	Perception and attitudes	Number	%
Emotion toward wild elephants	Hate	81	19.7
	Like	44	10.7
	Like if without problems	272	66.0
	Unknown	15	3.6
Coexistence with elephants	Cannot	201	48.8
	Far away from village	194	47.1
	Indifference	10	2.4
	Unknown	7	1.7
Causes of elephant-related problems	Shelter shortage	34	4.5
	Food shortage	165	21.7
	Excessive population	109	14.3
	Are fond of being close to village	182	23.9
	Are fond of crops	263	34.6
	Unknown	7	1.0
Approach to mitigate human–elephant conflicts	Inhabitants leave	88	19.7
	Elephants leave	249	55.8
	Reduction of rubber plantation	42	9.4
	Reduction of elephants	67	15.1

resources in southwestern China, as well as the powerful humans. Since then, more immigrants have moved into Xishuangbanna to develop tropical plantations.

The rubber plantation is one of the main factors worsening elephant-related problems in Xishuangbanna. Rubber was regarded as a key national raw material resource. The Chinese state council, in 1951, decided to develop rubber plantations in south and southwestern China. In Xishuangbanna, the most suitable areas (tropical rainforest) for rubber cultivation are where wild elephants preferred to live. Many rubber plantations developed by immigrants, followed by local people, in these areas destroyed the wild elephants' habitats. In 1976, forests covered approximately 70 % of Xishuangbanna, but by 2003 they covered less than 50 %. During this period, the area of tropical rainforests (where wild elephants preferred to live) was reduced by 67 % (Li et al. 2007). Wild elephants were driven to higher uplands. In addition, after the cold-resistant varieties of rubber trees were used, higher uplands with lower temperature were also cultivated with rubber trees (Chen and Wu 2009). The wild elephants' habitats decreased widely.

Nevertheless, local people preferred to have more rubber trees; 81.3 % of the 412 interviewees believed the expansion of cultivated land, especially rubber plantations, was reasonable; they would even plant more rubber trees if they had extra land. From data listed in Table 10.2, a positive relationship was found between cultivated areas and elephant-related problems in 16 villages; the Spearman's correlation coefficient was 0.635 (denotes 0.01 significance levels, $P = 0.008$), which implied that the larger was the cultivated area, the more serious were the elephant-related problems. Of 412 interviewees, 56.3 % understood clearly that wild food shortage and crop preference were the main causes for elephant-related problems, but only 9.4 % considered mitigating elephant-related problems by reducing rubber trees; 70.9 % hoped to reduce the elephant population or to move elephants away from areas where elephant-related problems happened (Table 10.3).

In fact, the rapid growth of rubber cultivation was only aimed at earning money. Local people, companies, and the government in Xishuangbanna were fond of sharing income from raw rubber products. They knew the negative impacts of rubber cultivation and tried to improve the situation, but became weak and inefficacious, because they never considered stopping the increasing growth of rubber cultivation. Facing the heavy cost of elephant-related problems, they had no choice but asked to for compensation from conservation departments of government.

To improve the social and economic situation in Xishuangbanna, highways of different grades were built, forming the dense highway net. The development of a highwaynet met human demands, but blocked the migration of wild elephants, reinforcing elephant-related problems (Fig. 10.3).

The Simao-Xiaomengyang-Mohan Highway, a part of the Kunming-Bangkok Highway, opened in 2006. It passed through Mengyang sub-reserve and cut off the movement of elephants from east to west; and it also stretched across Mengla and Shangyong sub-reserve and disrupted the movement of elephants between the two sub-reserves. In Mengyang sub-reserve, 2 overpasses and 23 underpasses were built for the elephants (Zhang et al. 2006); the elephants used 11 of the 25 underpasses sometimes but rarely used the other 14 passes (Chen et al. 2008). They climbed on the highway, or stopped and rushed blindly when they came to the highway from the east. Elephants were sequestered into two large groups after the highway opened in 2006, one in the western remote mountains, another in an area about 10 km wide on the eastern side of the highway, where elephant-related problems recently became serious. In Mengla and Shangyong sub-reserves, elephant movement was also blocked. A group of more than 30 elephants from Shangyong sub-reserves passed the highway to Mengla sub-reserve in January 2009 and did not return. Later, in January 2011, 21 elephants in Mengla sub-reserve fled westward to Mengban and Yaoqu, where elephants had not appeared in the past 30 years.

10.5 Discussion and Conclusion

We provided a case to discuss the reason why human–wildlife conflict became common and serious. The environment-friendly TEK, including a simple ecological view of forest priority, a biocultural rural landscape, and elephant culture, not only played an important role in indigenous daily life but also benefited the conservation of forests and all wildlife in the forests. The neglect of TEK and the rapid spread of exotic selfish consciousness increased elephant-related problems.

National and local government, the nature reserve, and experts made great efforts to reduce elephant-related problems in China (Chen et al. 2006); distinct achievements were noted (Wu 2008). Unfortunately, local people, companies, and governments at different levels chose what they wanted when they made decisions to balance conservation and development. People ignored or even hated elephants surviving in wildness; few of them were willing to make any concession. Our study indicated that 95.9 % of the 412 interviewees did not like coexisting with the wild elephants; 70.9 % of interviewees hoped to reduce the elephant population or to move elephants away from places where elephant-related problems happened (Table 10.3). In contrast, all over Xishuangbanna, in both urban and rural areas, elephant designs and sculptures, as local symbols, were commonly seen anywhere such as on roads, bridges, buildings, parks, clothes and personal adornment, scenic spots, handiwork, literature, and art, creating a richly atmospheric elephant culture. Such elephant culture showing enthusiastically eroded the environment-friendly TEK with elephant adoration and love. Therefore, it is necessary to consider cultural factors to mitigate elephant-related problems.

In conclusion, it could be understood clearly that uncivilized human behavior invaded elephant habitats, provoking the aggressive behavior of elephants. For the even more serious elephant-related problems, human population growth, habitat fragmentation, change of elephants' ecological behaviors, and diet and water availability were the ostensible reasons; the fundamental reasons were the neglect of environment-friendly TEK and rapid spread of exotic selfish consciousness. Therefore, it is necessary to refigure moderate and environment-friendly TEK to mitigate elephant-related problems and to build a sustainable biocultural landscape in Xishuangbanna.

Acknowledgments Foundation item: Under the auspices of National Natural Science Foundation of China (No. 30870431)

References

Chen MY, Wu ZL, Dong YH, Liu DJ, Chen YP, Yang SJ (2006) Research on Asian elephant in China. Science Press, Beijing

Chen MY, Li ZL, Dong YH (2008) Wild elephant in China. Yunnan Science and Technical Press, Kunming

Chen YF (2005) Looking for pacification with Asian elephant. Chin Wildl 3:9–10

Chen YP, Wu ZL (2009) Ecological issues and loss risk of cold resistant rubber germplasm resource in Xishuangbanna. Chin J Appl Ecol 20:1613–1616

Dao GD (1996) Story on the historical culture of Dai Nationality. Nationality Press, Beijing

Elvin M (2004) The retreat of the elephants: an environmental history of China. Yale University Press, New Haven and London

Feng LM, Wang ZS, Lin L, Yang SB, Zhou B, Li CH, Xiong YM, Zhang L (2010) Habitat selection in dry season of Asian elephant (*Elephas maximus*) and conservation strategies in Nangunhe National Nature Reserve, Yunnan, China. Acta Theriol Sin 30:1–10

Gao LS (1998) The honest natural ecological world view of Dai people. Ideological Front 2:45–50

He QC, Wu ZL (2010) Research advance on the current situation and management of human–wildlife conflicts in China. Sichuan J Zool 29:141–143

He QC, Wu ZL, Zhou W, Dong R (2011) Perception and attitudes of local communities towards wild elephant-related problems and conservation in Xishuangbanna, southwestern China. Chin Geogr Sci 21:629–636

Hoare RE, Du Toit JT (1999) Coexistence between people and elephants in African savannas. Conserv Biol 13:633–639

Lang XD, Peng MC, Wang CY, Li YJ, Duan HX, Li XH, Jiang WG (2008) Assessment of Asian elephant habitat quality status quo in Nangunhe Valley, China. J Yunnan Univ Nat Sci 30:415–423

Li HM, Aide TM, Ma YX, Liu WJ, Cao M (2007) Demand for rubber is causing the loss of high diversity rain forest in SW China. Biodivers Conserv 16:1731–1745

Liu YF (2005) Application research on the participating popularization in agricultural technology. Rev China Agric Sci Technol 7:68–71

Loarie SR, Van Aarde RJ, Pimm SL (2009) Fences and artificial water affect African savannah elephant movement patterns. Biol Conserv 142:3086–3098

Ma Y, Miao YH (1989) The comparative study of portion land system in Xishuangbanna and the land systems in the Western Zhou Dynasty. Yunnan People's Press, Kunming

Nyhus P, Tilson R (2004) Agroforestry, elephants, and tigers: balancing conservative theory and practice in human-dominated landscapes of Southeast Asia. Agric Ecosyst Environ 104:87–97

O'Connell-Rodwell CE, Rodwell T, Rice M, Hart LA (2000) Living with the modern conservation paradigm: can agricultural communities co-exist with elephants? A five-year case study in East Caprivi, Namibia. Biol Conserv 93:381–391

Shou ZH, Gao YT, Lu CK (1959) Elephants in South Yunnan. Chin J Zool 5:206–209

Sukumar R (2006) A brief review of the status, distribution and biology of wild Asian elephant. Int Zoo Yearbk 40:1–8

Wen HR (1995) Studies on the change of flora and fauna in historical phases of China. Chongqing Press, Chongqing

Wu ZL (1997a) Application of participation rural appraisal (PRA) method in the study of species and landscape variation. Chin J Appl Ecol 8(Suppl):89–94

Wu ZL (1997b) An ecological study on the Holy Hill Tradition of the Bulang in the Mengyang Reserve of Xishuangbanna Nature Reserve. Chin J Ecol 16:45–49

Wu ZL (2008) Management achievement evaluation of Xishuangbanna National Nature Reserve. Science Press, Beijing

Wu ZL, Peng MC, Yang ZB (1997) Settlement pattern in Mengyang reserve of Xishuangbanna. Chin J Appl Ecol 8(Suppl):23–37

Wu ZL, Xu FA, He QC, Zhou W (2007) The emerging wild elephant disturbance areas in Xishuangbanna—cases in two villages: Nanping and Hetu. In: Liu SY, Yang ZS, Zhao QG (eds) Studies on land use and human-land relation in mountainous regions of China. Chinese Science and Technical Press, Beijing

WWF (2008) Common ground: solution for reducing the human, economic and conservation cost of human wildlife conflict. Species Programme, WWF International, Gland

Xu YC, Jiang HQ, Quan F (1987) Comprehensive investigation report on Xishuangbanna Nature Reserve. Science and Technical Press, Kunming

Yan F (1990) Talking about the elephant culture of Dai nationality. In: Wang YZ (ed) The palm leaves Buddhism sutra culture treatises. Yunnan People's Press, Kunming

Yu YZ, Wang KL, Chen HS, Xu LL, Fu W, Zhang W (2009) Farmer's perception and response towards environmental migration and restoration plans based on participatory rural appraisal: a case study of emigration region in the karst of Southwestern China. Acta Ecol Sin 29:1170–1180

Zhang SL, Tian Y, Li W (2006) Leave corridor for wildlife—effects of highway on wildlife and engineering strategy in tropical rain forest region. China Highway 11:102–104

Chapter 11
Effects of Tropical Successional Forests on Bird Feeding Guilds

Eurídice Leyequién, José Luis Hernández-Stefanoni,
Waldemar Santamaría-Rivero, Juan Manuel Dupuy-Rada,
and Juan Bautista Chable-Santos

Abstract Previous studies have emphasized the importance of including not only the potential and costs of different land use/land cover alternatives on carbon sequestration but that also there is a need to study the impact of the resulting land cover changes on biodiversity. Tropical forests are undergoing rapid transformation as the result of human activities, which have created more than 600 million ha of secondary vegetation. In particular, tropical dry forests (TDF) are under great pressure caused by conversion to agriculture and other land uses, resulting in a heterogeneous landscape mosaic of secondary forest in different stages of succession or small forest remnants embedded in a matrix of agriculture. Changes in the landscape mosaic affect patterns of animal species abundance and distribution and, consequently, influence community composition. Despite the prevalence of successional forests, few studies have examined their influence on higher trophic levels such as bird communities. The aim of this study was to examine the relative influence of successional age, vegetation structure, and landscape structure on bird guild composition in a TDF region in the Yucatan Peninsula, an important area for migratory birds characterized by high avian endemism. Species composition of different bird feeding guilds was calculated for 274 plots of bird point counts, and vegetation structure was obtained from a vegetation survey in the same plots. We used a land cover thematic map, derived from a supervised classification of SPOT5 satellite imagery, to calculate landscape pattern metrics. Species composition of birds was related to structure of vegetation, landscape metrics of patch types, and principal coordinates of neighbor matrices (PCNM) variables using canonical correspondence analysis (CCA). Overall, bird feeding guilds were

---

E. Leyequién (✉) • J.L. Hernández-Stefanoni • W. Santamaría-Rivero • J.M. Dupuy-Rada
Centro de Investigación Científica de Yucatán, Mexico
e-mail: leyequien@cicy.mx

J.B. Chable-Santos
Universidad Autónoma de Yucatán, Mexico

N. Nakagoshi and J.A. Mabuhay (eds.), *Designing Low Carbon Societies in Landscapes*, Ecological Research Monographs,
DOI 10.1007/978-4-431-54819-5_11, © Springer Japan 2014

influenced by stand age, vegetation structure, and spatial structure of sampled data, and marginally by landscape composition and configuration, but varied in their response and susceptibility to habitat changes. Sound conservation and management should take into account forest specialist species, which require pristine or late secondary forests to persist, and should consider a possible decline in species that may occur in secondary forests but would otherwise use mature forests, as well as declines in species which may feed in a variety of habitats but may not necessarily reproduce in all habitat types.

Keywords Dry tropical forest • Feeding guilds • Mexico • Neotropical birds • Secondary forests • Yucatan Peninsula

11.1 Introduction

Old-growth forests throughout the tropics are increasingly reduced, fragmented, and degraded. Second-growth forests are increasing in extent and in importance for conservation and the provision of resources such as carbon and ecosystem services (Wright 2005). Previous studies have emphasized the importance not only of including the potential and costs of different land use/land cover alternatives on carbon sequestration but that also there is a need to study the impact of the resulting land cover changes on biodiversity (Mathews et al. 2002; Witt et al. 2011). A more comprehensive analysis of carbon sequestration by secondary forests in tropical dry forests would consider not only predictable profits but also the additional environmental benefits and costs associated with secondary forests.

Secondary forests are increasing in area around the world, and specifically in the tropics there is a high conversion rate from primary to secondary forests after abandonment of agriculture (Wright 2005). These forests are important for local, regional, and global processes, from the regulation of global carbon cycles (Achard et al. 2002) to the maintenance of species diversity (Brown and Lugo 1990). Particularly, tropical secondary forests are important in the conservation of animal species, particularly under conditions suitable for forest recovery (Dunn 2004). In tropical America there is an estimated area of 38 million hectares of secondary forests (ITTO 2002), where slash-and-burn shifting cultivation is one of the most widespread practices contributing to the conversion of natural forests to secondary forests (Raman et al. 1998). Tropical dry forests (TDF) in particular are among the most extensive land cover types in the tropics and the most threatened ecosystems in tropical America (Sanchez-Azofeifa et al. 2005; Quesada et al. 2009; Portillo-Quintero and Sánchez-Azofeifa 2010). Depending on their intensity, human activities produce a mosaic of secondary forest in different stages of succession or result in small forest remnants embedded in a matrix of agriculture or pasturelands.

Birds are widely known taxa with a large amount of available ecological information (Dunning 1992; Bird Life International 2011) and are commonly used in- conservation assessment and monitoring (Furness and Greenwood 1993; Mayer and Cameron 2003). Several studies have demonstrated that bird community structure in tropical rainforests is influenced by the changes in vegetation structure associated with succession (Raman et al. 1998; Raman and Sukumar 2002). In contrast and compared to other ecosystems, there is little ecological research on successional changes in bird communities in TDF (Leyequién et al., unpublished data). Moreover, successional vegetation patterns in TDF differ from those in humid forests in terms of species diversity, stem density, and convergence of plant species composition toward mature forest species (Lebrija-Trejos et al. 2008).

The rate at which animal communities recover during forest succession depends on landscape composition and configuration, which reflect habitat loss and fragmentation (Leyequién et al. 2010). In tropical forests, habitat loss and fragmentation affect bird communities through a decrease of suitable habitat and its degradation, the separation of habitat fragments by an anthropogenic matrix (e.g., croplands, pasturelands, urban areas), and by increasing edge effects (Arriaga-Weiss et al. 2008).

Feeding guilds are important to understand changes in animal community structure and functioning following a disturbance (Gray et al. 2007), such as the conversion of mature forests to agriculture and the subsequent forest succession after abandonment. There is evidence that different bird feeding guilds have dissimilar resilience to habitat disturbance (May 1982; Pearman 2002; Dunn 2004; Gray et al. 2007), and forest insectivores are especially at risk (Canaday 1996; Ford et al. 2001) because of their high sensitivity to habitat disturbance and fragmentation (Şekercioğlu et al. 2002). For example, many bark insectivores are forest dependent and/or specialists in their foraging strategy (e.g., army ant-following species such as *Dendrocincla*, or feeding often at bromeliads such as *Xiphorhynchus*); although other bark insectivores are facultative feeders, as are woodpeckers, the majority need large trees for foraging and are not present in heavily deforested areas (Neotropical Birds Online 2011).

In Mexico, only 27 % of TDF remains undamaged, and the most extensive contiguous areas of TDF occur in the Yucatan Peninsula, where natural and human disturbances have produced an intricate landscape mosaic. Despite the prevalence of successional forests, few studies have examined their influence on higher trophic levels such as bird communities. The Yucatan Peninsula is a region of high avian endemism as well as an important area for migratory birds. This study examined the relative influence of successional age, vegetation structure, and landscape structure on bird guild composition in a TDF region in the Yucatan Peninsula, and thus under different potential carbon sequestration scenarios. We put a special emphasis on bark insectivores as they are a group vulnerable to habitat degradation.

11.2 Methods and Materials

11.2.1 Study Area

The study area is located in the Yucatan Peninsula, México, in the South of *Ticul* hills (20°01′7″–20°09′36″ N, 89°35′59″–89°23′31″ W) (Fig. 11.1). The region encompasses TDF classified as Medium Stature Semi-Deciduous Forest where the majority of plant species (50–75 %) drop their leaves during the dry season (Miranda and Hernández 1963). The climate is tropical, warm, subhumid, with a mean annual temperature of 26 °C and mean annual precipitation of approximately 1,000–1,200 mm, concentrated between May and October, and a marked dry season between November and April. The landscape consists of limestone hills alternating with flat areas, and the elevation ranges from 60 to 160 m a.s.l. (Flores and Espejel 1994). The landscape is dominated by secondary forests with different age of abandonment, patches of agricultural areas (mainly under slash-and-burn shifting cultivation, i.e., *milpa*), and rural settlements (Dupuy et al. 2011).

11.2.2 Sampling Design

We used a hierarchical sampling design. First, we selected 23 landscape units of 1 km^2 considering the whole range of forest fragmentation; second, we located 12 plots within each landscape unit, using a stratified random design based on (1) stand age (determined using interviews with local residents who had inhabited the area >40 years and owned/practiced agricultural activities on the land), and (2) for plots with ≥15 years of abandonment, topographic position: hills or flat areas.

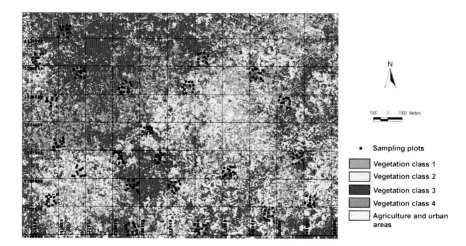

Fig. 11.1 Study area in the Yucatan Peninsula, Mexico

11 Effects of Tropical Successional Forests on Bird Feeding Guilds 181

This approach resulted in four vegetation classes (strata): (1) 3- to 8-year-old secondary forest, (2) 9- to 15-year-old secondary forest, (3) >15-year-old secondary forest on flat areas, and (4) >15-year-old secondary forest on hills (for details, see Dupuy et al. 2011).

11.2.3 Avifauna

We assessed bird guild species richness and composition on 274 point-counts with different stand age ranging from 3 to 70 years old [we excluded two point-counts from the total of 276 plots because they were in recently abandoned (<1 year) *milpa*]. Point-counts were located in the same location as for the vegetation sampling (see Sect. 11.2.4). The bird survey was carried out during the reproductive season (May 2009 and May 2010) and migratory season (November 2008 and November 2009). We used a modified double-observer method (Nichols et al. 2000; for full description see Leyequién et al. 2007), registering all birds seen or heard for a period of 12 min within a fixed radius of 60 m. Minimum distance among plots was 200 m to avoid double counting. The total sampling time was 203.2 h.

All birds were classified into feeding guilds based on their primary food source and their foraging strata using existing literature and field observations (Milesi et al. 2002; Gray et al. 2007; MacKinnon and Wood 2009, personal communication). We excluded some guilds based on the following criteria: (1) carnivores and full aerial birds were excluded because of their low number of records, and (2) omnivores were excluded because they are highly resilient (insensitive) to habitat disturbance and forest succession. We used only nine feeding guilds for the analysis (Appendix 11.1). Although we are aware that various species in the selected feeding guilds are facultative species that could shift to secondary resources depending on ad hoc conditions, we grouped bird species into one exclusive feeding guild to maintain sufficiently large sample sizes for each guild class (see review by Philpott et al. 2008; Şekercioğlu et al. 2004).

11.2.4 Vegetation Variables

A total of 274 plots of 200 m^2 were sampled in which only woody plants more than 5 cm dbh (diameter at breast, 1.3 m height) were registered, hereafter called adults. Woody plants 1–5 cm dbh, hereafter called saplings, were sampled in a nested 50 m^2 subplot. All individuals were taxonomically identified at species level or the closest possible taxon, and their height and the dbh of each stem were measured (see Dupuy et al. 2011 for more details). We calculated various vegetation variables but only ecologically relevant and statistically nonredundant variables were used in the analysis: stand age (AGE), total species richness (SPPT), species richness of adults (SPPa), species richness of saplings (SPPs), abundance of adult individuals (ABUNa), abundance of saplings (ABUNs), total basal area (ABT), total basal area of adults (ABa), total basal area of saplings (ABs), mean height of adults (Ha), total density (TD), mean height of saplings (Hs), density of adults (Da), density of saplings (Ds), and Simpson (SIMP) and Shannon (SHAN) diversity indices.

11.2.5 Landscape Variables

We used a classified map derived from a SPOT5 satellite imagery acquired on January 2005 to calculate the landscape metrics for the 23 landscape units, using the software Fragstats (McGarigal et al. 2002). The classified map had an overall accuracy of 75.6 % and Cohen's k statistic of 0.7 (Hernández-Stefanoni et al. 2011).

The landscape metrics considered for the analysis are based on patch-type indices and were selected considering variables that quantify different aspects of landscape configuration, that are ecologically relevant, and are commonly used in the landscape ecology literature for the taxonomic group of interest (Mazerolle and Villard 1999). Based on these criteria, six indices were selected in this study: percentage of land of each patch type (PLAND), patch density (PD), edge density (ED), mean area weighted shape index (SHAPE), mean area weighted Euclidean nearest neighbor distance (ENN), and total edge contrast index (TECI). (For details of how metrics were calculated, see Hernández-Stefanoni et al. 2011).

11.2.6 Spatial Structure of Data

A set of explanatory spatial variables was generated from the geographic coordinates of the sampling plots, using a principal coordinate of neighbor matrices (PCNM) analysis (Borcard et al. 2004). This set of variables (called PCNM vectors) represents a spectral decomposition of the spatial relationships among the sampling plots and provide a better representation of the spatial structure present in the sampling data than simple geographic coordinates (Borcard et al. 2004). PCNM vectors are also uncorrelated variables that can be used as predictors in a canonical correspondence analysis (CCA) because they are not subject to multicollinearity errors (Borcard and Legendre 2002). We obtained 60 PCNM vectors, among which 36 had positive values and a significant autocorrelation ($P < 0.05$, tested by Moran's I.)

11.2.7 Data Analysis

We calculated species richness for all feeding guilds jointly and for bark insectivores alone, and in both cases we used the number of species-by-point data. To compare observed species richness among forest classes, we computed sample-based rarefaction (MaoTau) using EstimateS V8.2 (Colwell 2005) and individual-based rarefaction for a comparable number of individuals (1,169 for all feeding guilds jointly and 64 for bark insectivores) using EcoSim700 (Gotelli and Entsminger 2001).

For the species guild composition analysis, we examined the strength of species–environment relationships, whereby we tested for significant effects of three sets of variables: (1) vegetation structure (including stand age), (2) landscape pattern metrics (composition and configuration), and (3) spatial structure of sampling plots (for detailed description, see Table 11.1). To test for significant effects we used constrained analysis (i.e., CCA).

Two different data sets were tested: (1) the number of individuals per feeding guild, and (2) the number of species per feeding guild. For the CCA we used a biplot scaling with interspecies distances, no data transformation, and a Monte Carlo permutation test with 9,999 permutations under the reduced model (to minimize type I error). We tested for the significance of the first axis and of all canonical axes. Finally, to separate the marginal (independent) and conditional (partial) effects of each set of variables we used a variance partitioning method (Borcard et al. 1992)

We also estimated the predictive capacity of stand age for explaining abundance and species richness for all feeding guilds using forward simple linear regressions (SPSS ver. 11.5); when needed, the dependent variable was log-transformed to fulfil the assumption of linearity.

11.3 Results

11.3.1 Bird Feeding Guild Abundance, Species Richness, and Diversity in Forest Classes

A total of 13,721 individuals of 128 species were included in the analysis; 11 species were excluded because they were raptors and fully aerials, or omnivores. For all feeding guilds, the observed, rarefied, and estimated species richness did not differ significantly among forest classes (Table 11.2). For bark insectivores (BI), observed species richness was significantly higher in >15-year-old forests on flat areas than in >15-year-old forests on hills; estimated species richness was higher in >15-year-old forests on flat areas compared to 3- to 8-year-old forests (Table 11.2).

Table 11.1 Description of biophysical variables grouped into three sets of variables: vegetation structure, landscape structure, and spatial structure of sampling sites

Set of variables	Description	Code
Vegetation set		
Stand age	Age after abandonment following slash-and-burn shifting cultivation	AGE
Total species richness	Total number of woody plant species per plot, including adults and saplings	SPPT
Species richness for adults	Number of woody plant species >5 cm dbh [diameter at breast (1.3 m) height] per plot	SPPa
Species richness for saplings	Number of woody plant species 1–5 cm dbh [diameter at breast (1.3 m) height] per plot	SPPs
Abundance of adult individuals	Number of individuals of adults per plot	ABUNa
Abundance of sapling individuals	Number of individuals of saplings per plot	ABUNs
Total basal area	Basal area (m^2/ha) of all woody plants per plot, including adults and saplings	ABT
Total basal area for adults	Basal area(m^2/ha) of all woody plants for adults per plot	ABa
Total basal area for saplings	Basal area (m^2/ha) of all woody plants for saplings per plot	ABs
Mean height of adults	Mean canopy height (m) for adults per plot	Ha
Mean height of saplings	Mean canopy height (m) for saplings per plot	Hs
Total density	Total number of individuals per hectare	TD
Density for adults	Number of adults per hectare	Da
Density for saplings	Number of of saplings per hectare	Ds
Simpson diversity index	Computed for all woody plants, including adults and saplings	SIMP
Shannon diversity index	Computed for all woody plants, including adults and saplings	SHAN
Landscape structure set		
Proportional abundance of each patch type	High values indicate dominance of certain successional patch type	PLAND
Patch density	High values indicate higher number of small patches per area unit	PD
Edge density at the class level	High values indicate higher edge density per successional class	ED
Overall shape complexity	High values indicate more irregular patch shape	SHAPE
Patch isolation		ENN
Edge contrast	High values indicate higher landscape fragmentation	TECI
Spatial structure set		
PCNM vectors	Represents a spectral decomposition of the spatial relationships among the sampling plots and provides a better representation of the spatial structure present in the sampling data than simple geographic coordinates	PCNM

Table 11.2 Changes in bird feeding guild species richness (observed and estimated) and diversity for all feeding guilds and for bark insectivores only: rarefaction for all guilds, 1,169 individuals; rarefaction for bark insectivores, 64 individuals. Different letters (a-b) indicate significant differences according to Tukey's post hoc test (P < 0.05)

Successional class	Jointly feeding guilds								Only bark insectivores						
	Total number of samples	Total number of individuals	Observed number of species	95 % CI (Mao Tau)	Number of species (rarefied)	95 % CI (Mao Tau)	Estimated number of species	95 % CI (Chao 1)	Total number of individuals	Observed number of species	95 % CI (Mao Tau)	Number of species (rarefied)	95 % CI (Mao Tau)	Estimated number of species	95 % CI (Chao 1)
3–8 year	46	1,461	93[a]	82.96–103.04	95.08[a]	91–97	94.23[a]	90.07–111.02	39	6[ab]	2.76–9.24	7.63[a]	6–8	6.38[a]	6.12–9.73
9–15 year	77	2,531	90[a]	80.36–99.64	92.52[a]	88–97	99.22[a]	96.11–113.94	84	9[ab]	6.26–11.74	8.90[a]	7–11	9.42[ab]	9.12–12.96
>15 year on flat	87	2,934	98[a]	89.54–106.46	91.93[a]	86–97	99.03[a]	96.84–111.05	162	10[b]	8.13–11.87	8.73[a]	7–11	10.95[b]	10.72–13.74
>15 year on hills	64	2,138	91[a]	81.05–100.95	92.93[a]	89–96	96.31[a]	93.38–110.58	119	7[a]	3.87–10.13	6.39[a]	5–8	8.45[a]	8.38–9.49

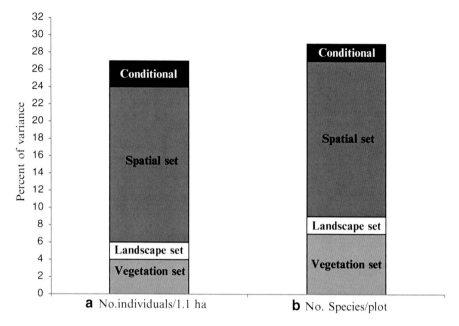

Fig. 11.2 Amount of variance explained by each set of variables (marginal effect), and the conditional effect explained by all sets of variables for (**a**) number of individuals and (**b**) number of species. Unexplained variance is not depicted

11.3.2 Guild Species Composition: Relationship to Stand Age, Vegetation Structure, Landscape Composition and Configuration, and Spatial Structure of Data

11.3.2.1 Number of Individuals per Feeding Guild

The percentage of explained variance by the first two axes, using the number of individuals per feeding guild, was 27 % (first axis: F ratio = 1.58, P value = 0.0001), where the marginal effect of the spatial structure of sampled data (PCNM vectors) explained the majority of variance (18 %), followed by the vegetation set (4 %), and last by the landscape set (2 %), whereas the conditional effect explained was 3 % (Fig. 11.2a).

The correlation biplot and correlation matrix resulting from the CCA yielded the following main results (Fig. 11.3a).

1. Overall, the first canonical axis had a strong positive correlation with the PCNM vectors, and with the vegetation variables, whereas there was a weaker association with landscape vegetation variables. The second canonical axis had a stronger positive correlation with stand age, spatial structure of data, structural vegetation variables (e.g., ABa, ABT, ABUNa, SPPT) and SIMP, whereas there was a weaker correlation with landscape variables (see Table 11.3).

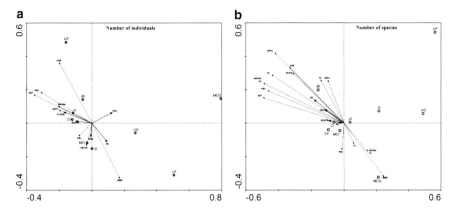

Fig. 11.3 Distribution of nine avian feeding guilds using (**a**) number of individuals and (**b**) number of species. Canonical correspondence analysis (CCA) ordination diagram with feeding guilds (*circle*) and environmental variables (*arrows*); first axis is horizontal, second axis vertical. Feeding guilds: *BI* bark insectivores, *CI* canopy insectivores, *N* nectarivores, *MCI* mid-canopy insectivores, *UI* understory insectivores, *CF* canopy frugivores, *UF* understory frugivores, *UG* understory granivores, *MCG* mid-canopy granivores. Vegetation set: *AGE* stand age, *SPPT* total species richness, *SPPa* species richness for adults, *SPPs* species richness for saplings, *ABUNa* abundance of adult individuals, *ABUNs* abundance of all sapling individuals, *ABT* total basal area, *ABa* total basal area for adults, *ABs* total basal area for saplings, *Ha* mean height of adults, *Hs* mean height of saplings, *TD* total density, *Da* density for adults, *Ds* density for saplings, *SIMP* Simpson diversity index, *SHAN* Shannon diversity index. Landscape structure set: *PLAND* proportional abundance of each patch type, *PD* patch density; for landscape configuration: *ED* edge density at the class level, *SHAPE* overall shape complexity, *ENN* patch isolation, *TECI* fragmentation. *PCNM* For display purposes the spatial structure vectors are not exhibited in the ordination diagram as well as the samples

2. Bark insectivores (BI) responded more strongly to the vegetation structure gradient than to the landscape structure, and it can be inferred that higher abundances of BI occurred in older age stands than in younger stands. The latter indicates that the number of individuals of BI increased with successional age (AGE), with the concomitant changes in vegetation structure (e.g., increase in total basal area, basal area, and abundance of adults); in contrast, the abundance of BI decreased with an increase of basal area and abundance of saplings. Moreover, abundances of BI increased as the values of PD decreased.
3. Canopy insectivores (CI) also responded more strongly to the vegetation structure gradient than to the landscape structure. And, similar to BI, the abundance of CI increased with an increase in total basal area, basal area, and abundance of adults, stand age, and moderately by an increase in PLAND, but presented a lower average abundance in comparison to BI.
4. The abundance of nectarivores (N) responded more strongly to the vegetation structure gradient (e.g., stand age, total basal area, basal area, and abundance of adults), rather than to the landscape structure, for which only PLAND had a moderate effect. In addition, the average abundance was lower compared to BI but higher compared to CI.

Table 11.3 Correlation matrix resulting from the canonical correspondence analysis (CCA) for individual-based and species-based analyses

Correlation matrix					
Individual-based analysis			Species-based analysis		
	Axis 1	Axis 2		Axis 1	Axis 2
ABa	−0.5091	0.3273	AB+5	−0.68596	0.26928
ABs	−0.1341	−0.1353	AB1-5	−0.01936	−0.22066
ABT	−0.5805	0.2997	ABT	−0.72776	0.21142
ABUNa	−0.3351	0.1818	ABUN+5	−0.72226	0.32978
ABUNs	−0.0573	−0.237	ABUN1-5	0.21846	−0.2321
AGE	−0.3273	0.6366	AGE	−0.48906	0.46068
Ha	−0.6582	0.2136	D+5	−0.72226	0.32978
SIMP	0.2778	−0.5766	D1-5	0.21846	−0.2321
SPPT	−0.3276	0.1377	DT	0.09152	−0.17248
PLAND	−0.0948	0.0397	H+5	−0.64878	0.39908
ENN	−0.0041	−0.0485	H-5	−0.21076	0.35816
PD	0.0516	−0.0664	SHAN	−0.4609	0.41338
SHAPE	−0.0549	0.0077	SIMP	0.36168	−0.4411
TECI	0.0693	0.0381	SPP+5	−0.62172	0.5808
PCNM1	−0.2319	−0.4026	SPP-5	−0.1452	0.35596
PCNM2	−0.387	0.5001	ED	−0.27258	0.19514
PCNM3	−0.15	−0.1218	PD	−0.05148	0.0022
PCNM4	0.3564	0.3807	SHAPE_AM	−0.14102	0.02046
PCNM5	0.0579	0.2292	TECI	−0.17534	0.1122
PCNM6	−0.2196	−0.327	PCNM1	0.0044	−0.43252
PCNM7	−0.2088	0.5346	PCNM2	−0.539	−0.15488
PCNM8	−0.0705	0.1947	PCNM3	−0.40304	0.44022
PCNM9	−0.1113	0.1113	PCNM4	0.10054	−0.2893
PCNM10	0.0633	−0.4392	PCNM5	0.02354	0.23914
PCNM11	0.1968	0.2316	PCNM6	−0.20196	0
PCNM12	0.1446	−0.4089	PCNM7	0.36608	0.16148
PCNM13	−0.6078	−0.2229	PCNM8	0.03234	0.52404
PCNM14	−0.4575	0.1188	PCNM9	−0.0803	−0.1441
PCNM15	−0.144	0.3168	PCNM10	−0.11594	−0.2255
PCNM16	0.4203	−0.3384	PCNM11	0.29326	0.18084
PCNM17	−0.0363	0.0321	PCNM12	0.05104	0.08778
PCNM18	0.2121	−0.0612	PCNM13	−0.31834	0.08668
PCNM19	0.3873	−0.1428	PCNM14	−0.23166	0.13706
PCNM20	0.4875	0.2253	PCNM15	−0.41382	−0.21098
PCNM21	0.1497	−0.3516	PCNM16	0.53064	0.42416
PCNM22	−0.0579	−0.3402	PCNM17	0.0319	0.16522
PCNM23	0.0684	0.1113	PCNM18	0.07106	−0.09746
PCNM24	0.3987	0.072	PCNM19	−0.09086	−0.03696
PCNM25	0.0351	−0.261	PCNM20	0.1738	−0.21252
PCNM26	−0.066	0.2058	PCNM21	0.25278	0.19206
PCNM27	−0.1914	−0.0318	PCNM22	0.11462	−0.07106
PCNM28	−0.2541	0.0225	PCNM23	−0.06512	0.02552

(continued)

11 Effects of Tropical Successional Forests on Bird Feeding Guilds

Table 11.3 (continued)

Correlation matrix					
Individual-based analysis			Species-based analysis		
	Axis 1	Axis 2		Axis 1	Axis 2
PCNM29	0.1056	0.0333	PCNM24	0.03564	−0.19976
PCNM30	0.2994	−0.3426	PCNM25	0.14102	−0.0176
PCNM31	0.0729	0.2817	PCNM26	−0.06292	0.0803
PCNM32	0.2421	−0.1872	PCNM27	0.03476	−0.09944
PCNM33	−0.285	−0.1725	PCNM28	0.0044	−0.07898
PCNM34	0.0012	0.0411	PCNM29	0.19602	−0.0286
PCNM35	0.0189	−0.2403	PCNM30	−0.0209	−0.0088
PCNM36	−0.2559	−0.0447	PCNM31	−0.15224	−0.05852
PCNM37	−0.0597	−0.0963	PCNM32	0.07392	−0.10076
			PCNM33	0.03256	−0.24772
			PCNM34	0.18084	0.18942
			PCNM35	−0.02574	−0.0979
			PCNM36	−0.20218	0.2167
			PCNM37	−0.11528	0.06204

5. Mid-canopy insectivores (MCI) also responded more strongly to the vegetation structure gradient, and increased with basal area and abundance of saplings, as well as with an increase in SIMP values. In contrast, abundance of this guild decreased with an increase of total basal area and basal area of adults. Specifically, the landscape variables that influenced MCI were PD and ENN, showing a positive relationship.

6. Understory insectivores (UI) showed patterns similar to MCI, responding more strongly to the vegetation variables (increased as basal area and abundance of saplings increased; and decreased with an increase of total basal area and basal area of adults). Landscape variables that influenced positively the abundance of UI were PD and ENN.

7. Canopy frugivores (CF) were positively influenced mainly by stand age and basal area (total and of adults). There was little evidence of quantifiable effects of landscape structure.

8. Understory frugivores (UF) showed an increase in abundance with an increase in the values of TECI and PD, and SIMP; and an increase in abundance with a decrease in basal area (total and of adults).

9. Understory granivores (UG) showed similar patterns to UF: increasing abundance with an increase in TECI and PD, and SIMP, and an increase in abundance with a decrease in basal area.

10. Mid-canopy granivores (MCG) presented higher abundance with an increase in the values of TECI.

11.3.2.2 Number of Species per Feeding Guild

The total amount of explained variance by the first two axes, using the number of species per feeding guild, was 29 % (first axis: F ratio $= 1.64$, P value $= 0.0001$); similarly to the individuals-per-feeding guild analysis, the spatial structure of sampling data (PCNM vectors) explained most of the variation (18 %), followed by the vegetation variables (including age) (7 %), and last by the landscape variables (2 %), whereas the conditional effect explained (2 %) (Fig. 11.2b).

The following are the main results from the correlation biplot, resulting from the CCA (Fig. 11.3b).

1. Overall, the first canonical axis had a stronger correlation with the spatial structure of sampling data, stand age, and vegetation variables, whereas in general landscape structure had a marginal negative correlation. The second canonical axis had a stronger correlation with the stand age and vegetation variables, whereas the spatial structure of sampling data and landscape structure had a marginal correlation (Table 11.3).

2. Bark insectivores (BI) were mainly associated with the gradient in vegetation age and structure, indicating that the number of BI species increased with successional age (AGE) with concomitant changes in vegetation structure (e.g., increasing tree height and basal area). In contrast, landscape structure played a marginal role in determining the number of BI species.

3. Increase in the number of canopy insectivore (CI) species was mainly associated with the vegetation gradient (e.g., abundance, density, basal area of adults) and stand age; in general, landscape structure variables (in both first and second axis) had a low overall correlation and therefore played a marginal role in explaining the number of CI species.

4. Increase in the number of nectarivore (N) species was predominantly influenced by an increase in the abundance, density, and basal area of saplings, and by an increase in SIMP, and showed a marginal influence of landscape structure.

5. Mid-canopy insectivores (MCI) responded more strongly to the vegetation structure gradient. The number of species of MCI was positively correlated with basal area and abundance of saplings, as well as with the SIMP values.

6. Understory insectivores (UI) responded to the vegetation structure variables and positively correlated to density and abundance of saplings, but negatively correlated to abundance, density, and basal area of adults.

7. Canopy frugivores (CF) were positively correlated with stand age and the concomitant changes in vegetation structure (e.g., increase in abundance and height of adults).

8. Understory frugivores (UF) showed a positive correlation with basal area of saplings and a fairly marginal correlation with landscape structure.

9. Understory granivores (UG) showed a positive correlation with basal area, abundance, and density of saplings, and a fairly marginal correlation with landscape structure.

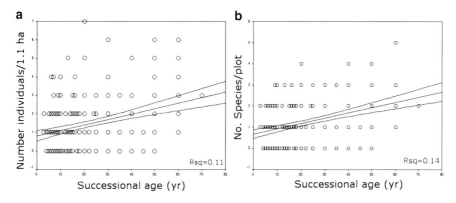

Fig. 11.4 Structural attributes of bark insectivores as a function of successional age: regression models for (**a**) number of individuals and (**b**) number of species

10. Mid-canopy granivores (MCG) were positively correlated to abundance, density, and basal area of saplings and SIMP, but negatively to abundance, density, and basal area of adults.

Stand age was positively related with the abundance (P value $= 0.001$, $R^2 = 11\,\%$) and species richness (P value $= 0.001$, $R^2 = 14\,\%$) of bark insectivores (Fig. 11.4.), whereas for the rest of the feeding guilds we did not find significant results.

11.4 Discussion

11.4.1 Trends in Bird Feeding Guild Species Richness and Diversity

Our results show that total species richness and diversity of bird feeding guilds in the current landscape do not differ significantly among vegetation age classes. These results differ from previous findings that species richness and diversity increase with successional age (May 1982; Raman et al. 1998), and may be partly attributed to the low level of deforestation and fragmentation of our study landscape, which is dominated by >15-year-old forests (nearly 60 % of the total area). Our results suggest that the avian community perceives the landscape as fairly homogeneous and may be able to cross the successional matrix. Additionally, in our study we did not consider seasonal variation as we used 1-year cycle data and used no discrimination between the reproductive and migratory seasons; hence, seasonal variation may be masked. In contrast, Santamaría-Rivero (2011) found, for the same study area, significant differences in species richness among vegetation age classes during the reproductive season, and it has been suggested that species mobility could decrease as a response of individuals protecting their breeding territory, constructing nests, or providing parental care (Volpato et al. 2009). In contrast, he found no differences during the migratory season; this pattern could be influenced by the presence of migratory birds that may exert pressure as potential competitors on feeding resources, thus forcing species with enough plasticity to occupy other habitats (Greenberg 1990).

Interestingly, in our study we found significant differences in species richness of bark insectivores among forest classes, which suggests that diversity of habitat specialists increases with succession and the associated changes in forest structure, that is, increase in tree size and hence in bark surface. The habitat characteristics associated mainly with late succession, such as large trees (Thompson et al. 1995) or climatic buffering (Greenberg 1992), may influence the habitat selection of bark insectivores (Smith et al. 2001). Additional research is needed to account for other specialized groups of forest-dependent species that are less tolerant of habitat disturbance.

11.4.2 Species Guild Composition: Relationship to Stand Age, Vegetation Structure, Landscape Composition and Configuration, and Spatial Structure of Data

The age of successional forests along with the related physiognomic and floristic changes in vegetation had perceptible effects on the feeding guilds. Interestingly, our results show that, in most cases, comparable patterns in bird feeding guild–environment relationships were found when using either the number of individuals per guild or number of species per guild. Other studies have found, when using abundance or presence/absence coded data to measure changes in bird communities, that the results for both approaches do not show strong differences and both data sets are highly correlated, indicating high similarity among their predictions (Bart and Klosiewski 1989).

In our study feeding guilds varied in their response and susceptibility to these factors, which suggests that this ecological trait is useful to detect differential responses to habitat disturbance, as has been found in other studies in wet tropical regions (Raman et al. 1998; Gray et al. 2007). Also, responses of bird species are presumably affected by their ecology and evolutionary history (Bennet and Owens 2002), such as phylogeny, body size, local population size, and geographic range, which were not taken into consideration in this study.

Overall, most of the variation in number of individuals and number of species per feeding guild was accounted for by the spatial structure of data and vegetation variables (including stand age), whereas landscape variables had a particular marginal influence in both analyses. The low marginal effect of landscape composition and configuration variables on bird feeding guilds suggests that, in this study, habitat characteristics play a greater role than landscape structure. In a tropical region, a comparable study found evidence that the bird community composition was more similar between the same age classes than with the adjacent vegetation regardless of distance between similar age class–vegetation patches (Raman et al. 1998), which stresses the importance of specific habitat attributes required by different bird feeding guilds regardless of the landscape structure. Also, the marginal effects of landscape structure found in this study were possibly affected by the landscape matrix, which is dominated by late succession (nearly 60 % corresponds to >15-year-old vegetation), suggesting that the avian community perceives the landscape as fairly homogeneous.

The most important set of variables influencing bird feeding guilds was the spatial structure of sampling plots; this result could imply that species composition is highly associated with avian dispersal (Haas 1995). However, it is possible that variation in space likely had a considerable environmental or biotic component that was not detected because some relevant environmental or biotic variables, such predation and competition, were not considered in our analysis

Another important element influencing bird feeding guilds is the relatively small influence of the landscape matrix that differentially affects the capacity of species to move among patches of dissimilar habitats, which in turn may provide foraging or breeding habitat and ultimately influence the local persistence and abundances of species (Renjifo 2001). In our study, the landscape matrix is relatively homogeneous, possibly providing a nonhostile landscape for dispersal that allows generalists or species capable of thriving in a variety of habitats to persist.

As expected, bark insectivores (BI), constituting mostly forest-dependent species or specialists in their foraging strategy (e.g., army ant-following species such as *Dendrocincla*, or those feeding often at bromeliads such as *Xiphorhynchus*) in this study, were positively correlated with stand age and its associated changes in vegetation structure (e.g., increase in basal area and tree height). Dependency of BI on higher trees and thicker stems has been previously documented (Arriaga-Weiss et al. 2008), giving support to the hypothesis that forest birds follow the recovery of forest vegetation (Raman et al. 1998). In contrast, BI were negatively correlated to the abundance and density of saplings, which offer a comparatively lower surface area for feeding; such negative effects have been reported in other studies about forest succession (Raman et al. 1998) and other tropical forests (Thiollay 1994). In addition, canopy insectivores showed consistent patterns for both individuals-based and number of species-based analyses in which tree basal area (i.e., total basal area and basal area of adults) and abundance of adult plants were important predictors that showed a positive correlation. For insectivores, the structural complexity in vegetation, associated with older succession, has been proposed as being influential in providing opportunities for foraging (Willson 1974; Holmes et al. 1979; Terborgh 1985). Thus, an increase in vertical strata, tree height, and canopy cover provides potential new foraging opportunities (Bowman et al. 1990; Brady and Noske 2009). On the other hand, nectarivores (a feeding guild represented only by hummingbirds in this study) showed contrasting results between individuals-based and number of species-based analyses. The number of individuals showed a positive correlation with abundance and basal area of adult individuals, whereas the number of species showed a positive correlation with abundance, density, and basal area of saplings. The response in abundance suggests the use of secondary resources as supplementary food; especially, *Amazilia candida* and *A. yucatanensis*, the most abundant species in the older succession, are reported as facultative feeders consuming insects as a secondary resource (Arizmendi et al. 2010). Additionally, higher abundance of insects has been reported as related to forest complexity, which is typically associated with older succession (Şekercioğlu et al. 2002). In contrast, the positive correlation of nectarivore species richness with abundance and density of saplings (associated with early succession) could be explained by the low sensitivity to disturbance of the hummingbird species grouped in this feeding guild; moreover, nectarivore species have been reported as frequent visitors of earlier successional stages (Robinson and Terborgh 1997) where flowering herbaceous and scrubby vegetation is more abundant. However, the observed trend could also be attributed to sampling bias, as in older successional stages these hummingbird species could be difficult to detect because of their small size and also the weak vocalizations of some of the species.

The abundance and species richness of mid-canopy insectivores (MCI) showed a positive correlation to basal area and abundance of saplings and the Simpson diversity index (dominance). A previous study in our research area found that the early successional stages are related to higher abundance and basal area of saplings where plant communities are dominated by few dominant species (Dupuy et al. 2011). The species composition of the MCI in this study includes mostly habitat generalists that reside or feed in open woodlands, plantations, clearings, scrubby vegetation, forest edge, or secondary forests (Shulenberg 2010); consequently, it is not surprising that MCI were associated with early succession. Also, understory insectivores (UI) presented similar patterns as the MCI, for both abundance and species richness, and similarly most of the species grouped as UI can inhabit second-growth forest and scrubby vegetation where the foliage is concentrated in the first meters above the ground and where food and microclimate conditions are suitable (Karr and Freemark 1983; Johns 1991).

Canopy frugivores (CF) showed a positive correlation with stand age and basal area, both for abundance and for species richness. The CF guild is composed of a wide array of species with unique natural history and ecology (Wiens 1989; Simberloff 1994); however, some species attributes appear to be related to vulnerability to habitat alteration. For example, body size may influence the response of CF, because large-bodied species (e.g., *Amazona albifrons*, and *Patagioenas flavirostris* with more than 130 g) tend to be more susceptible to habitat alteration (Raman and Sukumar 2002) as they may have smaller population sizes, lower reproductive rates, and larger home or geographic range requirements than small-bodied species (Gaston and Blackburn 1995). In addition, some species have specific reproduction strategies that link them to older succession; for example, species such as *Amazona albifrons*, *Aratinga nana*, and *Tityra semifasciata* require thicker and higher stems to construct their nests although they can forage in a wide range of habitats. In contrast, understory frugivores (UF) were positively correlated with early succession structural characteristics, which is not surprising as this guild was basically represented by *Crypturellus cinnamomeus*, a habitat generalist that inhabits secondary forests, thickets, and shrublands, where it usually hides under the dense understory (Bribiesca-Formisano et al. 2010). Moreover, the marginal influence of landscape structure in terms of patch density and higher fragmentation (TECI) could be partially explained by the low sensitivity to disturbance of *C. cinnamomeus*. In addition, understory granivores (UG) comprised mostly species that inhabit scrub or shrub vegetation, early successional forests, dense undergrowth, and for some species even open habitats. The majority of UG species also have low sensitivity to disturbance, which explains their positive correlation with early succession structural characteristics (e.g., abundance and density of saplings and a plant community dominated by one or few species), and the marginal positive correlation with patch density and fragmentation (TECI). It is noteworthy that the only two species with high sensitivity to disturbance, *Dactylortyx thoracicus* and *Meleagris ocellata*, occurred in low numbers in early succession or were very rare during the sampling. Finally, MCG were also composed of species that inhabit scrub vegetation, second-growth scrub, shrubby areas, and open woodlands, and commonly inhabit disturbed habitat; thus, these species are habitat generalists with no reported negative effect of habitat fragmentation. Two MCG species, *Zenaida asiatica* and *Cardinalis cardinalis*, have actually expanded their geographic range northward in past centuries, correlating with human habitation and agricultural activities (Shulenberg 2010).

The different responses among feeding guilds indicate that this trait may be a suitable predictor of avian susceptibility to habitat disturbance. Secondary forests are ubiquitous in the tropics and specifically in the TDF (Wright 2005; Sanchez-Azofeifa et al. 2005; Miles et al. 2006), and thus potentially play a key role in the conservation of animal communities as the bulk of tropical species will need to persist in degraded and secondary habitats (Bhagwat et al. 2008). However, it is important to remain cautious if sound conservation and management are to be done, because forest specialist species require pristine or late secondary forests to persist (Laurence 2007), species that may occur in secondary forests but would otherwise use mature forests may decline in secondary forests (Graham and Blake 2001), and species that possibly feed in a variety of habitats may not necessarily reproduce in all habitat types. Moreover, in coming decades the tendency in changes of forest cover is toward clearing, logging, fragmentation, or degradation (Rappole et al. 1992; Wright and Muller-Landau 2006), in which the type and rate of recovery could strongly affect species prone to extinction (Laurence 2007). Similarly, secondary forests are typically transient elements of the landscapes that undergo frequent reclearing, and which could go beyond survival thresholds of vulnerable species to disturbance. Finally, it is of great importance to account for the synergetic effect of both the temporal variability of secondary forests as a potential habitat and the landscape matrix in which secondary forests are immersed, especially in highly heterogeneous and fragmented landscapes.

In general, bark insectivores (BI) increased in species density and abundance with stand age. Although BI have been reported to adapt to microclimatic changes linked to disturbance (Johns 1991), our results and those of other studies show an increase in species richness in BI with tree height (Arriaga-Weiss et al. 2008), and consequently if the surface area of feeding sites increases (Thiollay 1994; Raman et al. 1998), which are features associated with late succession. Also, BI are associated with habitats that are structurally complex, which in turn provides a higher abundance of insects (Johnson 2000; Brady and Noske 2009). Our results highlight the importance of maintaining large forest tracts within the landscape matrix, if more specialized feeding guilds such as the BI are to be conserved.

11.5 Conclusion

Insofar as possible, when developing a national carbon sequestration strategy, the costs and benefits of different land use/land cover alternatives should be taken into account, which includes the potential effects on biodiversity. In this study, we provided important elements to assess the impact of a priori identified factors influencing bird diversity that may derive from carbon sequestration policies in a tropical dry forest: these include the variation in species diversity and composition of the bird communities in secondary forests of different ages. Cost–benefit analysis can be done when all relevant gains in a given land use/land cover scenario are included, such as the impacts on biodiversity.

Acknowledgments We thank James Callaghan and Kaxil Kiuic A.C. for logistic support. We also thank Rosalina Rodríguez Román, Filogonio May Pat, Fernando Tun Dzul, Víctor Marín Pérez, Ramiro Lara Castillo, Feliciano Pech Pinzón, Evelio Uc Uc, Mario Evelio Uc Uc, and Santos Armín Uc Uc for fieldwork and technical assistance. Funding for this research was provided by CICY, FOMIX-Yucatán (project YUC-2008-C06-108863) and CONACYT (CB-127800)

Appendix 11.1 Bird feeding guilds based on their primary food source and foraging strata, and presence of bird species in four forest classes

| Species | Feeding guild | Age of secondary forest | | | |
		3–8 years old	9–15 years old	>15 years old, flat land	>15 years old, hill
Crypturellus cinnamomeus	UF	P	P	P	P
Ortalis vetula	MCG	P	P	P	P
Meleagris ocellata	UG	A	P	P	A
Colinus nigrogularis	UG	A	P	P	A
Dactylortyx thoracicus	UG	P	P	P	P
Patagioenas cayennensis	CF	A	A	A	P
Patagioenas flavirostris	CF	P	P	P	P
Zenaida asiatica	MCG	P	P	P	P
Columbina passerina	UG	P	P	P	P
Columbina talpacoti	UG	P	P	A	A
Claravis pretiosa	UF	A	P	P	P
Leptotila verreauxi	UG	P	P	P	P

(continued)

11 Effects of Tropical Successional Forests on Bird Feeding Guilds 197

Appendix 11.1 (continued)

Species	Feeding guild	Age of secondary forest			
		3–8 years old	9–15 years old	>15 years old, flat land	>15 years old, hill
Leptotila jamaicensis	UG	P	P	P	P
Amazona albifrons	CF	P	P	P	P
Aratinga nana	CF	P	P	P	P
Piaya cayana	CI	P	P	P	P
Dromococcyx phasianellus	UI	P	P	P	P
Chordeiles acutipennis	UI	A	A	P	A
Campylopterus curvipennis	N	A	A	P	P
Anthracothorax prevostii	N	P	A	P	P
Chlorostilbon canivetii	N	P	P	P	P
Amazilia candida	N	P	P	P	P
Amazilia rutila	N	P	P	P	P
Amazilia yucatanensis	N	P	P	P	P
Trogon melanocephalus	MCI	P	P	P	P
Trogon collaris	MCI	A	P	A	P
Trogon caligatus	MCI	P	P	P	P
Momotus momota	MCI	P	P	P	P
Eumomota superciliosa	MCI	P	P	P	P
Melanerpes pygmaeus	BI	P	P	P	P
Melanerpes aurifrons	BI	P	P	P	P
Picoides scalaris	BI	A	P	P	A
Veniliornis fumigatus	BI	P	P	P	P
Colaptes rubiginosus	BI	P	P	P	P
Dryocopus lineatus	BI	P	P	P	P
Campephilus guatemalensis	BI	A	P	P	A
Dendrocincla anabatina	BI	A	A	P	P
Dendrocincla homochroa	BI	A	P	P	P
Sittasomus griseicapillus	BI	P	P	P	P
Xiphorhynchus flavigaster	BI	P	P	P	P
Thamnophilus doliatus	UI	P	P	P	P
Camptostoma imberbe	CI	P	P	P	P
Myiopagis viridicata	MCI	P	P	P	P
Elaenia flavogaster	CI	P	P	P	P
Oncostoma cinereigulare	MCI	P	P	P	P
Tolmomyias sulphurescens	CI	P	P	P	P
Platyrinchus cancrominus	UI	A	A	P	P
Contopus virens	MCI	P	A	P	A
Contopus cinereus	MCI	P	P	P	P
Attila spadiceus	CI	P	P	P	P
Empidonax flaviventris	MCI	P	P	P	P
Empidonax virescens	MCI	A	P	A	P
Empidonax minimus	MCI	P	P	P	P
Myiarchus yucatanensis	CI	P	P	P	P
Myiarchus tuberculifer	CI	P	P	P	P

(continued)

Appendix 11.1 (continued)

Species	Feeding guild	Age of secondary forest			
		3–8 years old	9–15 years old	>15 years old, flat land	>15 years old, hill
Myiarchus tyrannulus	CI	P	P	P	P
Pitangus sulphuratus	CI	P	P	P	P
Myiodynastes luteiventris	CI	A	A	P	A
Megarynchus pitangua	CI	P	P	P	P
Myiozetetes similis	CI	P	P	P	P
Tyrannus melancholicus	CI	P	P	P	P
Tyrannus couchii	CI	P	P	P	P
Tityra semifasciata	CF	P	P	P	P
Tityra inquisitor	CI	A	P	A	A
Pachyramphus major	CI	P	P	P	P
Pachyramphus aglaiae	CI	P	P	P	P
Saltator coerulescens	CF	P	P	P	P
Saltator atriceps	CF	P	P	P	P
Vireo griseus	MCI	P	P	P	P
Vireo pallens	MCI	P	P	P	P
Vireo flavifrons	CI	A	A	P	P
Vireo olivaceus	CI	P	P	A	A
Vireo flavoviridis	CI	P	P	P	P
Hylophilus decurtatus	CI	P	P	P	P
Cyclarhis gujanensis	CI	P	P	P	P
Cyanocorax yncas	MCI	P	P	P	P
Psilorhinus morio	CI	P	P	P	P
Cyanocorax yucatanicus	MCI	P	P	P	P
Pheugopedius maculipectus	UI	P	P	P	P
Thryothorus ludovicianus	UI	P	P	P	P
Uropsila leucogastra	MCI	P	P	P	P
Ramphocaenus melanurus	MCI	P	P	P	P
Polioptila caerulea	CI	P	P	P	P
Polioptila plumbea	CI	P	P	P	P
Hylocichla mustelina	UI	P	A	P	P
Turdus grayi	CI	P	P	P	P
Dumetella carolinensis	UI	P	P	P	P
Melanoptila glabrirostris	MCI	P	P	P	P
Vermivora cyanoptera	CI	A	A	A	P
Oreothlypis peregrina	CI	A	A	P	A
Parula americana	CI	P	P	P	P
Setophaga petechia	MCI	A	A	P	P
Setophaga magnolia	MCI	P	P	P	P
Setophaga caerulescens	CI	A	A	P	A
Setophaga virens	CI	P	A	P	P
Setophaga dominica	MCI	P	P	A	A
Mniotilta varia	MCI	P	P	P	P
Setophaga ruticilla	CI	P	P	P	P
Seiurus aurocapilla	UI	P	P	P	A

(continued)

11 Effects of Tropical Successional Forests on Bird Feeding Guilds

Appendix 11.1 (continued)

Species	Feeding guild	Age of secondary forest			
		3–8 years old	9–15 years old	>15 years old, flat land	>15 years old, hill
Geothlypis trichas	UI	P	A	A	A
Setophaga citrina	UI	P	P	P	P
Icteria virens	UI	P	A	A	A
Eucometis penicillata	UI	A	A	P	P
Cyanerpes cyaneus	CI	A	A	A	P
Volatinia jacarina	UG	P	P	A	A
Sporophila torqueola	UG	P	P	A	A
Tiaris olivaceus	UG	P	P	A	P
Arremonops rufivirgatus	UI	P	P	P	P
Arremonops chloronotus	UI	P	P	P	P
Habia fuscicauda	UI	P	P	P	P
Piranga roseogularis	CI	P	P	P	P
Piranga rubra	CI	A	P	A	P
Cardinalis cardinalis	MCG	P	P	P	P
Cyanocompsa parellina	UG	P	P	P	P
Passerina cyanea	UG	P	P	P	P
Granatellus sallaei	MCI	P	P	P	P
Dives dives	CI	P	P	P	P
Molothrus aeneus	UG	P	P	P	P
Icterus prosthemelas	CI	P	P	P	P
Icterus spurius	CI	A	A	P	A
Icterus cucullatus	CI	P	P	P	P
Icterus chrysater	CI	P	P	P	P
Icterus mesomelas	CI	P	P	A	A
Icterus auratus	CF	P	P	P	P
Icterus gularis	CI	P	P	P	P
Amblycercus holosericeus	UI	P	P	P	P
Euphonia affinis	CF	P	P	P	P
Euphonia hirundinacea	CF	P	P	P	P

P present, *A* absent

Feeding guilds: *BI* bark insectivores, *CI* canopy insectivores, *N* nectarivores, *MCI* mid-canopy insectivores, *UI* understory insectivores, *CF* canopy frugivores, *UF* understory frugivores, *UG* understory granivores, *MCG* mid-canopy granivores

References

Achard F, Eva HD, Stibig HJ, Mayaux P, Gallego J, Richards T, Malingreau JP (2002) Determination of deforestation rates of the world's humid tropical forests. Science 297:999–1002

Arizmendi MC, Rodríguez-Flores C, Soberanes-González C (2010) White-bellied Emerald (*Amazilia candida*). In: Schulenberg TS (ed) Neotropical birds online. Cornell Lab of Ornithology, Ithaca. http://neotropical.birds.cornell.edu/portal/species/overview?p_p_spp=252571 Accessed 5 Oct 2011

Arriaga-Weiss SL, Calmé S, Kampichler C (2008) Bird communities in rainforest fragments: guild responses to habitat variables in Tabasco, Mexico. Biodivers Conserv 17:173–190

Bart J, Klosiewski SP (1989) Use of presence/absence to measure changes in avian density. J Wildl Manag 53:847–852

Bennet PM, Owens IPF (2002) Evolutionary ecology of birds. Oxford University Press, Oxford

Bhagwat SA, Willis KJ, Birks HJB, Wittaker RJ (2008) Agroforestry: a refuge for tropical biodiversity? Trends Ecol Evol 23:261–267

Bird Life International (2011) Species factsheet. http://www.birdlife.org/ Accessed 5 Oct 2011

Borcard D, Legendre P (2002) All-scale spatial analysis of ecological data by means of principal coordinates of neighbour matrices. Ecol Model 153:51–68

Borcard D, Legendre P, Avois-Jacquet C, Tuomisto H (2004) Dissecting spatial structure of ecological data at multiple scales. Ecology 85:1826–1832

Borcard D, Legendre P, Drapeau P (1992) Partialling out the spatial component of ecological variation. Ecology 73:1045–1055

Bowman DMJS, Woinarski JCZ, Sands DPA, Wells A, McShane VJ (1990) Slash and burn agriculture in the wet coastal lowlands of Papua New Guinea: response of birds, butterflies and reptiles. J Biogeogr 17:227–239

Brady C, Noske R (2009) Succession in bird and plant communities over a 24-year chronosequence of mine rehabilitation in the Australian monsoon tropics. Restor Ecol 18 (6):855–864

Bribiesca-Formisano R, Rodríguez-Flores C, Soberanes-González C, Arizmendi MC, Behrstock B (2010) Thicket Tinamou (Crypturellus cinnamomeus). In: Schulenberg TS (ed) Neotropical Birds Online. Cornell Lab of Ornithology, Ithaca. http://neotropical.birds.cornell.edu/portal/species/overview?p_p_spp=59956 Accessed 5 Oct 2011

Brown S, Lugo AE (1990) Tropical secondary forests. J Trop Ecol 6:1–32

Canaday C (1996) Loss of insectivorous birds along a gradient of human impact in Amazonia. Biol Conserv 77:63–77

Colwell RK (2005) EstimateS: statistical estimation of species richness and shared species from samples. Version 7.5. User's Guide and application. http://purl.oclc.org/estimates. Accessed 7 April 2011

Dunn RR (2004) Recovery of faunal communities during tropical forest regeneration. Conserv Biol 18:302–309

Dunning JB (1992) CRC handbook of avian body masses. CRC, Boca Raton

Dupuy JM, Hernández-Stefanoni JL, Hernández-Juárez RA, Tetetla-Rangel E, Lopez-Martínez JO, Leyequién-Abarca E, Tun-Dzul F, May-Pat F (2011) Patterns and correlates of tropical dry forest structure and composition in a highly replicated chronosequence in Yucatan, Mexico. Biotropica 44:151–162

Flores S, Espejel I (1994) Vegetation types of the Yucatan Peninsula. Yucatan Etnoflora. UADY, México. vol 3, 135 p. (in Spanish)

Ford HA, Barret GW, Saunders DA, Recher HF (2001) Why have birds in the woodlands of Southern Australia declined? Biol Conserv 97:71–88

Furness RW, Greenwood JJD (1993) Birds as monitors of environmental change. Chapman & Hall, London

Gaston KJ, Blackburn TM (1995) Birds, body size and the threat of extinction. Philos Trans Biol Sci 347:205–212

Gotelli NJ, Entsminger GL (2001) Swap and fill algorithms in null model analysis: rethinking the knight's tour. Oecologia 129:281–291

Graham CH, Blake JG (2001) Influence of patch- and landscape-level factors on birds assemblages in a fragmented tropical landscape. Ecol Appl 116:1709–1721

Gray MA, Baldauf SL, Mayhew PJ, Hill JK (2007) The response of avian feeding guilds to tropical forest disturbance. Conserv Biol 21:133–141

Greenberg R (1990) Ecological plasticity, nephobia, and resource use in birds. Stud Avian Biol 13:431–437

Greenberg R (1992) Forest migrants in non-forest habitats on the Yucatan Peninsula. In: Hagan JM III, Johnston DW (eds) Ecology and conservation of neotropical migrant landbirds. Smithsonian Institution Press, Washington, DC, pp 273–286

Haas CA (1995) Dispersal and use of corridors by birds in wooded patches on an agricultural landscape. Conserv Biol 9:845–854

Hernández-Stefanoni JL, Dupuy JM, Tun-Dzul F, May-Pat F (2011) Effects of landscape structure and stand age on species richness and biomass of a tropical dry forest across spatial scales. Landsc Ecol 26:355–370

Holmes RT, Schultz JC, Nothnagle P (1979) Bird predation on forest insects: an exclosure experiment. Science 206:462–463

International Tropical Timber Organization (ITTO) (2002) Guidelines for the restoration, management and rehabilitation of degraded and secondary tropical forests. ITTO Policy Development Series No 13. ITTO, Yokohama, Japan, p 84

Johns AD (1991) Responses of Amazonian rain forest birds to habitat modification. J Trop Ecol 7:417–437

Johnson MD (2000) Effects of shade-tree species and crop structure on the winter arthropod and bird communities in a Jamaican shade coffee plantation. Biotropica 32:133–145

Karr JR, Freemark KE (1983) Habitat selection and environmental gradients: dynamics in the 'stable' tropics. Ecology 64:1481–1494

Laurence WF (2007) Have we overstated the tropical biodiversity crisis? Trends Ecol Evol 22:65–70

Lebrija-Trejos E, Bongers F, Pérez-Garciía EA, Meave J (2008) Successional change and resilience of a very dry tropical deciduous forest following shifting agriculture. Biotropica 40:422–431

Leyequién E, de Boer WF, Cleef A (2007) Influence of body size on coexistence of bird species. Ecol Res 22:735–741

Leyequién E, de Boer WF, Toledo VM (2010) Bird community composition in a shaded coffee agro-ecological matrix in Puebla, Mexico: the effects of landscape heterogeneity at multiple spatial scales. Biotropica 42:236–245

Mathews S, O'Connor R, Platinga AJ (2002) Quantifying the impacts on biodiversity of policies for carbon sequestration in forests. Ecol Econ 40:71–87

May PG (1982) Secondary succession and breeding bird community structure: patterns of resource utilization. Oecologia (Berl) 55:208–216

Mayer AL, Cameron GN (2003) Landscape characteristics, spatial extent, and breeding bird diversity in Ohio, USA. Divers Distrib 9:297–311

Mazerolle MJ, Villard MA (1999) Patch characteristics and landscape context as predictor of species presence and abundance: a review. Ecoscience 6:117–124

Mcgarigal K, Cushman SA, Neel MC (2002) FRAGSTATS: spatial pattern analysis program for categorical maps. Computer software program. University of Massachusetts, Amherst

Miles L, Newton AC, DeFries RS, Ravilious C, May I, Blyth S, Kapos V, Gordon JE (2006) A global overview of the conservation status of tropical dry forests. J Biogeogr 33:491–505

Milesi FA, Marone L, Lopez de Casenave J, Cueto VR, Mezquida ET (2002) Guilds as indicators of environmental conditions: a case study with birds and habitat perturbations in Monte central, Argentina. Ecol Austral 12:149–161

Miranda F, Hernández XE (1963) Los tipos de vegetación de México y su clasificación. Bol Soc Bot Méx 28:28–79

Nichols JD, Hines JE, Sauer JR, Fallon FW, Fallon JE, Heglund PJ (2000) A double-observer approach for estimating detection probability and abundance from point counts. Auk 117:393–408

Pearman PB (2002) The scale of community structure: habitat variation and avian guilds in tropical forest understorey. Ecol Monogr 72:19–39

Philpott SM, Arendt WJ, Armbrecht I, Bichier P, Diestch TV, Gordon C, Greenberg R, Perfecto I, Reynoso-Santos R, Soto-Pinto L, Tejeda-Cruz C, Williams-Linera G, Valenzuela J, Zolotoff JM (2008) Biodiversity loss in Latin American coffee landscapes: review of the evidence on ants, birds, and trees. Conserv Biol 22:1093–1105

Portillo-Quintero CA, Sánchez-Azofeifa GA (2010) Extent and conservation of tropical dry forests in the Americas. Biol Conserv 143:144–155

Quesada M, Sanchez-Azofeifa GA, Alvarez-Añorve M, Stoner KE, Avila-Cabadilla L, Calvo-Alvarado J, Castillo A, Espíritu-Santo MM, Fagundes M, Fernandes GW, Gamon J, Lopezaraiza-Mikel M, Lawrence D, Cerdeira-Morellato LP, Powers JS, Neves FS, Rosas-Guerrero V, Sayago R, Sanchez-Montoya G (2009) Succession and management of tropical dry forests in the Americas: review and new perspectives. For Ecol Manag 258:1014–1024

Raman TR, Sukumar R (2002) Responses of tropical rainforest birds to abandoned plantations, edges and logged forest in the Western Ghats, India. Anim Conserv 5:201–216

Raman TR, Rawat GS, Johnsingh AJT (1998) Recovery of tropical rainforest avifauna in relation to vegetation succession following shifting cultivation in Mizoram, north-east India. J Appl Ecol 35:214–231

Rappole JH, Morton ES, Ramos MA (1992) Density, philopatry, and population estimates for songbird migrants wintering in Veracruz. In: Hagan JM, Johnston DW (eds) Ecology and conservation of Neotropical migrant landbirds. Smithsonian Institution Press, Washington, DC, pp 337–344

Renjifo LM (2001) Effect of natural and anthropogenic landscape matrices on the abundance of subandean bird species. Ecol Appl 11:14–31

Robinson SK, Terborgh J (1997) Bird community dynamics along primary successional gradients of an Amazonian whitewater river. Ornithol Monogr 48:641–672

Sanchez-Azofeifa GA, Quesada M, Rodriguez JP, Nassar JM, Stoner KE, Castillo A, Garvin T, Zent EL, Calvo-Alvarado JC, Kalacska MER, Fajardo L, Gamon JA, Cuevas-Reyes P (2005) Research priorities for neotropical dry forests. Biotropica 37:477–485

Santamaría-Rivero W (2011) Efectos de la estructura del paisaje sobre la riqueza y abundancia de diferentes gremios de alimentación de aves, al sur de Yucatán, México. Centro de Investigación Científica de Yucatán, Mérida, Yucatán, México, p 96

Şekercioğlu CH, Ehrlich PR, Daily GC, Aygen D, Goehring D, Sandi RF (2002) Disappearance of insectivorous birds from tropical forest fragments. Proc Natl Acad Sci USA 99:263–267

Şekercioğlu CH, Daily GC, Ehrlich PR (2004) Ecosystem consequences of bird declines. Proc Natl Acad Sci USA 101:18042–18047

Shulenberg TS (2010) Neotropical birds online. Cornell lab of ornithology, Ithaca. http://neotropical.birds.cornell.edu/portal/ Accessed 5 Oct 2011

Simberloff D (1994) Habitat fragmentation and population extinction of birds. Ibis 137:S105–S111

Smith AL, Salgado-Ortiz J, Robertson RJ (2001) Distribution patterns of migrant and resident birds in successional forests of the Yucatan Peninsula, Mexico. Biotropica 33:153–170

Terborgh JW (1985) Habitat selection in Amazonian birds. In: Cody ML (ed) Habitat selection in birds. Academic, Orlando, pp 311–340

Thiollay JM (1994) Structure, density and rarity in an Amazonian rain forest bird community. J Trop Ecol 10:449–481

Thompson FR III, Probst JR, Raphael MG (1995) Impacts of silviculture: overview and management. In: Martin TE, Finch DM (eds) Ecology and management of Neotropical migratory birds: a synthesis and review of critical issues. Oxford University Press, New York, pp 201–219

Volpato G, López EV, Mendoca LB, Bocon R, Bisheimer MB, Serafini PP, Anjios LD (2009) The use of the point count method for bird survey in the Atlantic forest. Zoología 26:74–78

Wiens JA (1989) Spatial scaling in ecology. Funct Ecol 3:385–397

Willson MF (1974) Avian community organization and habitat structure. Ecology 55:1017–1029

Witt GB, Noël MV, Bird MI, Beeton RJS, Menzies NW (2011) Carbon sequestration and biodiversity restoration potential of semi-arid mulga lands of Australia interpreted from long-term grazing exclosures. Agric Ecosyst Environ 141:108–118

Wright SJ (2005) Tropical forests in a changing environment. Trends Ecol Evol 20:553–560

Wright SJ, Muller-Landau HC (2006) The future of tropical forest species. Biotropica 38:287–301

Part IV
Ecologies in Protected Areas

Chapter 12
Understanding Development Trends and Landscape Changes of Protected Areas in Peninsular Malaysia: A Much Needed Component of Sustainable Conservation Planning

Saiful Arif Abdullah, Shukor Md. Nor, and Abdul Malek Mohd Yusof

Abstract The establishment of protected areas in peninsular Malaysia was initiated during the British colonial period. The combination of political scenario and socio-economic development has influenced the planning and management of protected areas in peninsular Malaysia. As a result, some of them did not receive much attention and have been exposed to various human land use activities such as agriculture, urbanization, and building highways, a concern particularly since the rapid land development for agriculture in the 1950s and the 1960s. However, from the 1980s to recent years, urbanization and other similar types of development are emerging to affect the sustainability of protected areas. Assessing landscape element change or simply landscape change of protected areas and their link to its development trends is an urgent need to identify the main priorities for protection and conservation. Therefore, this chapter presents the development trends of protected areas, followed by some analysis on landscape changes both inside and outside of protected areas in three temporal years: 1988, 1996, and 2005. The degree of their impact on ecosystem of the protected areas is also presented. The objective is to provide understanding of the linkages between the development trends and landscape changes and their impact on the ecosystems of the protected areas.

S.A. Abdullah (✉)
Institute for Environment and Development (LESTARI), Universiti Kebangsaan Malaysia, 43600 Bangi, Selangor Darul Ehsan, Malaysia
e-mail: saiful@ukm.my; saiful_arif2002@yahoo.com

S.Md. Nor
School of Environment and Natural Resources, Faculty of Science and Technology, Universiti Kabangsaan Malaysia, 43600 Bangi, Selangor Darul Ehsan, Malaysia
e-mail: shukor@ukm.my

A.M.M. Yusof
Department of Wildlife and National Park, Km 10, Jalan Cheras, 51400 Kuala Lumpur, Malaysia
e-mail: malek@wildlife.gov.my

N. Nakagoshi and J.A. Mabuhay (eds.), *Designing Low Carbon Societies in Landscapes*, Ecological Research Monographs, DOI 10.1007/978-4-431-54819-5_12, © Springer Japan 2014

Keywords Biodiversity • Conservation planning • Land use policy • Sustainable development • Tropical landscape • Wildlife management

12.1 Introduction

Globally, the establishment of protected areas is an approach widely used to conserve and protect forests and habitats from further loss and degradation. Historically, the modern concept of protected areas began in the 1800s, but the development scenarios and characteristics differ between countries or regions (Brovko and Fomina 2008). The first modern designation of a protected area was in 1872: the declaration of Yellowstone National Park in the United States (Brovko and Fomina 2008).

In a broader perspective, the concept of protected areas is not restricted to conservation and protection of the forest and its flora and fauna, but also embraces marine ecosystems and historical and cultural sites with unique characteristics. This perspective has been adopted by the International Union for Conservation of Nature (IUCN) to define protected areas for recognition at the international level. Toward the end of the twentieth century, voices to develop models of protected area management became more prominent. As a result, the number and extent of protected areas increased. As of 2003, there are about 102,102 protected areas in the world with a total area of about 18.8 million km^2 (Chape et al. 2003).

In peninsular Malaysia, the establishment of protected areas was influenced by a combination of the country's political scenario and socioeconomic development. The initiative was started in the 1900s during the colonial period as an effort to protect and conserve wildlife from hunting activity and land use development (Aiken 1994). Many protected areas, however, did not receive much attention for various factors and constraints such as institutional overlapping, legislation, and land use policy. As a result, these areas have been exposed to a variety of land use activities such as agriculture, human settlements, urbanization, and road networks.

These trends may affect the landscape and habitat quality of protected areas and their surroundings. Thus, understanding landscape change of protected areas and its link to development trends is urgently needed and should be incorporated in conservation planning for sustainability. To understand the link, this chapter first presents the development trends of protected areas in peninsular Malaysia, followed by landscape change analysis both inside and outside of protected areas over three decades. Also, hemeroby analysis is used to measure how landscape changes have affected the naturalness of protected areas.

12.2 Development Trends

The development trends of protected areas in peninsular Malaysia (Fig. 12.1) can be divided into the colonial period and the national period. The colonial period refers to the era when the country was governed by the British whereas the national period is after the country received independence in 1957. The purpose of dividing into two periods is to compare the trend and efforts taken in development of protected areas between the two periods.

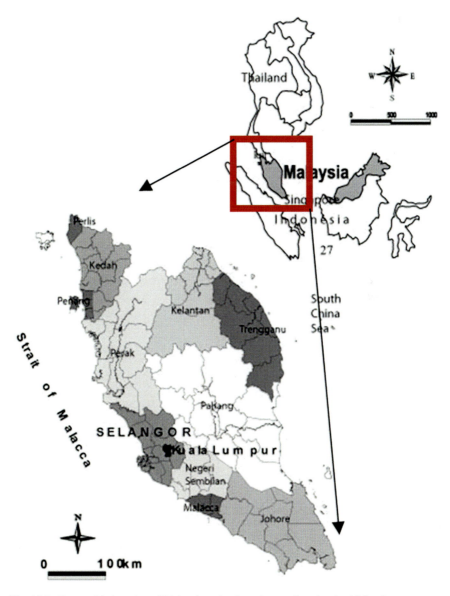

Fig. 12.1 Geographic location of Malaysia and enlarged area of peninsular Malaysia

12.2.1 The Colonial Period

Before the 1900s, peninsular Malaysia did not have any protected area. The first protected area in peninsular Malaysia was established in 1903 when Chior Wildlife Reserve was gazetted under the Birds and Wildlife Protected Ordinance, followed by the gazettement of Bukit Nanas Wildlife Reserve in 1906. By the 1920s, increasing the number and extent of protected areas became necessary to prevent further loss of forest and wildlife habitat resulting from the large-scale rubber plantations in the country (Jomo et al. 2004). At the time about 405,000 ha of primary forests was cleared, and by 1921 the total area of rubber plantations was approximately 800,000 ha (Berger 1990). As a result, certain animal species, particularly large mammals, had suffered serious threat from this activity (Hubback 1923, 1924, 1929 cited in Aiken 1994). For example, the populations of seladang (*Bos gaurus*) and Sumatran rhinoceros (*Dicerorhinus sumatrensis*) were depleted (Aiken 1991/1992). Other factors also contributed to this loss, such as poaching, overexploitation of commercial hunting, and trapping (Aiken, 1994).

In 1922 more protected areas were established: three in the state of Selangor, namely Bukit Kutu Wildlife Reserve, Fraser's Hill Wildlife Reserve, and Kuala Selangor Wildlife Reserve, and one each in the state of Perak (Sungkai Wildlife Reserve), the Federal Territory of Kuala Lumpur (KL Golf Course Bird Sanctuary), the state of Pahang (Krau Wildlife Reserve), and the state of Negeri Sembilan (Port Dickson Islands Reserve). Since then, the number increased gradually, and by the end of the 1930s the total number of protected areas in peninsular Malaysia was 18 (Fig. 12.2), which includes the gazettement of King George V National Park, now known as Taman Negara, in 1938–1939. However, the establishment of protected areas was halted during the 1940s, partly by World War II between 1942 and 1945. The effort then continued until 1956 when another four protected areas were established: two bird sanctuaries (Batu Gajah Bird Sanctuary and Four Island Reserve) and two wildlife reserves (Templer Wildlife Reserve and Pahang Tua Wildlife Reserve) (Fig. 12.2).

The total extent of protected areas increased significantly in the 1930s, mainly because of the establishment of Taman Negara in 1938–1939. Then, the total extent remained the same until 1957 (Fig. 12.3). At the same time, however, protected areas in peninsular Malaysia experienced degazettement and regazettement from pressure from land use and economic development. Thus, during the colonial period the total cumulative area was about 617,875 ha, or 756,716 ha if degazettement/regazettement is not taken into consideration (Fig. 12.3). In terms of number, the majority of protected areas were wildlife reserves, but national parks accounted for almost 70 % of the cumulative area (with degazettement/regazettement) (Table 12.1).

This in situ conservation can be considered as the main approach to protect wildlife and its habitats during the colonial period. Nevertheless, almost all protected areas, particularly wildlife reserves and national parks, were established at the remote, hilly, and undulating terrain of the main range of peninsular

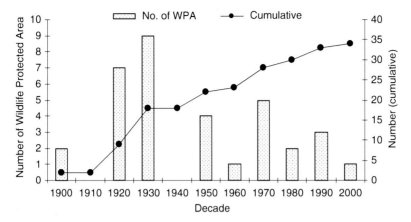

Fig. 12.2 Number of protected areas established in each decade and their cumulative number from the 1900s to 2000s. *Note*: Wang Pinang, Endau-Rompin National Park, Tasik Bera Reserve, and Tasik Chini Reserve are not included in this study because in terms of management there is some overlapping between the federal and state jurisdiction

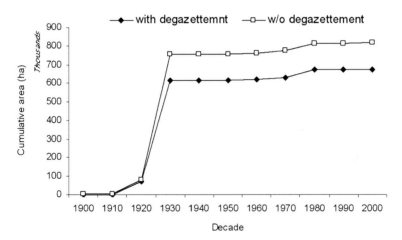

Fig. 12.3 Cumulative area of protected areas in peninsular Malaysia from 1900s to 2000s. *Note*: Area of several protected areas (Port Dickson Island BS, Four Island Reserve BS, Nine Island Reserve BS) not included because data are not available

Malaysia, the Titiwangsa Range. This area is relatively less favorable for in situ wildlife conservation compared to the lowland forests, which are considerably richer in wildlife species. Furthermore, only a small part covered the lowland forest, and the protected areas also suffered from insecurity of tenure as well as less representation of certain types of forests or ecosystems (Aiken 1994).

Table 12.1 Number and total extent of each type of protected area in peninsular Malaysia from the 1900s to 2000

Period	Decade	Wildlife reserve		National park		Wildlife sanctuary		Bird sanctuary	
		Number	Area (ha)	Number	Area (ha)	Number	Area (ha)	Number	Area (ha)
Colonial Period	1900	2	705						
	1910								
	1920	5	71,829					2	403[a]
	1930	6	110,460	3	432,173				
	1940								
	1950	2	2,301					2	4[a]
	Subtotal	15	185,295	3	432,173			4	407[a]
National Period	1960	1	4,330						
	1970	1	61			2	2	2	9,455[a]
	1980	1	40,197			1	21		
	1990					3	134		
	2000			1	1,200				
	Subtotal	3	44,588	1	1,200	6	157	2	9,455[a]
	Total (overall)	18	229,883	4	433,373	6	157	6	9,862[a]

[a]Area of several protected areas are not included because data are not available

12.2.2 The National Period

During the national period, the number and the total extent of protected areas increased slightly from the colonial period (Fig. 12.3). Between the 1960s and the 1980s, more wildlife reserves were established, including Sungai Dusun Wildlife Reserve in 1964, Tanjung Tuan Wildlife Reserve (1972) and Endau-Rompin Pahang Wildlife Reserve (1986); these are located in the states of Selangor, Negeri Sembilan, and Pahang, respectively, and the largest is Endau-Rompin Pahang (40,197 ha). As in the colonial period, the establishment of protected areas during the 1960s and the 1980s was also mainly driven by development of land for large-scale agriculture. However, at this time oil palm played the most crucial role in causing loss of forested areas in peninsular Malaysia compared to rubber and other cash crops (Abdullah and Hezri 2008).

The ex situ approach began to take place in wildlife conservation in the 1970s whereby two wildlife sanctuaries were established, that is, Bukit Pinang and Sidam Kanan in the state of Kedah: both are specifically for conservation of river terrapin. In the following decades, four wildlife sanctuaries were established: Zoo Melaka (1983), Bukit Palong Wildlife Sanctuary (1991), Bota Kanan Wildlife Sanctuary (1993), and Pulai Deer Wildlife Sanctuary (1993). During this period, parts of wildlife reserve areas were also allocated or developed as centers of ex situ conservation programs, particularly for large mammals, such as tapir (*Tapirus indicus*) and Sumatran rhinoceros (*Dicerorhinus sumatrensis*) at Sungai Dusun Wildlife Reserve. Other animals include seladang (*Bos gaurus*) in Jenderak Selatan and elephant (*Elephus maximus*) in Kuala Gandah, both located at the periphery of Krau Wildlife Reserve, and a deer conservation program at Sungkai Wildlife Reserve. The increase in the number of wildlife sanctuaries during this period was most likely related to the lack of suitable sites for in situ conservation areas. Rubber and oil palm plantations have caused clearance of a large proportion of lowland forests. As a result, many protected areas were relegated to the uneconomical land of remote areas (Aiken 1994) or, alternatively, several small areas were selected for ex situ conservation programs of threatened species. Degazettement and regazettement also occurred for some protected areas during this period. Nevertheless, the most significant was the degazettement of the entire 71,347 ha of Cameron Highlands Wildlife Reserve in 1962, 4 years after it was established in 1958. As it is located in a mountainous area, pressure from large-scale farming and an infrastructure for tourism might be the major factors that caused degazettement.

It took almost 20 years to set up another large tract of forest as a protected area in peninsular Malaysia. This area materialized in 2003 whereby a total of 1,200 ha of Pantai Acheh Forest Reserve was declared as Taman Negara Pulau Pinang (Penang National Park). This national park is currently the largest protected area in peninsular Malaysia, covering both terrestrial and marine ecosystems.

12.3 Landscape Changes

In this section, landscape changes at the inside and outside of 12 protected areas in 1988, 1996, and 2005 are presented: Chior, Bukit Kutu, Sungkai, Bukit Sungai Puteh, Endau-Kota Tinggi Barat, Endau-Kluang, Segamat, Klang Gate, Templer, Fraser's Hill, Sungai Dusun, and Endau-Rompin Pahang wildlife reserves. Data were based on satellite image analyses using a Geographic Information System (GIS). The 'outside' was defined as a 5-km zone surrounding each protected area. Based on the image analysis, there are seven categories of landscape elements in the protected areas, namely, built-up area, cleared land, forest, oil palm, rubber, shrubs, and water body.

Over the three decades, forest was the primary landscape element at all protected areas (Fig. 12.4a–l). However, three patterns were observed: (1) forest declined over the years, which is very obvious in Bukit Sungai Puteh and Segamat; (2) forest fluctuated, for example, in Klang Gate and Chior; and (3) forest remained similar (e.g., Sungai Dusun and Sungkai). Between 1988 and 2005 seven of the areas had forest cover of more than 80 %: Endau-Kota Tinggi Barat, Endau-Kluang, Bukit Kutu, Fraser's Hill, Sungkai, Sungai Dusun, and Templer. In Endau-Rompin Pahang, the proportion of forest remained above 80 % from 1988 to 1996 but became less than that in 2005. Except Sungkai and Sungai Dusun, all these areas are located in the hilly or mountainous areas of peninsular Malaysia.

Of most concern in terms of forest depletion and intensive human land use were Bukit Sungai Puteh and Segamat (Fig. 12.4a, g, respectively). In these reserves, the proportion of forest reduced tremendously and by 2005 only about 50 % remained in Segamat and 20 % in Bukit Sungai Puteh. Furthermore, both reserves are already fragmented, which could affect their effectiveness and reliability as protected areas; this is also true for Chior.

In Segamat, rubber and oil palm were the main threats to the reserve (Fig. 12.4g) whereas it was built-up areas in Bukit Sungai Puteh (Fig. 12.4a). Thus, the major human land use forming threat to wildlife protected areas depends on their location, either in remote areas or adjacent to urban regions. Elevation is also a factor that affects intrusion and the intensity of human land use. Relatively, protected areas in mountainous landscapes are less affected by human land use than those located in the lowlands (Butler et al., 2004). For example, in the three decades investigated here Bukit Kutu and Fraser's Hill suffered less intrusion by human land use than Templer, which is located at low altitude. However, this study also suggests that low altitude does not necessarily increase human land use intensity. At those locations, the size and the management of the protected area are also important in minimizing the activity, as obviously occurred in Sungkai and Sungai Dusun, which are located in lowlands and are larger than Templer. Furthermore, these areas have staff to manage and monitor the reserves whereas Templer has none.

Connectivity to other forested areas in the surroundings is also important to minimize further intrusion of human activities into protected areas, as well as to maintain the structure and function or ecological integrity of the landscape

(Rizkalla and Swihart 2007). Therefore, the distribution of landscape elements at the outside of protected areas is crucial to determine the changes at the inside. Studies by Nur Hairunnisa (2011) at Krau Wildlife Reserve in 1989, 1996, and 2007 revealed that the intensity of human land use at the inside was significantly correlated to that outside (Fig. 12.5a–c).

At the outside zones, Fig. 12.6a–l shows that for all areas, except Bukit Sungai Puteh and Segamat wildlife reserves, the proportion of forest was the highest during

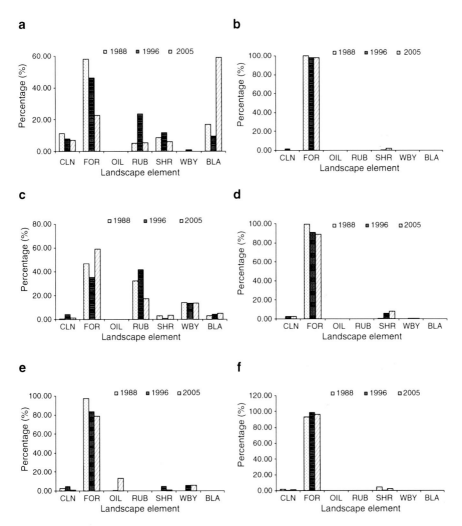

Fig. 12.4 Pattern change of landscape elements of protected areas in 1988, 1996, and 2005: Bukit Sungai Puteh (**a**), Endau-Kota Tinggi Barat (**b**), Klang Gate (**c**), Endau-Kluang (**d**), Endau-Rompin Pahang (**e**), Fraser's Hill (**f**), Segamat (**g**), Sungkai (**h**), Bukit Kutu (**i**), Sungai Dusun (**j**), Templer (**k**), and Chior (**l**). *CLN* cleared land, *FOR* forest, *OIL* oil palm, *RUB* rubber, *SHR* shrub, *WBY* water body, *BLA* built-up area

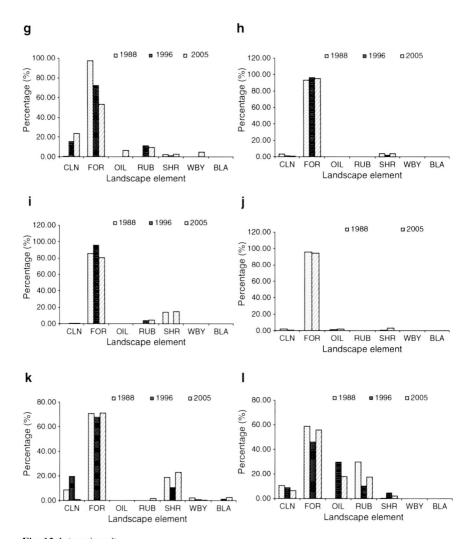

Fig. 12.4 (continued)

the three decades compared to other landscape elements. On one hand, this is because they are adjacent to or integral with permanent forest reserve (e.g., Sungkai, Bukit Kutu, Sungai Dusun, Templer, Endau Kota Tinggi Barat, Klang Gate, Fraser's Hill), whereas others such as Endau-Kluang and Endau-Rompin Pahang are bordering each other as well as Endau-Rompin National Park (Johor Site). Nevertheless, there were also cases in which the proportion of forest cover gradually reduced while human land use increased. In Bukit Sungai Puteh, the surrounding area was mostly covered by built-up area and by 2005 the proportion built up was more than 50 % (Fig. 12.6a).

A similar pattern also occurred at Klang Gate, where the proportion of built-up area is likely to increase if the scenario persists. The forest cover outside Segamat

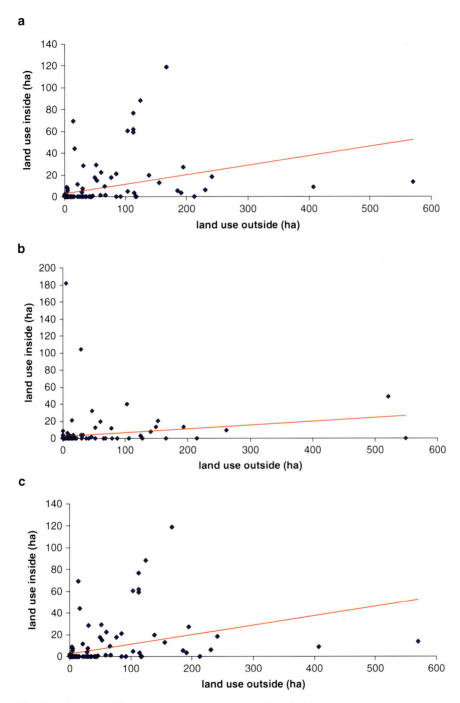

Fig. 12.5 Intensity of human land use inside was significantly influenced by the changes at the outside of Krau Wildlife Reserve in 1989 (**a**), 1996 (**b**), and 2007 (**c**). (From Nur Hairunnisa 2011)

Wildlife Reserve was reduced and its proportion became almost the same as that for rubber and oil palm land uses (Fig. 12.6g). The largest forest patch at the outside of this reserve is part of Endau Rompin National Park. However, conditions at the outside of Chior were much better than at Segamat where the proportion of human land use was less than 30 % (Fig. 12.6l), but there is potential for this intrusion to become more severe in the near future.

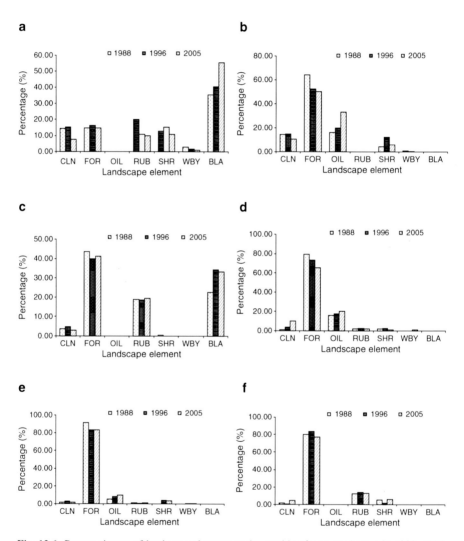

Fig. 12.6 Pattern change of landscape elements at the outside of protected areas in 1988, 1996, and 2005: Bukit Sungai Puteh (**a**), Endau-Kota Tinggi Barat (**b**), Klang-Gate (**c**), Endau-Kluang (**d**), Endau-Rompin Pahang (**e**), Fraser's Hill (**f**), Segamat (**g**), Sungkai (**h**), Bukit Kutu (**i**), Sungai Dusun (**j**), Templer (**k**), and Chior (**l**). *CLN* cleared land, *FOR* forest, *OIL* oil palm, *RUB* rubber, *SHR* shrub, *WBY* water body, *BLA* built-up area

12 Understanding Development Trends and Landscape Changes...

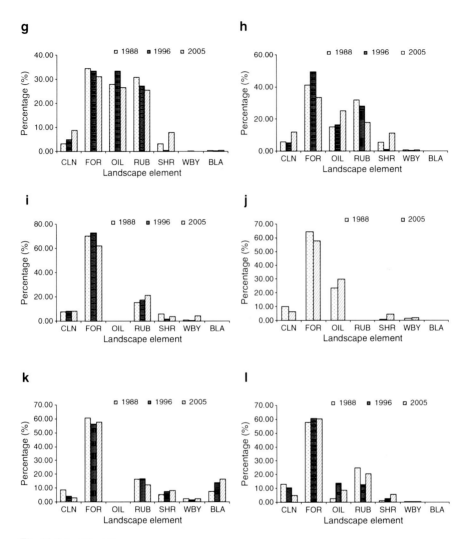

Fig. 12.6 (continued)

12.4 Hemeroby

The significance of the impact of human land use on naturalness of protected areas was measured using the Hemeroby Index (simply, hemeroby) at the inside and outside of each study area over the three decades (i.e., 1988, 1996, 2005). This index is an integrative measure for the degree of human impact on ecosystems in a particular area and the organisms that inhabit it (Sukopp 1976, cited in Steinhardt et al. 1999). In this study we follow Abdullah et al. (2009) to define the human

Table 12.2 Definition of the degree of human impact or hemeroby of the study

Degree of hemeroby[a]	Human impact[b]	Degree of naturalness[a]	Landscape element[b]
Metahemerobe	Vegetation cleared	Artificial	Built-up area
Polyhemerobe	Destruction for man-made ecosystems, land clearing, or abandoned with patchy and/or no vegetation	Strange to natural	Cleared land, water body
Oligohemerobe	Very limited removal of trees, managed logging scheme	Close to natural	Forest
Euhemerobe	Agricultural with application of fertilizers and drainage systems, managed cropland schemes	Relatively far or far from natural	Oil palm, rubber
Mesohemerobe	Once cleared and in succession process with occasional removal by humans	Seminatural	Shrub

[a]Based on Sukopp (1976, see Steinhardt et al. 1999)
[b]Based on Abdullah et al. (2009)

impact, which corresponds to naturalness as in Sukopp (1976, cited in Steinhardt et al. 1999) (Table 12.2). The following is the calculation of the index:

$$M = 100 \sum_{h=1}^{m} \frac{f_m}{m} h \qquad (12.1)$$

where m is the number of categories of hemeroby, f_m is the proportion of the area of the category m, and h is hemeroby linear factor (i.e., $h = 1$ for minimal and $h = m$ for maximum). M is equal to 100 if the whole area is classified as metahemeroby.

At the inside, the high degree of human impact was very obvious in Bukit Sungai Puteh and Klang Gate because of their adjacency to the urban conurbation of Kuala Lumpur. Nevertheless, the most severe was Bukit Sungai Puteh, where the hemeroby increased tremendously from about 30 % to almost 70 % by 2005 (Fig. 12.7a). In Klang Gate the degree of human impact fluctuated, ranging between 20 % and 40 % (Fig. 12.7c). The ecosystems in Segamat and Chior were also among those most impacted by human land use. The hemeroby in Segamat steadily increased from less than 5 % in 1988 to almost 40 % in 2005 (Fig. 12.7g), whereas it remained the same in Chior (between 30 % and 40 %) (Fig. 12.7l). In Endau-Kluang, Endau-Rompin Pahang, Endau-Kota Tinggi Barat, Fraser's Hill, and Bukit Kutu the impact was low, which was most probably attributable their large size and location at hilly sites. Nonetheless, of most concern are Endau-Kota Tinggi Barat and Endau-Kluang, because their hemeroby increased gradually. This pattern corresponds to the pattern of hemeroby in their surroundings, which is also true for Bukit Sungai Puteh. However, the surroundings of Bukit Sungai Puteh were the most degraded, with the hemeroby reaching almost 70 % in 2005.

At their outsides, the hemeroby of all protected areas, except Segamat, Chior, and Klang Gate, was slightly higher in 2005 than in 1988 and 1996. This observation suggests that over the decades the intensity of land development has spread toward the protected areas. In fact, protected areas with management plans and staff

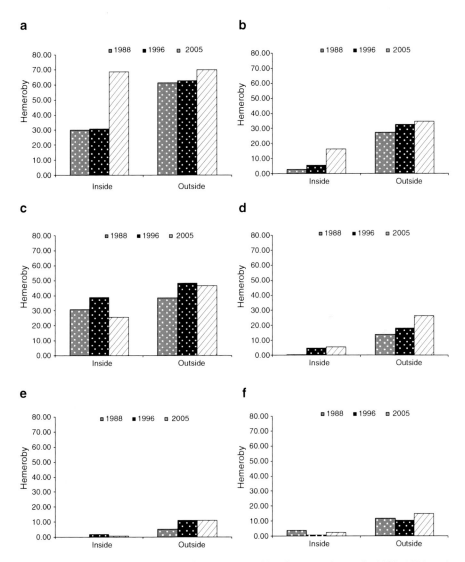

Fig. 12.7 Degree of hemeroby at the inside and outside of protected areas in 1988, 1996, and 2005: Bukit Sungai Puteh (**a**), Endau-Kota Tinggi Barat (**b**), Klang-Gate, Endau-Kluang (**d**), Endau-Rompin Pahang (**e**), Fraser's Hill (**f**), Segamat (**g**), Sungkai (**h**), Bukit Kutu (**i**), Sungai Dusun (**j**), Templer (**k**), and Chior (**l**)

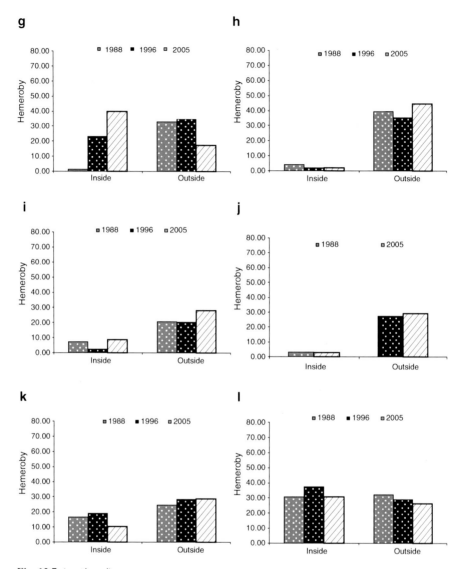

Fig. 12.7 (continued)

for monitoring also suffered the same situation, for example, Sungai Dusun. It is most alarming, obviously, if the protected area has no management plan and staff. Their location, either adjacent to forest reserves, oil palm, or rubber, also has an important role to these circumstances. In this context the most typical example is Bukit Sungai Puteh, where the demand for housing development has caused the clearance of many hectares of forest reserve and oil palm and rubber areas. In Segamat, the hemeroby on the inside was higher than on the outside by 2005, which suggests that land development (particularly for agriculture) became intense on the inside. In case of Chior this probably occurred earlier, because in 1988 the hemeroby on the inside and that outside were very similar.

12.5 Conclusion

The development of protected areas in peninsular Malaysia has been influenced by both the socioeconomic and political scenarios of the country. Generally, the development trends can be divided into the colonial period and the national period. In the colonial period, the in situ approach was mainly applied. During this period, most of the protected areas were established at remote and mountainous sites, driven by the intense land development of rubber and oil palm at the lowland areas. However, most of the establishment happened between the 1920s and the 1930s, and in terms of coverage it was primarily represented by Taman Negara. Since then, development was slowed down and halted in the 1940s, partly because of World War II. Afterward, the number and the extent of protected areas remained similar until the country received independence in 1957. The effort continued after that year, but the number and size of areas increased only slightly. In the national period, the ex situ conservation approach was started, possibly driven by the reduction in population of several animal species caused by forest fragmentation and habitat degradation.

Generally, over the three decades (1988, 1996, and 2005) all protected areas were exposed to various land use activities. Intensity of human land use inside the protected areas was apparently concomitant to what happened on the outside. The most severe were the changes in Bukit Sungai Puteh followed by Segamat, Chior, and Klang Gate. Except Klang Gate, at present, all these reserves are fragmented; for Bukit Sungai Puteh and Klang Gate, the urban sprawl of Kuala Lumpur City is the main threat to their sustainability, whereas rubber and oil palm plantations are the main threats for Segamat and Chior. Hemeroby analysis shows that Bukit Sungai Puteh, Segamat, Chior, and Klang Gate are vulnerable to any further land use development. Endau-Kota Tinggi Barat, Sungkai, Bukit Kutu, Sungai Dusun, and Templer are considered highly threatened; less threatened reserves include Endau-Kluang, Endau-Rompin Pahang, and Fraser's Hill. Nevertheless, other factors such as social, economic, physical (e.g., topography), and political (border) should be taken into consideration in future analysis to understand the link between development trends and landscape changes in wildlife protected areas, which is pivotal to improve conservation planning for reasons of both sustainability and sustainable development of the country.

Acknowledgments This study is part of a research project ScFund 04-01-02-SF0378 entitled "Landscape ecological assessment of protected areas in peninsular Malaysia for sustainable conservation planning" funded by the Ministry of Science, Technology and Innovation Malaysia.

References

Abdullah SA, Hezri AA (2008) From forest landscape to agricultural landscape in the developing tropical country of Malaysia: pattern, process and their significance on policy. Environ Manag 42:907–917

Abdullah SA, Yusof AMM, Nor SM (2009) Conservation status of wildlife protected areas in peninsular Malaysia: an assessment based on landscape approach. In: Ainsworth G, Garnett S (eds) RIMBA: sustainable forest livelihoods in Malaysia and Australia. LESTARI UKM, Bangi

Aiken SR (1991/92) The writing on the wall: declining fauna and the report of the Wild Life Commision of Malaya (1932). Wallaceana 66–67:1–6

Aiken SR (1994) Peninsular Malaysia's protected areas' coverage, 1903–92: creation, rescission, excision and intrusion. Environ Conserv 21:49–56

Berger R (1990) Malaysia's forest: a resource without a future. Packard, Chichester

Brovko PF, Fomina NI (2008) The history of establishment of the national park network in countries of the Asian-Pacific region. Geogr Nat Resour 29:221–225

Butler BJ, Swenson JJ, Alig RJ (2004) Forest fragmentation in the Pacific Northwest: quantification and correlations. For Ecol Manag 189:363–373

Chape S, Blyth S, Fish L, Fox P, Spalding M (compilers) (2003) 2003 United Nations List of Protected Areas. IUCN, Gland, Switzerland and Cambridge; UNEP-WCMC, Cambridge

Jomo KS, Chang YT, Khoo KJ (2004) Deforesting Malaysia: the political economy and social ecology of agriculture expansion and commercial logging. Zed Books, London

Nur Hairunnisa R (2011) Landscape ecological assessment of Krau wildlife reserve, peninsular Malaysia for sustainable conservation planning. Master's thesis, Universiti Kebangsaan Malaysia, Bangi

Rizkalla CE, Swihart RK (2007) Explaining movement decisions of forest rodents in fragmented landscapes. Biol Conserv 140:339–348

Steinhardt U, Herzog F, Lausch A, Müller E, Lehmann S (1999) Hemeroby index for landscape monitoring and evaluation. In: Pykh YA et al (eds) Environmental indices: system analysis approach. EOLSS, Oxford

Chapter 13
Land Use Trends Analysis Using SPOT 5 Images and Its Effect on the Landscape of Cameron Highlands, Malaysia

Mohd Hasmadi Ismail, Che Ku Akmar Che Ku Othman, Ismail Adnan Abd Malek, and Saiful Arif Abdullah

Abstract A large part of the steep mountain land in Peninsular Malaysia is covered by forests. Cameron Highlands is a mountainous region with a climate favorable to the cultivation of tea, subtropical vegetables, and flowers. Rapid economic growth and land use practices, however, have altered the environmental landscape of the area. Thus, this study was carried out to examine the rate of loss and pattern of fragmentation of the tropical mountain forests in Cameron Highlands. Temporal remotely sensed data (SPOT 5 images) of the years 2000, 2005, and 2010 were processed to develop a land use map of each year, which then analyzed their landscape fragmentation using GIS. Results showed that forest fragmentation occurred particularly in the period between 2005 and 2010. In 10 years the Cameron Highlands had lost about 2 % of its forested areas, mainly from agricultural activities. This study concludes that Cameron Highlands needs conservation efforts that should be focused on the management of the natural system and restoration project, and on management of the external influences, particularly on sustainable forest exploitation in the highland.

Keywords Remote sensing • Highland landscape • Land use/land cover change • Fragmentation

---

M.H. Ismail (✉) • C.K.A. Che Ku Othman • I.A. Abd Malek
Faculty of Forestry, Universiti Putra Malaysia, 43400 UPM Serdang, Selangor, Malaysia
e-mail: mhasmadi@upm.edu.my; cheku_85@yahoo.com; ismail@upm.edu.my

S.A. Abdullah
Institute for Environment and Development (LESTARI), Universiti Kebangsaan Malaysia, 43600 Bangi, Selangor, Malaysia
e-mail: saiful@ukm.edu.my

N. Nakagoshi and J.A. Mabuhay (eds.), *Designing Low Carbon Societies in Landscapes*, Ecological Research Monographs, DOI 10.1007/978-4-431-54819-5_13, © Springer Japan 2014

13.1 Introduction

Monitoring and management of natural resources requires timely, synoptic, and repetitive coverage over a large area. Coverage from various spatial scales helps in assessing temporal and spatial changes. Remote sensing provides up-to-date, reliable, spatial data at regular time intervals that are useful for land cover analysis. This method has prompted research using remote sensing where detection of a wide variety of changes is based upon spectral values from remotely sensed data. The analysis of landscape pattern, which utilizes spatial information, presents a great opportunity for the generation of ecological information, processes, and knowledge. Global environmental change is a result of land cover change (Skole et al. 1997). The ability to monitor land cover change at a variety of scales provides essential information required to assist in sustainable land management. In recent years, land management has moved toward a landscape approach that reflects a mix of social, environmental, and economic values. In landscape ecology, landscape scale is divided into ecological processes and human use through developed infrastructure, ownership, and management of resources (Turner et al. 2001).

Our landscape is continuously changing in response to both natural and human disturbances. Landscape changes often occur gradually over time as a series of small, localized events. The structure and function of a landscape can be perceived differently at different scales, and it is important for the observer to decide on appropriate scales for a study (Turner 1989). The relationship between human behavior and forest changes poses a major research challenge for development projects, policy makers, and environmental organizations that aim to improve forest management.

Landscape mapping is often the first step in many remote sensing projects (Watson and Wilcock 2001; Zha et al. 2003). Many landscape metrics used in remote sensing change detection are based on ecology, and these metrics have been developed for quantifying landscape structure (Turner et al. 2003; Cohen and Goward 2004; Kerr et al. 2001). Landscape metrics is a number or indices that describe the landscape configuration and composition to formulate and analyze either individual patches or the whole landscape. Landscape metrics are very important to detect the pattern of change that is not readily visible to the human eye or easily detectable by a human analyst. The metrics can be used to assess ecosystem health or as variables for models that support environmental assessment and planning efforts (Herzog et al. 2001; Patil et al. 2001).These metrics fall into two general categories: those that quantify the composition of the map without reference to spatial attributes, and those that quantify the spatial configuration of the map, requiring spatial information for their calculation (McGarigal and Marks 1995; Gustafson 1998). Using satellite imagery such as SPOT or Landsat, aerial photography, and geographic information systems, landscape ecologists are able to examine how the landscape has changed over time and how it is likely to change in the future. Once landscape changes are identified or predicted, the causes and the ecological and societal consequences of such changes can be examined.

Roy and Joshi (2002) clearly state that changing the landscape pattern through fragmentation can disrupt ecological processes that depend on movement within the landscape. Tropical mountain forests are among the most fragile and highly threatened of all tropical forest ecosystems (Bruijnzeel 2001). Forest landscape models have benefited greatly from technological advances, including increased computing capacity, the development of the Geographic Information System (GIS), remote sensing, and software engineering. The forest ecological processes and their interactions in forest landscape models can be represented by well-designed computer software (He et al. 2000).

Previous studies reported that the Cameron Highlands face various environmental problems caused by human activities such as agriculture, urbanization, infrastructure development, and deforestation, which contribute to degradation of the highland landscape and severe upland soil erosion (Aminuddin et al. 2005; Che Ku Akmar and Mohd Hasmadi 2010). To date there are limited studies on landscape pattern or changes in mountain areas in Malaysia. In this chapter, the rate of forest loss and pattern of landscape fragmentation in the tropical mountain forest of Cameron Highlands were examined by comparing temporal SPOT 5 images in years 2000, 2005, and 2010. The landscape structure changes were assessed based on their spatial configuration over time using selected landscape metrics or indices. The information obtained may be directly or indirectly useful to the management and development strategies for environmental sustainability of the highlands.

13.1.1 Remote Sensing and Landscape Interaction

Conceptually, remotely sensed images and landscape interaction exist where imagery can provide critical information on components in a landscape study. Size, shapes, perimeter, connectivity, orientation, presence of corridors, visibility, or diversity of patches are variables critical for describing the landscape mosaic. Remote sensing has made important contributions to biodiversity conservation planning through the measurement of deforestation rates, habitat fragmentation, habitat degradation, and isolation of protected areas in human altered landscapes. By using remote sensing surveying techniques, landscape-level patterns and change are regularly monitored. From this point resource managers and planners have growing expectations of remote sensing science and landscape modeling to provide better predictions of future environmental conditions and of the impacts that environmental and socioeconomic changes may have on ecosystems and the services they provide. Such models involve the integration of multiple component models, including those predicting species distributions and habitat requirements (Turner et al. 2003).

In the first place, forest cover is of great interest to many researchers for scientific reasons and land management purposes, for which forest cover is a basic source of information for habitat modeling and predicting and mapping forest health, including landscape degradation. The spatial pattern of a changing

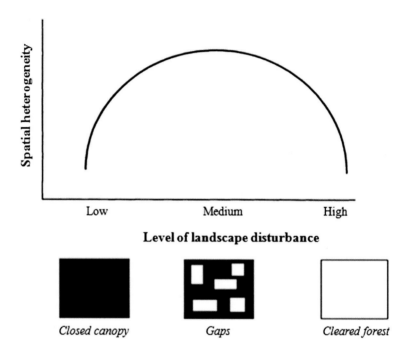

Fig. 13.1 The expected relationship between the level of landscape disturbance and the spatial heterogeneity of the landscape measured on spatial data at a spatial resolution of a few dozen meters. (From Eric 1999)

landscape on forest cover contains information on the processes of land cover change. Of key importance for monitoring forest degradation is the integration of information from spectral, spatial, and temporal resolution. The relationship between landscape-level disturbance and the spatial heterogeneity of the landscape measured on spatial data at a spatial resolution is shown in Fig. 13.1.

Quantifying forest degradation is more challenging than quantifying deforestation. Deforestation is "direct" mapping but forest degradation involves forest "pixels," which are complex mixtures of environment of different earth cover materials. Assessing the linking and interaction of satellite data and landscape ecology has become the essence of landscape fragmentation studies. A multiscale approach can be used to investigate changes of spatial landscape in a variety of levels between grain size and spatial extent. In a remote sensing image, grain size is equivalent to the spatial resolution of pixels, whereas spatial extent represents the total area examined (Hay et al. 2002). Remotely sensed data have different grain size, depending on their type or spatial extent, to facilitate landscape analysis and mapping according to objectives of the research. Remote sensors detect the radiance associated with a given pixel, and this information is then often converted to the spectral reflectance. Different cover types each have their own spectral reflectance. As an example of spectral reflectance in a satellite image, chlorophyll in vegetation primarily absorbs radiation in blue and red wavelengths and reflects radiation in green wavelengths. Thus, a pixel can be classified as deciduous forest, water, barren soil, clouds, and other. All this classification depends on the spectral reflectance for each satellite type. Remote sensing is now being used in the currently emerging integrated regional ecosystem consultations. Satellite remote sensing and sensor-related technological advancements have made it possible to acquire precise vegetation classification and mapping.

The development of remote sensing and GIS technologies facilitates accurate monitoring of vegetation change as less time consuming, cost effective, and highly efficient. Remote sensing and GIS have offered vast opportunities for investigation at scales larger than in the past for contemporary research into processes and landscape patterns. Many researchers believe that the integration of remote sensing techniques within analysis of environmental change is essential if ecologists are to meet the challenges of the future, specifically issues relating to global change; however, in practice, this integration has so far been limited (Griffiths and Mather 2000). The role of remote sensing and its interaction with landscape patterns are increasing in many parts of the world. The imagery, in conjunction with geospatial modeling, provides a platform for landscape characterization. Using satellite imagery, for example, Landsat TM, SPOT, IKONOS, aerial photography, and geographic information systems (GIS), landscape ecologists are able to examine how the landscape has changed over time and how it is likely to change in the future. Once landscape changes are identified or predicted, the causes and the ecological and societal consequences of such changes can be examined. In addition, the direct linkage of technology such as GIS with remote sensing and landscape ecology research allows us to integrate spatial land cover patterns and ecological processes in a manner that is essential to understanding the processes of change (Forman 1995). The advantage of using GIS is the ability to incorporate different source data into change detection applications.

13.2 Methodology

13.2.1 Description of the Study Area

The study area covers the western region of the Cameron Highlands district, State of Pahang, Peninsular Malaysia (Fig. 13.2). The area is located between 4°35′55.40″N and 101°29′07.05″E, and about 200 km from Kuala Lumpur City and only 120 km from Ipoh City in Perak state. The study area covers an area of about 27,009.8 ha from the total area of the Cameron Highlands district (71,225 ha). Elevation in the study area ranges between 1,070 m and 1,830 m above mean sea level. The highland has steep slopes; 66 % is more than 20° slope. The mean

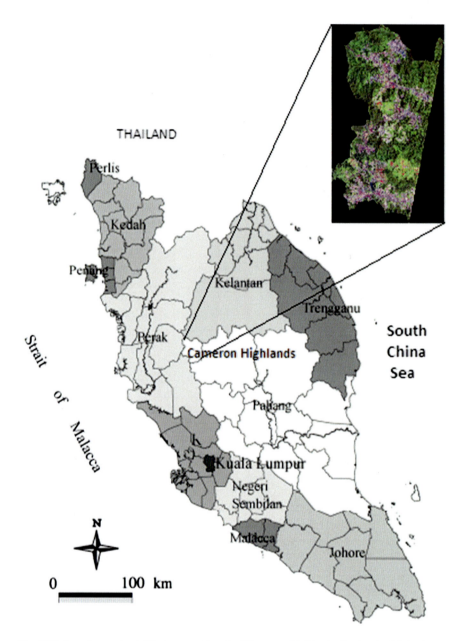

Fig. 13.2 Location and satellite imagery (*inset*) of Cameron Highlands, Peninsular Malaysia

temperature is about 24 °C in daytime and 14 °C at night. The average annual rainfall is 2,660 mm with two peaks in May and October. Cameron Highlands is drained by three main rivers, namely, Sg. Telom, Sg. Bertam, and Sg. Lemoi. Two of the main economic activities in Cameron Highlands are tourism and agriculture.

13 Land Use Trends Analysis Using SPOT 5 Images and Its Effect... 229

Fig. 13.3 The Cameron Highland's view of forests, farms, terraces, homes, and tea plantation

Popular as an agrotourism spot in Peninsular Malaysia, the landscapes of the highlands are covered by natural forests, orchards, vegetable farms, flower terraces, and nurseries (Fig. 13.3). Some of the flora and fauna in these highlands are considered as rare or endemic and are totally different from lowland vegetation types. The forests above 1,500 m are known as the upper mountain forests and are found on mountain summits and ridges on the western borders of this area. The lowland dipterocarp forest ranges from 300 m and above and mainly on the eastern part of the district. This highland is known to contain a diverse array of plant species and is very biodiverse, with many species that are restricted to mountains and highlands. Thus, the ecology of Cameron Highlands is extremely fragile and sensitive to disruption.

13.2.2 Data Acquisition and Methods

Three SPOT 5 satellite imageries were used, obtained from the Agency of Remote Sensing Malaysia (ARSM). The descriptions of the images are as presented in Table 13.1; multitemporal SPOT data are shown in Fig. 13.4. The cloud covers for these datasets are less than 5 %, and quality is good.

Analysis of SPOT 5 multitemporal data utilizes several processes. The summary of steps is outlined to meet the objective of the study. The processes were employed from various datasets to extract the information and gain results. Figure 13.5 shows a workflow of all the steps taken as well as providing various elements of tasks to gain results.

Table 13.1 Satellite data description

Data source	Date captured	Spatial resolution (m)	Band description
SPOT 5 269/341 (path/row)	8 July 2000	20	Four bands
SPOT 5 268/341 (path/row)	19 April 2005	10	Four bands
SPOT 5 268/341 (path/row)	20 February 2010	10	Four bands

Fig. 13.4 SPOT 5 images of 2000, 2005, and 2010 (RGB: 4-3-1)

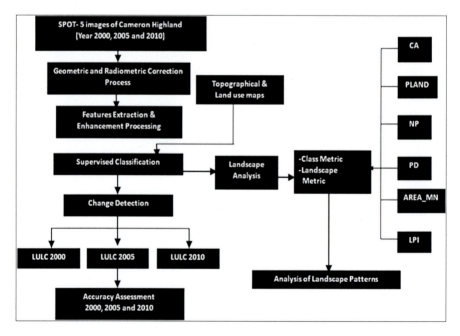

Fig. 13.5 Workflow for the research methodology. *CA* class area, *PLAND* percent of landscape, *NP* number of patches, *PD* patch density, *AREA_MN* mean patch area, *LPI* largest patch index

13.2.3 Land Cover Classification

Three SPOT 5 image series with 20-m raster grid resolution were acquired for the years 2000, 2005, and 2010, respectively. Each image was geometrically, atmospherically, and topographically corrected. The images were analyzed using ERDAS Imagine 9.1, ArcGIS 9.3 software. The land cover in Cameron Highlands were first defined into five classes: (1) water body, (2) tea plantation, (3) secondary forest/shrubs, (4) mixed agriculture/residential/road, and (5) primary forest.

The classification was automatically generated using a supervised maximum-likelihood classifier. Supervised image classification is a method in which the analyst initially defines small areas, called training sites, on the image that are representative of each desired land cover category (Kucukmehmetoglu and Geymen 2008). The classifiers then recognize the spectral values or signatures associated with these training sites. After the signatures for each land use/land cover category have been defined, the software then uses these signatures to classify the remaining pixels. Land use/land cover classification in each image was generated using combined bands of four, three, and one. Using AOI (area of interest), the spectral signature and spectral separability among classes were selected. The land cover classes then were verified following ground verification or truthing. The classified images were finally filtered by using 3×3 of median statistical filtering approach to reduce pixel overlaying of minor or isolated classes.

13.2.4 Accuracy Assessment

Accuracy assessment is an essential component of the investigation to quantify whether the data quality of one classification method is superior over others (Sader et al. 1995). Accuracy assessment is the process in which the image is partitioned into a set number of groups (classes) based on the values of the pixels in one or more image channels. The accuracy was set at more than 80 % as the threshold for accurate values. A total of 255 reference pixels per class were selected in a stratified random sampling approach for each class to assess the accuracy. Accuracy assessment involves identifying a set of sample locations (ground verification points) that would be visited in the field. Then the land cover identified in the field was compared to that mapped in the image during supervised classification for the same location by means of error or confusion matrices (Jensen 1986; Stehman 1996). Based on the confusion matrices, different accuracy measures were calculated: producer's accuracy, user's accuracy, and overall accuracy. To summarize the classification results, overall accuracies with 95 % confidence intervals were also generated. KAPPA analysis yields (Khat statistic) were also calculated to measure agreement or accuracy as suggested by Congalton (1991).

13.2.5 Landscape Pattern Analysis

There are more than a hundred indices for quantifying landscape pattern. A group of selected metrics (or indices) can be useful to interpret the landscape changes and considered relative to the type of patches (Apan and Peterson 1998). Landscape pattern metrics were run on images for the year studied. This analysis is designed to generate spatiotemporal indicators of landscape pattern as part of the scale-pattern-process paradigm (Walsh et al. 1998). Six landscape metrics were chosen in this study: class area (CA), percent of landscape (PLAND), number of patches (NP), patch density (PD), mean patch area (AREA_MN), and largest patch index (LPI) (Table 13.2). The spatial metrics were generated by FRAGSTATS version 3.3, a software package that calculates a number of spatial structures at three levels: entire landscape, class, and patch levels (McGarigal and Marks 1995; McGarigal et al. 2002).

13.3 Results and Discussion

13.3.1 Landscape Pattern Analysis

The trends of land cover change between years 2000, 2005, and 2010 in Cameron Highlands were varied for each class. Changes in land cover (Fig. 13.6) were derived from area estimates using land cover maps (Fig. 13.7).

Comparing three classification maps in general, the landscape change is not significant in the study area. Primary forest had increased slightly from year 2000 to year 2005, but decreased after that toward year 2010. This change demonstrated the recovery process of the secondary forest/shrubs in Cameron Highland and also in some part of the abandoned mixed agriculture area. The expansion of tea plantations is expected as shown in this study. Cameron Highland is the most famous tea producer in Malaysia where more than 2,000 ha of the highland is covered by tea plantations. According to Jamilah et al. (2006), the tea plantation landscape is the most preferred scene in Cameron Highlands. The scenic tea plantation view has been known ever since tea was introduced to Cameron Highlands in 1929. So, this land use has been maintained as the main scenic icon of Cameron Highlands and had increased in aerial extent gradually. It is noted that the other mixed agriculture/residential/roads are showing a decreased trend in the study periods because of the transformation of this class into other land cover types such as secondary forest and tea plantations. Water body area showed a minor increase (less than 100 ha) from years 2000 to 2010.

Table 13.2 Selected landscape metrics used in this study

No.	Landscape metrics	Abbreviation	Unit	Description
1	Class area	CA	Hectares (ha)	Sum of areas of all patches of the corresponding patch type
2	Percent of landscape	PLAND	Percentage (%)	Equals the percentage of the landscape composed of the corresponding patch type %LAND $= (CA/TLA) \times 100$ $-$TLA (total landscape area) $$\text{PLAND} = P_i = \frac{\sum_{j=1}^{n} a_{ij}}{A}(100)$$ by patch type (class) i $a_{ij} =$ area (m^2) of patch ij $A =$ total landscape area (m^2)
3	Number of patches	NP	No	The number of patches for each individual class. The higher NP indicates greater fragmentation NP $= N$ $N =$ total number of patches in the landscape
4	Patch density	PD	Number per 100 ha	Equals the number of patches in the landscape, divided by total landscape area (m^2), and multiplied by 10,000 and 100 (to convert to 100 ha) $N =$ total number of patches in the landscape $A =$ total landscape area (m^2)
5	Mean patch area	AREA_MN	No	Equals means of patch areas (ha)
6	Largest patch index	LPI	Percentage (%)	Equals the area (m^2) of the largest patch of the corresponding patch type divided by total landscape area (m^2), multiplied by 100 (to convert to a percentage) $a_{ij} =$ area (m^2) of patch ij $A =$ total landscape area (m^2)

13.3.2 Accuracy Assessment

The overall accuracies of the supervised classification of the satellite images accounted for 94 % (for year 2000), 91 % (for year 2005), and 88 % (for year 2010), respectively (Table 13.3). The kappa statistics were 0.90, 0.80, and 0.79, respectively for the years 2000, 2005, and 2010 classifications.

The KHAT statistic was calculated to determine the statistical significance or classification accuracy between the years. The computed KHAT statistic is a better indicator of percentage because the values approach 1. The overall image accuracies for each date, following verification from field data, were well above the

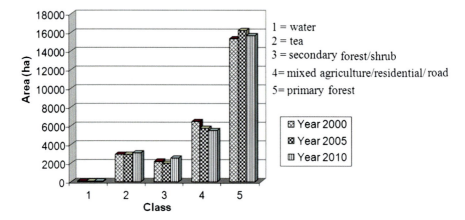

Fig. 13.6 Estimated area of land cover types in years 2000, 2005, and 2010 in the Cameron Highlands

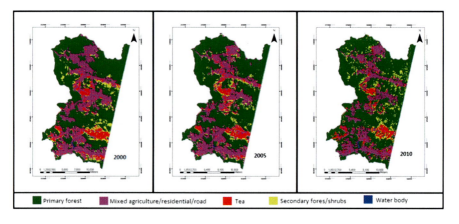

Fig. 13.7 Spatial variations of land use/land cover in the Cameron Highlands in years 2000, 2005, and 2010

Table 13.3 Accuracy assessment and kappa statistic for land cover classification (years 2000, 2005, and 2010)

Image	Accuracy assessment (%)	Kappa (K) statistic
SPOT 5 Year 2000	94	0.9
SPOT 5 Year 2005	91	0.84
SPOT 5 Year 2010	88	0.79

generally accepted 85 % standard of accuracy for image classifications (Foody 2003), with kappa statistics above 0.75 (Nagendra et al. 2003). The generally accepted overall accuracy level for land use maps is 85 %, which is approximately equal in accuracy for most categories (Jensen 1986; Mohd Hasmadi and Kamaruzaman 2008).

13 Land Use Trends Analysis Using SPOT 5 Images and Its Effect... 235

Table 13.4 Landscape structure of the Cameron Highlands in 2000, 2005, and 2010

Landscape metric/year	Land cover type				
	Primary forest	Secondary forest and shrubs	Mixed agriculture, residential, roads	Tea plantations	Water bodies
CA					
2000	15,384.4	2,185.6	6,468.1	2,939.9	31.8
2005	16,240.4	2,021.9	5,782.6	2,924.4	40.5
2010	15,691.8	2,553.5	5,546.8	3,121.5	96.2
PLAND					
2000	56.9122	8.1951	23.8588	10.9076	0.1263
2005	60.2359	7.281	21.4083	10.9299	0.1449
2010	58.1235	9.3544	20.5035	11.6637	0.3549
NP					
2000	499	1,306	564	1,010	10
2005	450	1,619	530	1,159	40
2010	545	1,802	437	1,441	110
PD					
2000	1.8474	4.8351	2.088	3.7392	0.037
2005	1.666	5.9938	1.9622	4.2908	0.1481
2010	2.0177	6.6713	1.6179	5.3349	0.4072
AREA_MN					
2000	30.8068	1.6949	11.4264	2.9171	3.4124
2005	36.1563	1.2147	10.9106	2.5473	0.9786
2010	28.8069	1.4022	12.6732	2.1863	0.8714
LPI					
2000	34.4468	0.3957	13.5959	2.0065	0.1003
2005	38.8871	0.2341	4.9977	1.568	0.0539
2010	40.6688	0.3753	5.9545	1.7055	0.1152

CA class area, *PLAND* percent of landscape, *NP* number of patches, *PD* patch density, *AREA_MN* mean patch area, *LPI* largest patch index

13.3.3 Analysis of Landscape Patterns

Spatially, changes in landscape of the Cameron Highlands are constantly dynamic. The changes are less than 5 % across the study period. Table 13.4 shows the landscape metric patterns in years 2000, 2005, and 2010, respectively. The total number of forest fragments slightly decreased from 1.8474 in 2000 to 1.666 in 2005, and then increased to 2.0177 in 2010. Areas of reforestation were significantly larger than areas of deforestation, across all dates. Patch size was a good indicator of economic activity.

PLAND is increased for tea plantation and water body from 10 % to 11 % and from 0.12 % to 0.35 %, respectively. Mixed agriculture, residential use, and roads decreased over the years (from 23 % in 2000 to 20 % in 2010). CA and PLAND are good indicators of landscape composition. PLAND is an important characteristic in a number of ecological applications. Both metrics have the same characteristic when spatial extent does not change or is opposite. These indices relate to numbers/total areas for land use and cover (LULC) classes. NP defined total patches for every class, and the major patch value in the study is from mixed secondary forest and shrubs, being about 1,802. These patch numbers depend on the pixel/window size from satellite image resolution. In the study, a low patch number is exhibited by the water body class (ten patches only). The highest PD shown in the study was by secondary forest and shrubs, ranging from 4.8 to 6.6. LPI measures the largest patch of landscape fragmentation in percentage values, and the largest patches shown in this study was primary forest class (40). As the performance of LPI is generally not sensitive to varying spatial aggregation, this index can serve for landscape composition but is not suitable for indication of fragmentation.

13.4 Conclusion

This research provides evidence for the usefulness of multitemporal remote sensing approaches and landscape metrics to monitor the rate of loss and pattern of fragmentation in the tropical mountain forest of Cameron Highlands. It can be concluded that, in general, changes in land use occurred over the study period. Increase in the number of forest patches was above 2 % between year 2000 and 2010, and substantial decrease was shown in the mixed agriculture/residential/road patch: the latter finding indicates improvement in forest sucession in Cameron Highlands. The positive trends in forest cover changes and expansion provide some evidence of the ecological sustainability of the area. The decreasing number of patches, from 23 % to 21 % in mixed agriculture/residential/road during the study period has, however, raised some questions regarding the agricultural practices in the highland, although this study has provided important insights into the dynamics of land cover in a forested area and other major land uses of Cameron Highands. Quantitatve information from this study reinforces understanding of the relationship between goverment policies and forest conditions. It is urgent to define a political attitude and conservation plan that minimizes landscape degradation in the near future in the Cameron Highlands. This study was only a preliminary step toward understanding the landscape and properties of the Cameron Highlands environment. Land use planning in the highland should be prepared in accordance with a protection strategy. Adaptation of the protection strategy to this area requires certain, urgent, and short-term measures for protection of the precious landscape and biodiversity. Any changes in land use types that can result in irreversible changes should be kept under control by a responsible authority. Further studies are needed to improve scientific knowledge in developing sustainable agriculture and tourism development strategies in Cameron Highlands, as well as the ability to quantify the biogeochemical and hydrological processes changes resulting from landscape changes in the highlands. The use of multitemporal remotely sensed data to conduct landscape trends analysis represents an exciting opportunity, not only to conduct change detection analysis but also to advance the disciplines of both landscape ecology and remote sensing in a spatial context.

Acknowledgments We gratefully acknowledge financial support from the Ministry of Higher Education Malaysia through Fundamental Research Grant Scheme FRGS-5523434 and Malaysian Remote Sensing Agency for providing satellite data (SPOT 5).

References

Aminuddin BY, Ghulam MH, Wan Abdullah WY (2005) Sustainability of current agricultural practices in the Cameron highlands, Malaysia: water, air, and soil pollution. Focus 5:89–101

Apan AA, Peterson JA (1998) Probing tropical deforestation: the use of GIS and statistical analysis of georeferenced data. Appl Geogr 18:137–152

Bruijnzeel LA (2001) Hydrology of tropical montane cloud forests: a reassessment. Land Use Water Resour Res 1:1–18

Che Ku Akmar CKO, Mohd Hasmadi I (2010) Land use in Cameron highlands: analysis of its changes from space. In: Proceedings of the world engineering congress: geometrics and geographical information science, Grand Margherita Hotel, Kuching, Sarawak, Malaysia, 2–5 Aug 2010, pp 190–195

Cohen WB, Goward SN (2004) Landsat's role in ecological applications of remote sensing. BioScience 54(6):535–545

Congalton RG (1991) A review of assessing the accuracy of classifications of remotely sensed data. Remote Sens Environ 37:35–46

Eric FL (1999) Monitoring forest degradation in tropical regions by remote sensing: some methodological issues. Glob Ecol Biogeogr 8:191–198

Foody GM (2003) Remote sensing of tropical forest environments: towards the monitoring of environmental resources for sustainable development. Int J Remote Sens 24(20):4035–4046

Forman RT (1995) Land mosaics: the ecology of landscapes and regions. Cambridge University Press, Cambridge

Griffiths GS, Mather PM (2000) Remote sensing and landscape ecology: landscape patterns and landscape change. Int J Remote Sens 21:2537–2539

Gustafson EJ (1998) Quantifying landscape spatial pattern: what is the state of the art? Ecosystems 1:143–156

Hay GJ, Dubé P, Bouchard A, Marceau DJ (2002) A scale-space primer for exploring and quantifying complex landscapes. Ecol Model 153(1-2):27–49

He HS, DeZonia BE, Mladenoff DJ (2000) An aggregation index (AI) to quantify spatial patterns of landscapes. Landsc Ecol 15(7):591–601

Herzog F, Lausch A, Muller E, Thulke HH, Steinhardt U, Lehmann S (2001) Landscape metrics for assessment of landscape destruction and rehabilitation. Environ Manag 27:91–107

Jamilah O, Ahmad MA, Manohar M, Zaliah S (2006) Consideration of visual aesthetics quality for landscape management decisions. In: Proceedings of the 2nd seminar on the environmental, centre for built environment, Kulliyyah of architecture and environmental design, IIUM, pp 206–301

Jensen JR (1986) Introductory digital image processing: a remote sensing perspective. Prentice Hall, Englewood Cliffs

Kerr JT, Southwood TRE, Cihlar J (2001) Remotely sensed habitat diversity predicts butterfly species richness and community similarity in Canada. Proc Natl Acad Sci USA 98:11365–11370

Kucukmehmetoglu M, Geymen A (2008) Measuring the spatial impacts of urbanization on the surface water resource basins in Istanbul via remote sensing. Environ Monit Assess 142:153–169

McGarigal K, Marks B (1995) FRAGSTATS: spatial pattern analysis program for quantifying landscape structure, vol 2.0. General technical report PNW-GTR-351. Forest Science Lab, Oregon State University, Corvallis, USDA Forest Service, Pacific Northwest Research Station

McGarigal K, Cushman SA, Neel MC, Ene E (2002) FRAGSTATS: spatial pattern analysis program for categorical maps. Computer software program produced by the authors at the University of Massachusetts, Amherst. www.umass.edu/landeco/research/fragstats/fragstats.html. Accessed 7 July 2009

Mohd Hasmadi I, Kamaruzaman J (2008) Satellite data classification accuracy assessment based from reference dataset. Int J Comput Inf Sci Eng 2(2):96–102

Nagendra H, Southworth J, Tucker C (2003) Accessibility as a determinant of landscape transformation in Western Honduras: linking pattern and process. Landsc Ecol 18:141–158

Patil GP, Brooks RP, Myers WL, Rapport DJ, Taillie C (2001) Ecosystem health and its measurement at landscape scale: toward the next generation of quantitative assessments. Ecosyst Health 7:307–316

Roy PS, Joshi PK (2002) Forest cover assessment in North-East India–the potential of temporal wide swath satellite sensor data. Int J Remote Sens 23:4881–4896

Sader SA, Ahl D, Wen-Shi L (1995) Accuracy of Landsat TM and GIS rule-based methods for forest wetland classification in Maine. Remote Sens Environ 53:133–144

Skole D, Justice C, Townshend J, Janetos A (1997) A land cover change monitoring program: strategy for an international effort. Mitig Adapt Strat GL 2:157–175

Stehman SV (1996) Estimation of kappa coefficient and its variance using stratified random sampling. Photogramm Eng Remote Sensing 26:401–407

Turner MG (1989) Landscape ecology: the effect of pattern on process. Annu Rev Ecol Syst 20:171–197

Turner MG, Gardner RH, O'Neill RV (2001) Landscape ecology in theory and practice: pattern and process. Springer, New York

Turner W, Spector S, Gardiner N, Fladeland M, Sterling E, Steininger M (2003) Remote sensing for biodiversity science and conservation. Trends Ecol Evol 18(6):306–314

Walsh SJ, Butler DR, Malanson GP (1998) An overview of scale, pattern, process relationships in geomorphology: a remote sensing and GIS perspective. Geomorphology 21(3-4):183–205

Watson N, Wilcock D (2001) Pre-classification as an aid to the improvement of thematic and spatial accuracy in land cover maps derived from satellite imagery. Remote Sens Environ 75(2):267–278

Zha Y, Goa J, Ni S (2003) Use of normalized difference built-up index in automatically mapping urban areas from TM imagery. Int J Remote Sens 24(3):583–594

Chapter 14
The Relationship Between Land Use/Land Cover Change and Land Degradation of a Natural Protected Area in Batang Merao Watershed, Indonesia

Rachmad Firdaus, Nobukazu Nakagoshi, Aswandi Idris, and Beni Raharjo

Abstract Land degradation in Batang Merao is regarded as one of the major environmental problems that can affect the sustainability of this natural protected area in Kerinci Seblat National Park. The aim of this study was to determine the linkages between land use/land cover (LULC) changes and land degradation using multitemporal Landsat data from 1990, 2000, and 2010. Based on a maximum-likelihood algorithm of the supervised classification method, images were classified into six classes: forest, mixed plantation, tea plantation, shrub/bush, agricultural land, and settlement. The results showed that during the past two decades, two major changes took place. Forest decreased at rates of 330.85 ha year^{-1} (period of 1990–2000) and 145.25 ha year^{-1} (period of 2000–2010); on the other hand, agricultural land, mixed plantation, and settlement have shown increments. Concerning land degradation, Batang Merao Watershed exhibited potential soil degradation where the mean annual potential land degradation was 128.03 tons ha^{-1} year^{-1} in 1990, 144.68 tons ha^{-1} year^{-1} in 2000, and 194.14 tons ha^{-1} year^{-1} in 2010. Based on statistical analysis (Pearson's correlation coefficient), this study reveals a relationship between LULC change and land degradation in that land cover type plays an important role in protecting soil from

---

R. Firdaus (✉)
Graduate School for International Development and Cooperation, Hiroshima University, 1-5-1 Kagamiyama, Higashi-Hiroshima 739-8529, Japan

Regional Development Planning Board of Jambi Province (Bappeda), Jl RM Nur Atmadibrata No. 1 Telanaipura, Jambi 36124, Indonesia
e-mail: r.firdaus2010@gmail.com

N. Nakagoshi • B. Raharjo
Graduate School for International Development and Cooperation, Hiroshima University, 1-5-1 Kagamiyama, Higashi-Hiroshima 739-8529, Japan

A. Idris
Fakultas Pertanian-PPMDAS, Jambi University, Jl. Raya Jambi-Muara Bulian KM 15, Mendalo Darat, Jambi 36123, Indonesia

N. Nakagoshi and J.A. Mabuhay (eds.), *Designing Low Carbon Societies in Landscapes*, Ecological Research Monographs, DOI 10.1007/978-4-431-54819-5_14, © Springer Japan 2014

land degradation in this watershed. Therefore, areas in an extremely high level of land degradation should be recommended as important for a conservation program.

Keywords Conservation • Environmental • Land cover • Land use • Protected area • Sustainability

14.1 Introduction

Studying land use/land cover (LULC) change is a critical requirement for the assessment of potential environmental impacts and the development of effective land management and planning strategies (Leh et al. 2013). Knowledge of the nature of LULC change and its configuration across spatial and temporal scales is consequently indispensable for sustainable environmental management and development (Turner et al. 1995). LULC is always dynamic when it constantly changes in response to the dynamic interaction between underlying drivers and proximate causes (Lambin et al. 2003). Change in LULC is a key driver of environmental changes (Lambin et al. 2003) on all spatial and temporal scales (Turner et al. 1994), and it can be a major threat to biodiversity (Verburg et al. 1999). Monitoring LULC change in the landscape of a watershed is becoming an important issue across various fields of development and sustainable management. Landscape changes include not only damage by agriculture, but also degradation of historic values and land conservation functions (Ohta and Nakagoshi 2011).

During the past few decades, LULC change and its impacts have become major problems and serious threats to environmental conditions. Many watersheds today suffer from several detrimental problems such as severe soil erosion, flood, drought, and declining land productivity or land degradation. Land degradation, a synonym for soil degradation (Kertész 2009) that implies soil functions have been damaged by climate or human activities (Maitima et al. 2004), is a critical environmental problem in many countries (Ouyang et al. 2010), especially in developing countries (Ananda and Herath 2003). About 85 % of land degradation in the world is associated with soil erosion (Oldeman et al. 1991), such as in the Citarik, West Java, about 94–103 tons ha^{-1} year^{-1} (Kusumandari and Mitchell 1997). Furthermore, land degradation has major implications for society in economic, social, and environmental perspectives.

Information on LULC dynamics and its impacts is very important for landscape management because it creates key environmental information for many resource management and policy purposes. Therefore, it is very useful for planners and policy makers to initiate remedial analysis on LULC change and land degradation. Furthermore, to strengthen the conservation and protection of the ecological environment, comprehensive planning is necessary with considerations that include balancing the social, safety, ecology, and landscape values and treating the whole watershed as a management unit (Wu and Feng 2006).

Batang Merao Watershed was selected for this research because it is the most important watershed around Kerinci Seblat National Park (KNSP), the biggest natural protected area in Sumatera. Batang Merao Watershed has a prominent role as a buffer zone for KNSP. Land degradation in Batang Merao Watershed will affect the sustainability of the conservation function of KNSP and downstream areas such as Sungai Penuh City and Merangin Regency. The purpose of this study was to analyze LULC changes, land degradation, and the relationship between them in three different years (1990, 2000, and 2010).

14.2 Methods

14.2.1 Study Area

Batang Merao Watershed is located in the northwest of Jambi Province: its geo-location is at $01°42'19''–02°08'14''$S, $101°13'11''–101°32'20''$E (Fig. 14.1). The elevation ranges from 766 to 3,236 m above sea level. The total area of the study site is 67,874.48 ha. This area covers ten subregencies and 124 villages. It is situated in a tropical zone where the annual mean precipitation is 2,495 mm year^{-1}, based on the past 20 years, and the annual mean temperature is 23.1 °C, based on the past 10 years.

14.2.2 Data

The basic data required to meet the objectives of this study were Landsat image data, Aster Global Digital Elevation Model (Aster GDEM), digital land use maps, climate data (annual rainfall), and Universal Soil Loss Equation (USLE) data. For supporting image analysis, some ancillary data were used including ground truth data acquired through a field survey on September 10–15, 2011, a digital administrative map of Jambi Province provided by the Gespatial Information Agency of Indonesia, and a digital watershed boundary map of Jambi Province published by the Ministry of Forestry of Indonesia. All the supplementary data were used to assist in image classification and to collect reference data for accuracy assessment, as described in Table 14.1.

14.2.3 Analysis

This research started with analyzing LULC change during the study period. Supervised classification, the most widely used technique for quantitative analysis of remote sensing image data (Pôças et al. 2011), was used to perform image classification. To evaluate the accuracy of the classified images, an accuracy assessment

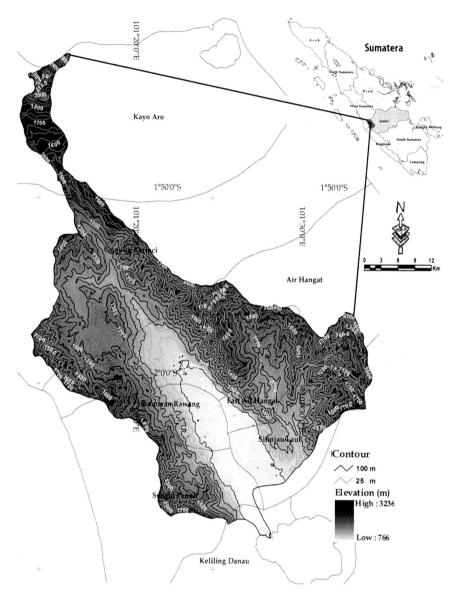

Fig. 14.1 Batang Merao watershed showing the elevation at 25-m and 100-m contour intervals

or confusion contingency matrix was implemented. The error matrix compared the linkages between known reference data (ground truth) and the corresponding results of an automated classification (Congalton 1991). The kappa coefficient, the value for estimating how accurate the remotely sensed classification is in the reference data (Jensen 2004), was used for accuracy assessment. The kappa (*Khat*) is computed as follows:

14 The Relationship Between Land Use/Land Cover Change and Land Degradation... 243

Table 14.1 Data collection and its description

Data	Description	Source
Landsat TM	Path 126/row 61	http://glovis.usgs.gov/
	July 13, 1990	
	May 5, 2000	
	June 18, 2010	
DEM	Aster GDEM 30 m	http://demex.cr.usgs.gov/
Administrative	Jambi Province	Geospatial Information
	Kerinci regency	Agency of Indonesia (BIG)
	Watershed boundary	Planning Agency of Jambi
Soil series map	Soil type 1990, 2000, 2010	Planning Agency of Kerinci
Rainfall	Monthly rainfall and	Forestry office of Batanghari (BPDAS)
	distribution 1990–2010	
Land use planning map	Regional land use planning	
Ground truth	Ground truth 2011	Field survey

$$\text{Khat} = \frac{N \sum_{i=1}^{r} - \sum_{i=1}^{r} (x_{i+} * x_{+i})}{N^2 - \sum_{i=1}^{r} (x_{i+} * x_{+i})}, \tag{14.1}$$

where r is the number of rows in the error matrix, x_i is the number of observations in row i and column i, x_{i+} and x_{+i} are the marginal totals for the row i and column i, respectively, and N is the total number of pixels. Value greater than 0.80 represent strong agreement or good classification performance, values between 0.40 and 0.80 indicated moderate agreement, and values less than 0.40 represent poor agreement (Jensen 2004).

LULC classification was modified from the LULC categories of the Indonesian National Standard No. 7645:2010 specified by the National Standard Agency of Indonesia, which refers to the FAO's land cover classification system and ISO 19144-1 (BSN-National Standardization Agency of Indonesia 2010). Because of differences in scale, accuracy, and type of LULC categories, synchronization and generalization were performed, resulting in six categories: forest, mixed plantation, tea plantation, shrub/bush, agricultural land, and settlement (Table 14.2).

Potential soil erosion was determined using the USLE method (Wischmeier and Smith 1978). This is the most widely used method and the simplest model for predicting soil erosion and still remains the best known because of its scientific basis, ease of use, low cost, and direct applicability in watershed (Sharma et al. 2011). It can be described in the following equation:

$$\mathbf{A} = \mathbf{R} \times \mathbf{K} \times \mathbf{L} \times \mathbf{S} \times \mathbf{C} \times \mathbf{P}, \tag{14.2}$$

where $\mathbf{A}$ is the annual soil loss (tons ha^{-1} $year^{-1}$), $\mathbf{R}$ is the rainfall/erosivity factor (mm $year^{-1}$), $\mathbf{K}$ is the soil erodibility factor (t J^{-1} mm^{-1}), $\mathbf{L}$ and $\mathbf{S}$ are slope length and steepness factors, respectively, and $\mathbf{C}$ and $\mathbf{P}$ are the crop management and conservation factors.

Table 14.2 Land use/land cover (LULC) classification and its general description

LULC classification	General description
Forest	Areas covered by dense trees with relatively darker green color
Tea plantation	An intensively managed plantation. It is characterized by a homogeneous canopy structure with single dominant species and regular spatial network and clear-cut boundaries with neighboring vegetation
Mix plantation	Areas covered by a combination of several woody and fruity plantations such as rubber, *Cynnamomun*, *Mango* sp., *Durio* sp., etc. It is indicated by irregular patterns
Shrub/bush	Areas covered by herbs, grass, and nonwoody herbs. These areas usually correspond to recently opened areas, the first phase of land conversion into both mixed plantation and tea plantation
Agricultural land	Areas dominantly cultivated by paddy field and potatoes, characterized by inundating of fields from irrigation or rainfall
Settlement	Areas occupied by houses or buildings including road network and other facilities

Source: The Indonesian National Standard No. 7645:2010 by the BSN-National Standardization Agency of Indonesia 2010

Statistical analysis with Pearson's correlation coefficient was applied to identify the relationship between LULC change and land degradation. In this analysis, the land use category served as the independent variable and the rate of land degradation served as the dependent variable. Because of the difference in scale between land use type (nominal scale) and land degradation class (ordinal scale), the land degradation class needed to be aggregated using the statistics method "summated rating scale." The next step was bivariate statistics correlation analysis between land use degradation to identify the relationship between LULC change and land degradation.

All image processing, classification, and change detection were performed using ERDAS Imagine 8.7; the GIS analysis was carried out using ArcGIS 10.1. Statistical analyses were conducted using PSAW SPSS Statistics ver. 18.0 for Windows.

14.3 Results

14.3.1 Changes in LULC

The accuracy of LULC change along with the overall accuracy and the Khat coefficient are summarized in Table 14.3. This table shows that the user's accuracy of individual category ranged from 50 % to 100 % and the producer's accuracy ranged from 68 % to 100 %. The overall accuracy of image classification was 81.93 % and the kappa coefficient was 0.776. The kappa coefficients indicated that the classified images showed moderate classification performance or moderate agreement.

Table 14.3 Accuracy assessment for supervised classification of land use/land cover (LULC)

LULC classification	Reference data						Total	User's accuracy (%)
	F	MP	TP	SB	AL	S		
Forest (F)	19						19	100.00
Mixed plantation (MP)	1	17		1	1		20	85.00
Tea plantation (TP)		1	8				9	88.89
Shrub/bush (SB)		2		7	1		10	70.00
Agricultural land (AL)		2		1	12		15	80.00
Settlement (S)		3		1	1	5	10	50.00
Total	20	25	8	10	15	5	83	Overall accuracy 81.93 %
Producer's accuracy (%)	95.00	68.00	100.00	70.00	80.00	100.00		Kappa coefficient 0.776

The distribution of LULC and its changes for 1990, 2000, and 2010 are summarized in Table 14.4. The spatial distribution of LULC over time is clearly visible in Fig. 14.2. The dynamic patterns of LULC change are also represented in Fig. 14.3. Throughout the period of study, several LULC categories had increased such as settlement, agricultural land, and mixed plantation. For example, settlement areas have increased from 1,150.29 ha in 1990 to 1,530.51 ha in 2000 and 1,587.48 ha in 2010. Mixed plantation areas have also shown an increase of 36.60 % (period 1990–2000) and 3.44 % (period 2000–2010). On the other hand, forest decreased from 20,297.58 ha in 1990 to 16,989.12 ha in 2000, and to 15,536.66 ha in 2010.

Table 14.5 and Fig. 14.4 show a clear pattern of changes characterized by potential land degradation levels. There was an increase in the total area with very high levels of land degradation, namely, 19.55 % in the period of 1990–2000 and 10.17 % in the period of 2000–2010. At the same time, the area with very low level of land degradation decreased by -5.70 % and -16.54 % in 1990–2000 and 2000–2010, respectively. Batang Merao Watershed also exhibited potential land degradation as the mean annual land degradation increased from 128.03 tons ha^{-1} $year^{-1}$ in 1990 to 144.68 tons ha^{-1} $year^{-1}$ in 2000, and to 194.14 tons ha^{-1} $year^{-1}$ in 2010.

14.3.2 Linkages Between LULC Change and Land Degradation

The result of Pearson correlation analysis is summarized in Table 14.6. The results depicted a negative correlation between several types of LULC and the rate of land degradation: forest (-0.86), mixed plantation (-0.74), and shrub/bush (-0.49),

Table 14.4 LULC change in 1990, 2000, and 2010

| LULC classification | Area (ha) | | | Change | | | | Average rate of change | | | |
| | 1990 | 2000 | 2010 | 1990–2000 | | 2000–2010 | | 1990–200 | | 2000–2010 | |
				ha	%	ha	%	ha year^{-1}	%. year^{-1}	ha year^{-1}	% year^{-1}
Forest	20,297.58	16,989.12	15,536.66	−3,308.46	−16.30	−1,452.46	−8.55	−330.85	−1.63	−145.25	−0.85
Mixed plantation	11,177.02	15,267.43	15,792.12	4,090.41	36.60	524.69	3.44	409.04	3.66	52.47	0.34
Tea plantation	931.46	1,078.82	997.71	147.35	15.82	−81.10	−7.52	14.74	1.58	−8.11	−0.75
Shrub/bush	21,587.39	19,734.09	20,154.28	−1,853.29	−8.59	420.18	2.13	−185.33	−0.86	42.02	0.21
Agricultural land	12,730.75	13,274.52	13,806.23	543.77	4.27	531.71	4.01	54.38	0.43	53.17	0.40
Settlement	1,150.29	1,530.51	1,587.48	380.22	33.05	56.98	3.72	38.02	3.31	5.70	0.37
Total	67,874.48	67,874.48	67,874.48								

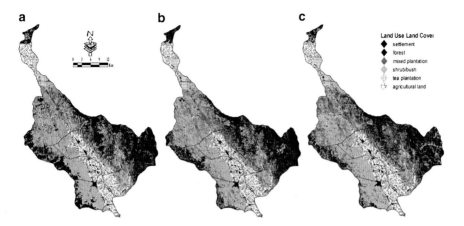

Fig. 14.2 Land use/land cover maps of the Batang Merao watershed in 1990 (**a**), 2000 (**b**), and 2010 (**c**)

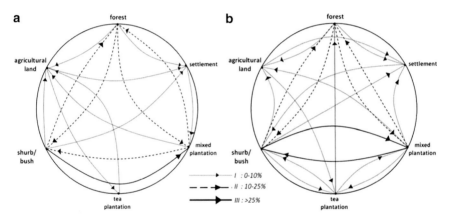

Fig. 14.3 Dynamic patterns of land use/land cover change in Batang Merao watershed in the periods 1990–2000 (**a**) and 2000–2010 (**b**)

respectively. The negative correlation implied that the larger are forest area, mixed plantation, and shrub/bush, the lower is the land degradation. However, there were some positive correlations between low-vegetation land use type and nonvegetation type to the rate of land degradation: agricultural land (0.95), settlement (0.75), and tea plantation (0.15), respectively, the larger. The positive correlation implied that the larger are agricultural land, tea plantation, and settlement, the higher is the level of land degradation in the watershed.

Table 14.5 Distribution of potential land degradation in Batang Merao watershed, Sumatera

Level of land degradation	Potential soil loss (tons ha^{-1} year^{-1})	Area (ha)			Change				Average rate of change			
					1990–2000		2000–2010		1990–2000		2000–2010	
		1990	2000	2010	ha	%	ha	%	ha year^{-1}	% year^{-1}	ha year^{-1}	% year^{-1}
Very low	<5	32,156.14	30,322.57	25,306.09	−1,833.57	−5.70	−5,016.48	−16.54	−183.36	−0.57	−501.65	−1.65
Low	5–10	4,584.36	1,616.81	3,837.67	−2,967.55	−64.73	2,220.86	137.36	−296.76	−6.47	222.09	13.74
Moderate	10–30	1,743.60	1,220.02	1,681.19	−523.58	−30.03	461.17	37.80	−52.36	−3.00	46.12	3.78
High	30–60	2,800.14	2,926.87	2,028.86	126.74	4.53	−898.01	−30.68	12.67	0.45	−89.80	−3.07
Very high	>60	26,590.24	31,788.20	35,020.68	5,197.96	19.55	3,232.47	10.17	519.80	1.95	323.25	1.02
Total		67,874.48	67,874.48	67,874.48								

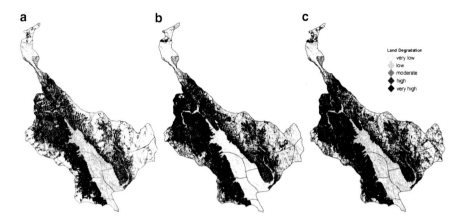

Fig. 14.4 Land degradation maps of the Batang Merao watershed in 1990 (**a**), 2000 (**b**), and 2010 (**c**)

Table 14.6 Pearson's correlation between LULC type and land degradation

LULC type	Statistical indicator	Land degradation
Forest	Pearson correlation	−0.86
	Significance (two-tailed)	0.34
Mixed plantation	Pearson correlation	−0.74
	Significance (two-tailed)	0.47
Tea plantation	Pearson correlation	0.15
	Significance (two-tailed)	0.91
Shrub/bush	Pearson correlation	−0.49
	Significance (two-tailed)	0.67
Agricultural land	Pearson correlation	0.95
	Significance (two-tailed)	0.21
Settlement	Pearson correlation	0.75
	Significance (two-tailed)	0.46

14.4 Discussion and Conclusion

In general, the results of the study disclosed that the Batang Merao Watershed had been under continual LULC changes from 1990 to 2010. Deforestation caused by agricultural activities (expansion of plantations and paddy fields) and increasing demand for settlement have imposed a threat of land degradation. This information is essential to preserve the natural protected areas. The results also showed that the areas were dominated by high levels of land degradation. The mean annual potential land degradation amounted to 128.03 tons ha^{-1} year^{-1} in 1990, 144.68 tons ha^{-1} year^{-1} in 2000, and 194.14 tons ha^{-1} year^{-1} in 2010.

According to the statistical analysis, this research showed that there was a relationship between the type of LULC and land degradation. Vegetation has an important role in protecting soil from erosion loss and is the key factor affecting land degradation in the watershed (Zhou et al. 2008). In addition, vegetation cover affected land degradation because of its litter production, organic matter accumulation, and plant roots (Wijitkosum 2012).

The results also concluded that changes in LULC and its dynamics were closely associated with human activities in the region such as the expansion of paddy fields and settlements. The results of the study clearly indicated that dynamic changes in LULC had been a driver of extensive land degradation in the study area. It showed areas with a high potential of land degradation as indicated by the mean annual land degradation in the study period. Among the USLE factors, the value of the LULC factor was dynamic over time and increased the total value and level of potential land degradation. The wide range of land degradation was subject to agricultural activities (agricultural land and mixed plantations).

Assessment of LULC dynamics and its linkages is an essential part in sustainable land management that can help people optimize land use and minimize environmental impacts such as land degradation. Therefore, to prevent an extremely high level of land degradation in these areas, the wise use of a land cover and soil conservation program is highly recommended to be widely implemented in the tea plantation and agricultural land.

Acknowledgments The authors acknowledge the Center for Development, Education and Training of Indonesian Planner (Pusbindiklatren-Bappenas RI), Regional Development Planning Board of Jambi Province, and the Global Environmental Leaders (GELs) Program of Graduate School for IDEC, Hiroshima University. We thank to Mr. Aris Rusyiana (a staff member of Indonesian Statistics Agency) for his statistical assistance.

References

Ananda J, Herath G (2003) Soil erosion in developing countries: a socioeconomic appraisal. J Environ Manag 68:343–353

BSN-National Standardization Agency of Indonesia (2010) Land cover classification. SNI 7645. Bahasa, Indonesia

Congalton RG (1991) A review of assessing the accuracy of classifications of remotely sensed data. J Remote Sens Environ 37:35–46

Jensen J (2004) Introductory digital image processing: a remote sensing perspective, 3rd edn. Prentice-Hall, Upper Saddle River

Kertész A (2009) The global problem of land degradation and desertification. Hung Geogr Bull 58:19–31

Kusumandari A, Mitchell B (1997) Soil erosion and sediment yield in forest and agro-forestry area in West Java, Indonesia. Soil Water Conserv 52:376–380

Lambin EF, Geist HJ, Lepers E (2003) Dynamics of land use and land cover change in tropical regions. Annu Rev Environ Resour 28:205–241

Leh M, Bajwa S, Chaubey I (2013) Impact of land use change on erosion risk: an integrated remote sensing, geographic information system and modeling methodology. Land Degrad Dev 24:409–421

Maitima J, Reid RS, Gachimbi LN, Majule A, Lyaruu H, Pomery D, Mugatha S, Mathai S, Mugisha S (2004) Regional synthesis paper: the linkages between and land use change, land degradation and biodiversity across East Africa. LUCID working papers 42, pp 1–58

Ohta Y, Nakagoshi N (2011) Analysis of factors affecting the landscape dynamics of islands in Western Japan. In: Hong SK et al (eds) Landscape ecology in Asian cultures. Springer, Tokyo, pp 169–185

Oldeman LR, Hakkeling RTA, Sombroek WG (1991) World map of the status of human-induced soil degradation: an explanatory note. GLASOID Report

Ouyang W, Skidmore AK, Hao F, Wang T (2010) Soil erosion dynamics response to landscape pattern. Sci Total Environ 408:1358–1366

Pôças I, Cunha M, Marcal ARS, Pereira LS (2011) An evaluation of changes in a mountainous rural landscape of Northeast Portugal using remotely sensed data. Landsc Urban Plan 101:2–9

Sharma A, Tiwari K, Bhadoria PBS (2011) Effect of land use land cover change on soil erosion potential in an agricultural watershed. Environ Monit Assess 173:789–801

Turner BL, Meyer WB, Skole DL (1994) Global land-use/land cover change: towards an integrated study. Ambio 23:91–95

Turner BL, Skole DL, Sanderson S, Fischer G, Fresco L, Leman R (1995) Land-use and land cover change: science/research plan. IGBP Report No. 1 and HDP Report No. 7

Verburg PH, Veldkamp TA, Bouma J (1999) Land use change under conditions of high population pressure: the case of Java. Global Environ Chang 9:303–312

Wijitkosum S (2012) Impacts of land use changes on soil erosion in Pa Deng sub-district, adjacent area of Kaeng, Krachan National Park, Thailand. Soil Water Res 7:10–17

Wischmeier WH, Smith DD (1978) Predicting rainfall erosion losses: a guide to conservation planning. In: Agricultural handbook No. 537. Department of Agriculture, Washington, DC, pp 1–69

Wu HL, Feng ZY (2006) Ecological engineering methods for soil and water conservation in Taiwan. Ecol Eng 28:333–344

Zhou P, Luukkanen O, Tokola T, Nieminen J (2008) Effect of vegetation cover on soil erosion in a mountainous watershed. Catena 75:319–325

Chapter 15
Ecotourism Activities for Sustainability and Management of Forest Protected Areas: A Case of Camili Biosphere Reserve Area, Turkey

Mustafa Fehmi Türker, İnci Zeynep Aydın, and Türkan Aydın

Abstract Ongoing altering interests of communities and diversified attitudes and behavior regarding forest resources have affected basic forest management policies. In this context, it seems necessary to provide sustainable economic and social development without endangering forest resources for today and future generations. "Sustainable forest management" is defined as operating forests and the biological diversity, fertility, ecological, social, and economic functions of forest lands in a sustainable way. Protected areas cover special, precious, and rare examples and species that are of utmost importance for science and education. For target management of protected areas, it is necessary to select an approach that includes sustainability and conservation of biological diversity. An important sustainability issue arises here. In addition, it is seen in most forest lands that villagers create a meaningful whole with protected areas. It should not be forgotten that the most important group with which Turkish forestry must communicate is forest villagers within the framework of sustainable forestry (Geray, 1998). Ecotourism is a kind of tourism that aims to raise the life standard of forest villagers, contribute to the national economy, and increase awareness of the structure of natural and cultural values. Ecotourism also provides sustainability of environmental resources and excludes activities aiming for short-term profits (Alkan et al., 2010), (Akıllı, 2004), (Porsuk, 2000).

In this study, conducted in Camili Biosphere Reserve in Artvin; suggestions including the use of ecotourism and the importance of public relations for the development of forest villagers have been put forth to explain the relationship between protected areas and sustainable forestry management and to solve problems in the sustainability process of the protected areas. Forest villagers were selected according to a full-count method. Questionnaires included such variables

---

M.F. Türker • İ.Z. Aydın • T. Aydın (✉)
Department of Forest Economic, Artvin Çoruh University, Artvin, Turkey
e-mail: mft@artvin.edu.tr; iza_5561@hotmail.com; aturkanal@hotmail.com

N. Nakagoshi and J.A. Mabuhay (eds.), *Designing Low Carbon Societies in Landscapes*, Ecological Research Monographs, DOI 10.1007/978-4-431-54819-5_15, © Springer Japan 2014

as forest villagers' personal information of every age group, socioeconomic effects of ecotourism activities, and destruction of the environment. These variables have been measured in terms of before and after ecotourism. In the process of evaluating the results, percentage, chi-square, paired t tests, and Wilcoxon analyses have been used (Aydın, 2010), (Demirayak, 2006).

Keywords Ecotourism • Forest protected areas • Forest villagers • Forestry management • Public relations • Sustainability

15.1 Introduction

Increased and diversified social demands because of rapid population growth, urbanization, and industrialization are changing people's attitudes about environment and natural resources. As a result of this process, the importance of ecological and environmental functions of nonwood forest resources is increasing. This change, in turn, brings about the development of a contemporary philosophy of utilization of forest resources in accordance with principles of sustainability and multidimensional utilization.

Forest ecosystems are in the focal point of the sustainable development process. The basic element of this process is human, and all other productions and utilities, especially raw wood materials, depend on humans.

The changes occurring in economic, political, cultural, technological, and environmental fields throughout the world have necessitated a review of national development policies and required a multidimensional approach such as ecotourism to develop rural areas by taking changed tourism demands into consideration.

Ecotourism is a type of tourism activity that safeguards environmental protection and prioritizes the well-being and prosperity of the region's people. As opposed to the generally perceived tourism definition, it is regarded as a sustainable and most effective way of natural conservation. Other types of tourism activities can also be evaluated as sustainable and fundamentally linked to societies whose main agenda is to keep green environments as is; however, the main object of ecotourism should be to preserve biological diversity.

Dasenbrock (2002), in his article titled "The Pros and Cons of Ecotourism in Costa Rica," described ecotourism as the world's largest and fastest growing service industry. However, he also mentioned some ups and downs of ecotourism in developing countries.

Ecotourism is an ideal type of touristic activity to help bolster both economic development and environmental protection. However, there are numerous examples around the world disclosing the situation is well outside what has long been intended or expected to be. Those examples that have not been properly managed

15 Ecotourism Activities for Sustainability and Management of Forest... 255

Fig. 15.1 View of Camili Village in research area. (Photograph from Aydın 2010)

can unintentionally yield more adverse effects than those of mass tourism activities (Keszi 1998).

The study area of this study, Camili Basin, is 35 km from the district center on the Georgia border of Borçka district of Artvin province. In the north, all streams in the Camili Basin, which is surrounded by the Turkey–Georgia border, flow toward the other side of the border. The abundance of water resources attracts attention in the entire Camili basin (villages, hamlets, plateaus). As agricultural areas are limited and the area is mostly rainy, irrigated farming is not usual. High mountains surrounding the basin from the east, south, and west provide a natural border and isolate the region geographically. In the basin, which has a hilly structure, the elevation increases from 350 m (Camili Village) to 3,414 m (climax of Kaçkar). Camili is reached through Cankurtaran Strait, which is at 1,870 m. Camili Village located in the center, and five villages in the nearby basin are together called Macahel. Other villages included in the project along with Camili Village are the Düzenli, Efeler, Kayalar, Maral, and Uğur villages (Fig. 15.1).

This study is based on the GEF-II Ecotourism Project that is being applied in Camili Biosphere Reserve, an area in the Borçka district of Artvin Province that is adjacent to Georgia (Fig. 15.2).

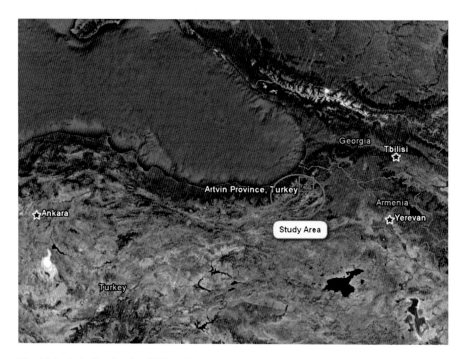

Fig. 15.2 Artin-Borçka-Camili biosphere reserve area, Turkey

15.2 Materials and Methods

15.2.1 Materials

The effects of ecotourism activity carried out by the GEF-II Project on the societal values (social, economic, educational, cultural, technological), environmental values, and communication values of the villagers living in this village and other villages in the Camili Basin in Borçka district of Artvin Province, adjacent to Georgia, were examined. Face-to-face interviews were performed to examine the socioeconomic effects in the study area. These acquired data, related to the socioeconomic characteristics of the countrymen, provided the primary information for this study.

Secondary information includes Camili GEF-II Project records, the studies conducted by Turkish Foundation of Struggle with Erosion, Forestation and Protection of Natural Beings (TEMA), interviews held with GEF-II and TEMA managers, and other resources related to ecotourism activities. Other resources are all kinds of books, articles, conference papers, and theses written or published on such subjects as rural development, forest villagers, tourism, and ecotourism.

To examine the socioeconomic impacts on the study area, face-to-face interviews were held with villagers who were determined with the method of complete inventory. In the interview forms distributed to the countrymen who took part in ecotourism activities, we tried to measure such variables as personal information of the interviewees, their views and levels of information about ecotourism activities, the phenomenon of migration from the country to the city, road intensity, production levels in stockbreeding for pre- and post-ecotourism activities, its impact on honey production, the amount in annual average agricultural products, change in land size, change in local handcrafts, annual levels of income, changes in living standard, and change in fuel consumption, etc., in terms of their level both before and after ecotourism.

15.2.2 Method

15.2.2.1 The Reason for Choosing This Area for the Study

This study is based on the GEF-II Ecotourism Project that is being applied in Camili Biosphere Reserve Area in the Borçka district of Artvin Province, which is adjacent to Georgia.

The main reasons for choosing this area for the study are the ecotourism activity in the area and its cultural and ecological richnesses. It was declared as the first Biosphere Reserve Area of Turkey on June 29th, 2005, to the "Human and Biosphere Network" of UNESCO. It is a protected area that is home to natural aged forests.

15.2.2.2 Preparation for Study Surveys and the Application Technique

We tried to show the contributions made by ecotourism activity for the villagers in the study area in socioeconomic, cultural, technological, and environmental terms by comparing pre- and post-ecotourism activities. Survey questions were prepared by taking this purpose into consideration. To ensure correct and reliable answers, questions were chosen as related to the subjects relevant for people who were a part of ecotourism activities.

15.2.2.3 Determination of the Number of Interviewees on Whom the Survey Would Be Applied

Persons were identified with the complete inventory method. Then, with direction from these people, 10 participants from Maral village, 14 participants from Uğur village, 11 participants from Düzenli village, 11 participants from Camili village, 12 participants from Efeler village, and 3 participants from Kayalar village were identified.

15.2.2.4 Statistical Evaluations

To realize the purpose of the study, in the analysis stage of the data obtained with the survey, percentage, chi-square, matched t test, and Wilcoxon analysis were used.

15.3 Results and Discussion

In this study conducted in the Camili Biosphere Reservation Area of Artvin Province, suggestions were developed to explain the relationship of sustainable forestry management with protected areas, and that ecotourism activities should be evaluated as a tool for the resolution of the conflicts that appear in the sustainability stage of protected area management and the importance of public relations activity for development of forest countrymen.

In the questionnaires provided to the villagers who were chosen based on the complete inventory method, these variables were measured: pre- and post-ecotourism periods; the impact of ecotourism on societal values, the changes it initiated in communication, and its effect on environmental values.

15.3.1 Impact on Societal Values

15.3.1.1 Residence of Households

Before ecotourism activities, 59 % of households in the area resided permanently and 41 % resided temporarily; after ecotourism activities, 65.6 % of households in the area were permanent residents permanently and 34.4 % were temporary residents. It can be stated that, depending on ecotourism activities, permanent residence in villages increased and temporary residence in villages decreased.

15.3.1.2 Size of Households

Before ecotourism activities, the number of household members was 4.79 on average; it was found to be 4.80 after ecotourism. The H_0 hypothesis was accepted (Table 15.1) for the significance test of the difference between two variables to see whether ecotourism had an impact on the change in household size. The result was 95 % reliability, which means that there was no statistically significant difference (t calculation $= -299$, $p > 0.05$).

15 Ecotourism Activities for Sustainability and Management of Forest... 259

Table 15.1 t test results applied to determine the difference in size of household between pre-ecotourism and post-ecotourism periods

	Dual differences							
				95 % matched two-example reliability interval				
	Average	SD	Average SE	Lower	Higher	t	Degrees of freedom (df)	Significance level
Pre-ecotourism Post-ecotourism	−0.016	0.428	0.055	−0.126	0.093	−0.299	60	0.766

Source: Aydın (2010)

15.3.1.3 Vocational and Educational Status

As a result of the conducted survey, a chi-square compliance test conducted to see whether villagers were distributed evenly among vocational groups (Table 15.2); accordingly, the H_0 hypothesis was rejected with 95 % statistical reliance and it was determined that the answers revealed a significant difference among vocational groups (chi-square calculation $= 41,787$, $p < 0.05$): 42.6 % were farmers, 11.5 were public servants and housewives, 4.9 % were workers and self-employed, 9.8 % were retired and shopkeepers, and 6.6 % were students.

In addition, the chi-square compliance test was conducted to see whether countrymen were distributed evenly among educational status groups (Table 15.3). The H_0 hypothesis was rejected with 95 % statistical reliance, and it was determined that educational status distributions were not even (chi-square calculation $= 41,787$, $p < 0.05$); it was found that 3.3 % were illiterate, 3.3 % were just literate, 34.4 % were elementary school graduates, 21.3 % were secondary school graduates, 26.2 % were high school graduates, 3.3 % held 2-year college degrees, and 8.2 % held graduate degrees.

According to these results, it is understood that frequencies related to educational status were not distributed evenly among all classes.

15.3.1.4 Measurement of Socioeconomic Impacts of Ecotourism

Percentages of the problems faced in the region before ecotourism activities were listed as the lack of transportation in winter months and obligation to use routes through Georgia (86.9 %), education/health problems (52.5 %), power blackouts (24.6 %), insufficiency of income (23 %), insufficiency of village roads (11.5 %), scarcity of livestock and agricultural products (6.6 %), and communication

Table 15.2 Chi-square results of distribution of participants in ecotourism activity according to vocational groups

Vocation	Observed value	Expected value	Residual
2	2	10.2	−8.2
3	3	10.2	−7.2
4	4	10.2	−6.2
6	12	10.2	1.8
7	14	10.2	3.8
26	26	10.2	15.8
Total		61	
Chi-square		41.787	
Degrees of freedom		5	
Significance level		0	

Source: Aydın (2010)

Table 15.3 Chi-square results of distribution of participants in ecotourism activity according to educational status groups

Educational status	Observed value	Expected value	Residual
Not graduate	6	12.2	−6.2
University	5	12.2	−7.2
Secondary	13	12.2	0.8
High school	16	12.2	3.8
Elementary	21	12.2	8.8
Total	61		
		Educational status	
Chi-square		52.194	
Degrees of freedom		4	
Significance level		0	

Source: Aydın (2010)

problems (1.6 %); after ecotourism activities, the problems mentioned were the lack of transportation in winter months and obligation to use routes through Georgia (85.2 %), education/health problems (49.2 %), power blackouts and insufficiency of income (9.8 %), and insufficiency of village roads (6.6 %).

H_0, which is the significance test of the difference between two matches (Table 15.4), was accepted to test for statistical difference between these values; it was determined with 95 % reliability that there was no difference between the problems faced in the region before and after ecotourism activities (t calculation $= 2,497$, $p > 0.05$).

The question "Was the initiation of ecotourism activities effective in the solution of problems faced in your area?" was answered as "yes" by 26.2 % of the villagers, "partially" by 42.6 % of villagers, and "no" by 31.1 % of the villagers.

15 Ecotourism Activities for Sustainability and Management of Forest... 261

Table 15.4 *t* test results for comparison of the problems faced in the region before and after ecotourism activities

| | Dual differences | | | | | | | |
| | Average | SD | Average SE | 95 % matched two-example reliability interval | | | Degrees of freedom (*df*) | Significance level |
				Lower	Upper	*t*		
Pre-ecotourism Post-ecotourism	57.016	178.343	22.834	11.34	102.69	2.497	60	0.015

Source: Aydın (2010)

Table 15.5 *t* test results for comparison of the change in the motives for migration from the country to the village before and after ecotourism activities

| | Dual differences | | | | | | | |
| | Average | SD | Average SE | 95 % matched two-example reliability interval | | | Degrees of freedom (*df*) | Significance level |
				Lower	Upper	*t*		
Pre-ecotourism Post-ecotourism	209.426	630.251	80.695	48.011	370.841	2.595	60	0.012

Source: Aydın (2010)

15.3.1.5 The Reasons for Migration from the Country to the City

The question "What are the reasons for migration from the country to the city?" before ecotourism was answered by the villagers as follows: 29.3 % for ensuring better education opportunities for their children, 65.3 % for problems faced in transportation, 30.7 % for limitations in health facilities, 18.7 % for more job opportunities, 12 % for higher income prospects, 6.7 % for better living standards, 1.3 % for scarcity and nonfertility of farmland, the roughness of terrain, and limited utilization of forests; the same question after ecotourism activities was answered by the villagers as 36.1 % for ensuring better education opportunities for their children, 82 % for problems faced in transportation, 26.2 % for limitations in health facilities, 6.6 % for more job opportunities, 3.3 % for higher income prospects, and 1.6 % for better living standards.

It can be stated that the reason for the results provided here is not ecotourism but that education, health, and transportation are still the most important motives for migration from the country to the city.

H_0, which is the significance test of the difference between two matches (Table 15.5), was rejected to test whether there was a statistical difference between these values; it was determined with 95 % reliability that there was a difference

Table 15.6 t test results for determination of the difference between the meaning of ecotourism before and after ecotourism activities

	Dual differences							
				95 % matched two-example reliability interval				
	Average	SD	Average SE	Lower	Upper	t	Degrees of freedom (df)	Significance level
Pre-ecotourism Post-ecotourism	4.934	3.628	0.465	4.005	5.864	10.623	60	0

Source: Aydın (2010)

between the motives for migration from the country to the city before and after ecotourism activities (t calculation $= 2{,}595$, $p < 0.05$).

This finding means that the significance level of the conducted t test (less than 0.05) indicates there is a statistical difference between the motives for migration from the country to the city between the pre- and post-ecotourism activities periods.

15.3.2   Changes Made in Communication

15.3.2.1   Level of Information About Ecotourism

The percentages of the answers given to the question "What did ecotourism mean to you?" before ecotourism activities were as follows: 78.7 % "I had no idea," 8.2 % "the tourism that understands and explains nature, 3.3 % "tourism that provides opportunities for higher income," "tourism that reflects the cultural values of the local people," and "tourism that protects forest resources," 1.6 % "environmentally friendly tourism"; the percentages of the answers given to the question "What does ecotourism mean to you?" after ecotourism activities were as follows: 26.2 % "the tourism that both protects nature and observes the welfare of the local people," 45.9 % "tourism that provides opportunities for higher income, 9.8 % "tourism that reflects the cultural values of the local people," 8.2 % "tourism that protects forest resources," 6.6 % "tourism that allows for more production," 1.6 % "tourism that understands and explains nature" and "tourism that ensures higher income for both myself and the organizers."

H_0, which is the significance test of the difference between two matches (Table 15.6), was rejected to test for statistical difference between these values; it was determined with 95 % reliability that there was a difference between the answers given to the question about the meaning of ecotourism before and after

15 Ecotourism Activities for Sustainability and Management of Forest... 263

ecotourism activities (t calculation $= 10,623$, $p < 0.05$). This result means that "significance of the difference between two matches test" (less than 0.05) indicates there is statistical difference between the answers given to the questions "what did ecotourism mean" before and after participation in ecotourism activities.

15.3.2.2 Participation in Ecotourism Activities

The answers given to the question "What are the most important three purposes for participation in ecotourism?" are as follows: 77 %, to increase my level of income, 27.9 %, for better evaluation of my products, 13.1 %, for better meeting the feed needs of my livestock and catering needs of myself, 34.4 %, for contribution to the cultural and social development of my area, 29.5 %, for contribution to the development of tourism, 50.8 %, as it allows for creation of new job opportunities, 31.1 %, as it could be a solution to transportation problems, 29.5 %, as it could solve education and health problems, 1.6 %, as it could contribute to the protection of natural, archeological. and historical areas, 29.5 %, as it could contribute to the protection of forest resources, 1.6 %, as it could contribute to the protection of the diversity of the flora, 3.3 %, for becoming more socialized, and 1.6 %, for enjoyment. However, the important point here is that people interviewed within the Project cannot obtain sufficient income from stockbreeding, which is the basic source of living, and for this reason they preferred ecotourism activity projects and the additional revenue created.

The question "Whether or not the purpose in their participation in ecotourism activities was realized" was answered as "yes" by 32.8 %, "partially" by 50.9 %, and "no" by 16.3 %. It can be stated that by participating in ecotourism activities villages can generate sufficient additional income and thus embrace new job opportunities.

15.3.3 Impact on Environmental Values

15.3.3.1 Relevance of Ecotourism with Forests and Other Resources

The Most Negative Impact of Ecotourism Activities on the Habitat

As a result of the survey, the following were indicated as the reason by the villagers: destruction of the forest and land use problems (8.2 %), destruction of farmland and forbiddance of hunting (3.3 %), pollution of streams (41 %), damage to the soil and the flora (11.5 %), and waste materials problem (23 %).

As a result of the chi-square compliance test, conducted to see whether the frequencies of the classes created as regards the results of the classification of the

Table 15.7 Chi-square results of the distribution of negative impacts of ecotourism activities on the habitat

Negative impacts	Observed value	Expected value	Residual
Hunting	2	12.2	−10.2
Destruction	10	12.2	−2.2
Soil damage	8	12.2	−4.2
Waste materials	16	12.2	3.8
Stream pollution	25	12.2	12.8
Total	61		
		Negative impacts	
Chi-square		24.984[a]	
Degrees of freedom		4	
Level of significance		0	

Source: Aydın (2010)

[a] O cells, (,0%) have expected frequencies less than 5. The minimum expected cell frequency is 12,2.

most important negative impact of ecotourism activities on the environment, it can be seen from Table 15.7 that the H_0 hypothesis is rejected with 95 % statistical reliability and that there is significant difference between the answers (chi-square calculation $= 24,984$, $p < 0.05$).

The Reasons for Short Periods of Ecotourist Visits

The answers provided by the villagers who took part in the survey to the question "What are the reasons for shortness of staying periods of ecotourists that have visited the region so far?" were as follows: the lack of cooperation, collaboration, and solidarity between the segments of ecotourism activity (27.9 %), the contracts of tour agencies that were concluded with ecotourists (24.6 %), lack of sufficient publication and communication services (16.4 %), ineffectiveness of promotion activities (14.8 %), that activities are concentrated in only certain months of the year (11.5 %), insufficient social activities in the region (3.3 %), and that vacation concepts of the ecotourists who visit the region differ (1.6 %).

As a result of the chi-square compliance test conducted to see whether the frequencies of the classes created about the results of the classification of the reasons for shortness of the stay of ecotourists, it is seen in Table 15.8 that the H_0 hypothesis is rejected with 95 % statistical reliability and that there is a significant difference between the answers (chi-square calculation $= 24,951$, $p < 0.05$) and that answers were not distributed homogeneously.

According to this result, it can be understood that frequencies of visit periods of the ecotourists is not distributed evenly between classes.

Table 15.8 Chi-square results of the distribution of findings as concerns the shortness of stay periods of ecotourists

Reasons	Observed values	Expected value	Residual
Few social activities	1	8.7	−7.7
Vacation understanding of ecotourists	2	8.7	−6.7
Concentration in certain months of the year	7	8.7	−1.7
No promotion	9	8.7	0.3
No communication	10	8.7	1.3
Tour agencies	15	8.7	6.3
No solidarity between segments	17	8.7	8.3
Total	61		
		Reasons	
Chi-square		24.951[a]	
Degrees of freedom		6	
Significance level		0	

Source: Aydın (2010)
[a] O cells, (,0%) have expected frequencies less than 5. The minimum expected cell frequency is 8,7.

15.3.3.2 The Impact of Ecotourism Activities on Land Use

The Impact of Ecotourism Activities on the Type of Utilization from Forests

The question "What is the primary mode of utilization from forests before and after ecotourism?" was answered by the villagers as "meeting the need for winter fuel wood" (93.4 %) and grazing fields for the livestock (6.6 %); the same villagers gave the following answers after ecotourism activities: "meeting the need for winter fuel wood" (75.4 %) and grazing fields for the livestock (6.6 %) and as raw material for wood-carving work (18 %).

H_0, the significance test of the difference between two matches (Table 15.9), was rejected to test whether there was a statistical difference between these values; it was determined with 95 % reliability that there was difference between the answers given for the priorities for utilization of the forests before and after ecotourism activities (t calculation $= -3,536, p < 0.05$).

This finding means that the smaller than 0.05 value for the significance level indicates there is statistical difference between the answers given regarding the primary types of utilization of forests before and after ecotourism activities.

15.3.3.3 Impact of Ecotourism Activities on the Type and Amount of Fuels

Impact on Annual Fuelwood

Before ecotourism activities the amount of wood used as fuel per household on an annual basis was 13.93 sters, whereas today the amount of wood used as fuel per household on an annual basis as fuel is 13.52 sters.

Table 15.9 t test results applied for determination of the difference between the most prominent mode of utilization from forests before and after ecotourism

	Dual differences							
				95 % matched two-example reliability interval			Degrees of freedom	Significance
	Average	SD	Average SE	Lower	Higher	t	(df)	level
Pre-ecotourism Post- ecotourism	−0.361	0.797	0.102	−0.565	−0.157	−3.536	60	0.001

Source: Aydın (2010)

H_0, which is the significance test of the difference between two matches (Table 15.10), was accepted to test whether there was a statistical difference between these values; it was determined with 95 % reliability that there was difference between the answers given to the question for the amount of wood consumed per household on an annual basis before and after ecotourism activities (t calculation $= 882, p > 0.05$).

This result means that there is no statistical difference between the two values. The decrease in the usage of wood today compared to the period before ecotourism activity considerably reduced the utilization by villagers of forests within the study area. However, this situation does not necessarily mean that villagers do not exert any pressure on the forests.

Impact on Annual Liquid Petroleum Gas (LPG) Usage

On the other hand, the average amount of LPG used per households each year was 2.02 before ecotourism activity and it has been determined that today the average amount of LPG used per household is 3.51 each year.

H_0, the significance test of the difference between two matches (Table 15.11), was rejected to test whether there was a statistical difference between these values; it was determined with 95 % reliability that there was a difference between the amount of LPG used per households each year before and after ecotourism activities (t calculation $= -6,631, p < 0.05$).

This result means that there is a statistically significant difference between two values. The question "Do you think that ecotourism has an impact on the type and amount of fuels?" was answered as "yes" by 31.1 %, "partially" by 26.2 %, and "no" by 42.6 %. According to these answers, it can be said that ecotourism activity had considerable impact on the change in the type and usage of fuels compared to the pre-ecotourism activity. To sum up, the usage of the amount of fuel changed compared to the period before ecotourism activity but it did not have any impact on the change of type of fuel.

15 Ecotourism Activities for Sustainability and Management of Forest...

Table 15.10 t test results conducted to determine the amount of wood consumed per household before ecotourism activity and

	Dual differences							
			Average	95 % matched two-example reliability interval			Degrees of	Significance
	Average	SD	SE	Lower	Higher	t	freedom (df)	level
Pre-ecotourism Post-ecotourism	0.41	3.63	0.465	−0.52	1.34	0.882	60	0.381

Source: Aydın (2010)

Table 15.11 t test results conducted to determine the amount of liquid petroleum gas (LPG) used per households each year before ecotourism activity and amount of LPG used per households each year after ecotourism activity

	Dual differences							
			Average	95 % matched two-example reliability interval			Degrees of freedom	Significance
	Average	SD	SE	Lower	Higher	t	(df)	level
Pre-ecotourism	−1.492	1.757	0.225	−1.942	−1.042	−6.631	60	0
Post-ecotourism								

Source: Aydın (2010)

15.4 Conclusion

In the evaluation, when the number of households and household sizes are considered, it is seen that although there is not a clear difference in statistical terms, as ecotourism activities begin to be satisfactory values began to change in a positive way. As a result, it can be said that as a function of the ecotourism activity conducted in the study area migration has ceased; even so that depending on the activities reverse migration has started.

In the test conducted, it has been determined that frequencies related to vocational fields are not distributed evenly among all groups. Thus, the uneven distribution of vocational groups appearing on the question form is a statistical indicator that the intensity among vocational groups is in farmer groups. Almost half the villagers interested in ecotourism are involved with farming, and their educational status is at elementary level.

There has been no change witnessed in problems before and after ecotourism activity. However, the amount of problems decreased after ecotourism activity, meaning that ecotourism activities had an impact on decreasing these problems as income increased as a result of ecotourism activities and thus interest arose for these activities, which in turn reduced the number of current problems.

Explicit difference has been found in statistical terms in "causes for migration from the country to the city" before and after ecotourism activity. However, these reasons are mostly the lack of transportation to the region and insufficient health and education opportunities. Explicit increase has been identified in statistical terms in the level of information about ecotourism activity before and after ecotourism activity. It can be understood that until today information has been provided on the projects and studies conducted on ecotourism. However, it is understood that information related to ecotourism was sufficient in the beginning, but later, parallel to the increase in awareness among people, sufficient information was not provided despite the increase in interest. As a result, as both Project managers and people within the scope of the activity witness a sufficient impact of ecotourism on generating additional income recourses, ecotourism activities are not at a sufficient level (as they do not have sufficient information despite the change in their opinions).

The purpose of villagers in participating in ecotourism activities has been determined that villagers who earn their living under rather disadvantaged conditions see the implementation of ecotourism activity as a way to avoid these harsh conditions. By participating in ecotourism activity, it can be claimed that villagers generate sufficient additional income and thus embrace new job opportunities.

It has been statistically determined that the most important negative effects of ecotourism activities on the environment are generally concentrated on waste materials and stream pollution. Although the study examines the economic dimension of ecotourism, its environmental dimension is also important. Financial support for the development of rest house management has proved to be insufficient. As sewage systems fit for rest house management have not been structured, pollution of streams and waste materials come to the fore as important problems. The statistical evaluation performed for the reasons of shortness of staying periods of ecotourists revealed that the most important reason is the lack of solidarity among segments. Therefore, villagers in the entire Camili basin must be unified under a single central unit, so that local people can use the right to self-determination and self-management and make contribution to the development of ecotourism. More support has to be provided for the promotion of the region so that ecotourists do not come via agencies. By this manner, new incentives can be developed, promotion of the culture can be allowed, and thus new job areas can be generated and diversity can be produced.

Statistical difference has been determined in the types of utilization of forests after ecotourism activities, which changed the types of utilization of forests. After ecotourism activities decrease has occurred in utilization in the form of winter fuelwood and an increase has been identified in utilization as raw material for carving activities.

Although no statistically significant difference has been witnessed in the amount of annual fuelwood used after ecotourism activities, a relative decrease has occurred in the amount of annual fuelwood utilization after activities. If the decrease in fuelwood utilization today is compared to the pre-ecotourism activity period, utilization of forests in the study area by villagers has considerably lessened.

Usage of LPG as wood raw material has increased after ecotourism activities as the need for fuel. In addition, according to the pre-ecotourism activity period, the increase in LPG usage amount is noted as the result of increase in revenue as ecotourism activities become more functional.

This conclusion means that people who had no idea about ecotourism before ecotourism activities (78 %) participated in the activities depending on the increase in their income and enjoyed more opportunities for obtaining higher income from these activities. It can be concluded that the awareness for environmental protection has been increased in ecotourism activities conducted in the study area.

References

Aydın IZ (2010) Ekoturizmin Türkiye-Orman Köyleri Kalknmaları Üzerindeki Sosyo-Ekonomik Etkilerinin Ölçümü: "(Artvin- Camili Biyosfer Rezerv Alanı Örneği)" Karadeniz Tekbnik Üniversitesi Fen Bilimleri Enstitüsü Yüksek Lisans Tezi, Trabzon, Türkiye

Akıllı H (2004) Ekoturizmin Sosyokültürel, Ekonomik, Yönetsel ve Çevresel Etkileri Bakımından İrdelenmesi; Antalya Köprülü Kanyon Milli Parkı Örneği, Akdeniz Üniversitesi KamuYönetimi Anabilim Dalı Yüksek Lisans Tezi, Antalya

Alkan H, Korkmaz M, Eker M (2010) Sürdürülebilir OrmanYönetiminde Yaşanan Gelişmeler, Karşılaşılan Sorunlar ve Çözüm Önerileri: Isparta Orman Bölge Müdürlüğü Örneği. Süleyman Demirel Üniversitesi Orman Fakültesi Orman Mühendisliği Bölümü 32260, Isparta, Türkiye

Dasenbrock J (2002) The pros and cons of ecotourism in Costa Rica, TED Case Studies Number 648, January

Demirayak F (2006) Türkiye'de Korunan Alanlar İçin Yeni Bir Yaklaşım Ortaklaşa Yönetim Ankara Üniversitesi Sosyal Bilimler Enstitüsü Kamu Yönetimi ve Siyaset Bilimi (Kent ve Çevre Bilimleri) Anabilim Dalı, Ankara, Türkiye

Geray U (1998) Orman Kaynakları Yönetimi, T.C. Başbakanlık DPT Ankara, Türkiye

Keszi J (1998) Formulatıon of an Ecotourısm Policy Framework for Manitoba. Master Thesis, Winnipeg, Manitoba, Canada

Porsuk T (2000) Sürdürülebilir Ormancılık Ölçütleri ve Türkiye'deki Durumun Belirlenmesi İstanbul Üniversitesi Fen Bilimleri Enstitüsü, Yüksek Lisans Tezi İstanbul, Türkiye

Chapter 16
Community Aspects of Forest Ecosystems in the Gunung Gede Pangrango National Park UNESCO Biosphere Reserve, Indonesia

Nobukazu Nakagoshi, Heri Suheri, and Rizki Amelgia

Abstract Indonesia's forest has high biodiversity in flora and fauna. The government has tried to maintain this biodiversity through managing the forest in accordance with the existing functions and conditions. Population is increasing, which affects demands for land and wood consumption. Statistics of the Ministry of Forestry (2010) recorded several disturbances of forest areas. The disturbances include land occupation by the community, which had reached 52,972 ha of forest ecosystems. This research has the main objectives to assess the level of dependency on forest products and to identify which factors are affecting the level of dependency on forest products in the transition zone of Gunung Gede Pangrango UNESCO Biosphere Reserve. Detailed demographic and economic household information was collected using the stratified random sampling method from 210 respondents in six villages. Statistics reveal that 361,002 people live around the area (Mulcahy and Mc Carthy, 2010). The statistical results show that about 58 % of the population respondents depend on forest products for about 0–20 % of their total monthly income and only 10 % of the respondents highly depend on forest products (more than 80 % of total monthly income). In average, the community level of dependency on forest products was 28 % of their total income. From nine independent variables, only five variables such as gender, household size, policy, off-forest income, and elevation, are mainly responsible for determining the level of dependency. It is statistically proven that off-forest income and elevation of the household have a strong effect related to household forest dependency;

N. Nakagoshi (✉) • H. Suheri
Graduate School for International Development and Cooperation, Hiroshima University,
1-5-1 Kagamiyama, Higashi-Hiroshima 739-8529, Japan
e-mail: nobu@hiroshima-u.ac.jp

R. Amelgia
Directorate General of Forest Utilization, Ministry of Forestry, Manggala Wanabakti Building,
7th Block, 5th Floor, Jalan Jendral Gatot Subroto, Jakarta 10270, Indonesia

N. Nakagoshi and J.A. Mabuhay (eds.), *Designing Low Carbon Societies in Landscapes*, Ecological Research Monographs, DOI 10.1007/978-4-431-54819-5_16, © Springer Japan 2014

gender and policy are significant at 5 % level and household size at 10 % level. Simply, the household with low off-forest income, large number of family members, and located in a remote area is more dependent on forest products compared to other households.

Keywords Biodiversity • Biosphere reserve • Forest ecosystem • Forest product • Indonesia • National park

16.1 Introduction

Indonesia's tropical forests are the third largest forests in the world. The forests play important roles to our environment, economy, and society. Indonesia's forest has high biodiversity in flora and fauna with about 12 % of the world's mammals, 16 % of its reptiles and amphibians, and 17 % of all bird species (Ministry of Forestry 2010).

The forest serves ecological functions such as habitat, carbon sink, carbon source, and also richness of biodiversity (Azizia and Sulistyawati 2009). The forest has crucial roles in the economic and social expansion of humankind. These habitats not only provide an essential source of food, fuel, and materials for housing but also perform a vital part in protecting and maintaining the stability of our natural environment (Riswan and Hartanti 1995).

Forest management in Indonesia is divided into four categories: (1) production forest, intended for selective logging as part of the permanent forest estate; (2) conversion forest, destined to be clear felled for agriculture, settlements, or non-forestry uses; (3) protection forest, mainly on high land or steep slopes, is to be retained for watershed protection; and (4) protected forest, which includes national parks and nature reserves where forest exploitation is forbidden (Ministry of Forestry 2006, 2008, 2010).

The national park is a protected area with large natural areas to protect large-scale ecological processes, along with the complement of species and ecosystem characteristic of the area, which provides a foundation for environmentally and culturally compatible spiritual, scientific, educational, and recreational opportunities. The primary objective of this area is to protect natural biodiversity together with its underlying ecological structure, support environmental processes, and furthermore promote education and recreation (IUCN 2008).

Additionally, Indonesia is the center of plant diversity for several genera and is one of the world centers of species diversity of hard corals and many groups of reef-associated flora and fauna. Hence, forest management needs to be professional and well planned and can be used optimally, outside the ability of the forest to produce sustainable benefits at the local, national, regional, and international levels. The government has tried to maintain biodiversity through managing the forest in accordance with the existing functions and conditions (Nakagoshi and Amelgia 2010).

The forest is a multifunctional natural resource in support of human life, not only as a place for the conservation of biodiversity and maintenance of ecosystem functions, but also to produce goods and services for the community (Yohanes 2010). Many studies have determined that forest resources are destroyed by overexploitation by concessionaires (private company or big estate plantations), squatters, shifting cultivators, a burgeoning population, and forest fires. The population is also increasing, and it affects demands of land for development and wood consumption.

Until 2010, Indonesia's Bureau of Statistic Center identified that the total population was around 237 million, that nearly 48.8 million are living in and around forest areas, and that 10.2 million are in poor economic conditions. Around 6 million people are directly dependent on forest products (wood and non-wood) and 3.4 million others are working in the forest industry. In 2009, the Ministry of Forestry recorded several disturbances of forest area. These disturbances include land occupation by the community, which has reached 52,972 ha of forest. The disturbance is caused by several factors, such as low economic income and few opportunities to obtain an occupation of good quality. This situation also occurs in Gunung Gede Pangrango National Park (GGPNP), one of the five oldest national parks in Indonesia (GGPNP 2009) and a core zone of Cibodas Biosphere Reserve.

In the present day, the park is under significant pressure caused by several threats, not only from the impact of development but also from the local community who exist in the surroundings of park and on the forest boundary (Indra Exploitasia 2010). One adjunct of the community aspect is forest dependency: people are still collecting several forest products to support their daily basic income, which is expected to weaken the existence of park ecosystems. Hence, it is of concern to identify the affecting factors.

The task of this study is to contribute a better understanding of local community current conditions related to their needs for forest products. More specifically, the objectives of this study are (1) to identify forest products utilization in the local community surrounding the GGPNP, (2) to assess the level of forest products dependency, and (3) to identify factors that affect the forest product dependency level. The level of dependency is evaluated from the proportion of forest income to total income.

Identification of the local community who live surrounding the GGPNP will help the government to develop strategies for collaboration with the local community to minimize disturbance of the forests.

16.2 Materials and Methods

16.2.1 Study Areas

Gunung Gede Pangrango Biosphere Reserve, better known as GGPNP, is a protected area that functions as a core zone of Cibodas Biosphere Reserve (Wiratno et al. 2003). It is one of the seven Biosphere Reserves in Indonesia, recognized by

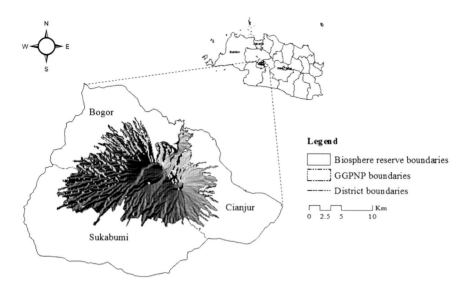

Fig. 16.1 Location of Gunung Gede Pangrango National Park (GGPNP) Biosphere Reserve

UNESCO in 1977, and one of the five oldest national parks in Indonesia, established in 1980. The GGPNP is administratively located in three districts of Cianjur, Sukabumi, and Bogor in the Province of West Java (Fig. 16.1). Sub-montane, montane, and sub-alpine forests are represented, situated between an altitude of 1,000 m and 3,019 m. The park is dominated by two volcanoes: Mount Gede and Mount Pangrango. The current total area of the GGPNP is 22,851 ha.

This park is very important as it is among the few remaining montane habitats and ecosystems of Java. More than 1,000 species of higher/flowering plants have been recorded, composing about 120 plant families. The park includes more than 200 species of orchids, about one third of all the orchid species found in Java. The vegetation in GGPNP is divided into three vegetation zones based on altitude: the Sub Montane Zone (1,000–15,000 m above sea level) dominated by *Schima walichii, Altingia axalsa, Castanopsis argentea*, and *Quercus* spp.; the Montane Zone (1,500–2,400 m above sea level) with *Podocarpus imbricatus, Quercus* spp., *Podocarpus neriifolius, Castanopsis* spp., and *Altingi exelsa*, and the seedlings in this area include *Strobilanthes cermuis, Begonia* spp., and *Melastoma* spp.; and the Sub Alpine Zone (>2,400 m above sea level) dominated by *Isachne pangrangensis, Anaphalis javanica, Vacinium varingifolium, Leptospernium flanescens, Symplocos javanica*, and *Eurya acuminata*.

The birds species excite particular interest, attracting ornithologists from all over the world; this is not surprising considering that more than half of Java's bird fauna can be seen here, including most of the "Javan endemic birds." Threatened or endangered mammals, which include the leopard, the Javan gibbon, and the Javanese leaf monkeys, are preserved in this area. This area is also the habitat of 245 species of Javanese birds and many species of fauna such as *Panthera tigris*

16 Community Aspects of Forest Ecosystems in the Gunung Gede Pangrango... 275

Table 16.1 Population and wide area surrounding Gunung Gede Pangrango National Park (GGPNP) Biosphere Reserve

| | Total population | | |
District	2003	2009	Area (ha)
Cianjur	60,741	150,556	187,068.00
Sukabumi	91,780	173,661	777,421.71
Bogor	59,852	160,000	4,464.63
Total	212,373	484,217	968,954.33

Resource: GGPNP Biosphere Reserve Management Board

sondaica, Bos javanicus, and *Hylobates moloch.* From the aspect of biodiversity, GGPNP is highly significant; therefore, in 1977 the UNESCO recognized GGPNP as a biosphere reserve.

For decades the GGPNP have been important sources for livelihood of local people and sometimes have contributed to national economic development. This park has contributed significantly to the reduction of poverty and sustainable development (Wiratno et al. 2003). The park carries out many functions that are beneficial to community and welfare, which include biodiversity, watershed protection, tourism, forest products, carbon, water, research, and cultural values. The way these functions are transformed into benefits for people, including the poor rural inhabitants living around the park, depend heavily on the proper management of the park to translate policy on sustainable land use into actions within the biosphere reserve.

The easy accessibility of the park makes it a popular recreation area. However, its location within West Java makes it all the more susceptible to the pressures of the ever-increasing populations surrounding it. With the three large cities of Jakarta, Bandung, and Bogor located not more than 80 km from this park, the GGPNP Biosphere Reserve is one of Indonesia's popular resorts for family picnics and camping. Greater efforts are required to prevent local populations and visitor pressures from becoming a threat to ecosystem conservation. This biosphere reserve is divided into three zones according to UNESCO definition: (1) the core zone; (2) the buffer zone; and (3) the transition zone.

Recently, human activities have been increasing because of high population density and economic pressure in the surrounding areas. Local communities use several products from the GGPNP to support their lives, using these products for household purposes. There are 66 villages around the transition zone of the park with 46 villages located close to the forest boundaries. Table 16.1 shows the alteration of total population in the transition zone. There were 212,373 people living in the transition zone in 2003, increasing up to 484,217 people in 2009. The total area is 968,954.33 ha, divided into three districts: Sukabumi, Cianjur, and Bogor. About 66 % of local people in transition zone have low education and 12 % of them have better education (Fig. 16.2). The major ethnic group in this area is Sudanese with agriculture as the main source of livelihood.

Population pressure has forced the transition-zone farmers to engage in the buffer zone. Table 16.2 shows the types of forest disturbances in GGPNP Biosphere Reserve in 2003–2009.

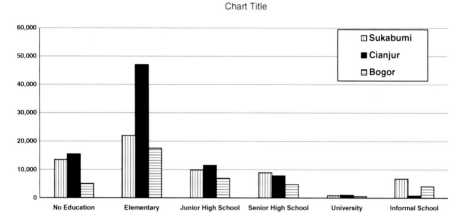

Fig. 16.2 Total population with their education level in surroundings of the GGPNP Biosphere Reserve

Table 16.2 Forest disturbances in GGPNP Biosphere Reserve

		Year						
No.	Type of disturbance	2003	2004	2005	2006	2007	2008	2009
1.	Illegal agricultural field (ha)	9.93	–	–	–	–	–	7.48
2.	Illegal logging (m^3)	2,076.00	–	–	12.75	9.00	18.00	13.25
3.	Illegal hunting (cases)	16.00	22.00	–	73.00	22.00	21.00	136.00
4.	Forest fire (ha)	–	0.23	–				–

Resource: The GGPNP (2009)

Worried by these conditions, the GGPNP Biosphere Reserve Management Board has several social programs to mitigate this interference. The social programs are an implementation of The Law of Republic of Indonesia Number 5/1990 about conservation of living resources and ecosystem. The aims of the social program was to improve the wealth of the local community in the supporting area and to increase environmental awareness. Therefore, through community awareness it is expected that forest disturbances will be reduced.

In 1993–1994, the GGPNP introduced the village economic development program. This program includes the multipurpose tree species seeds system, rabbit and sheep breeding system, and capital for entrepreneur groups system (Siarudin and Mile 2005). In working on the integration of conservation objectives into development of land use planning, since 1994, a multistakeholder partnership as a consortium of 14 institutions (which consists of NGOs, private sectors, local community, and government bodies) was established to encourage cooperation and to develop sustainable land use surrounding the park. The consortium is known as GEDEPAHALA, which stands for Gede-Pangrango-Halimun-Salak, the name of the mountains that connect the national park. Halimun and Salak mountains are now managed as a national park as well. Accordingly, activities in the park are carried out collaboratively between the park authority, international organizations, local NGOs in the local community, local government, and the private sector. Multi-stakeholder partnerships will bring a big impact on conservation. The private sector's full support and involvement are very much needed to support sustainable land use. On December 2007, the GEDEPAHALA then launched one of its programs named Adoption Trees, a program that aims to implement conservation and management to support restoration of degraded lands in accordance with sustainable land use by local people and supported by wider communities.

In 2006, the Ministry of Forestry launched another social program, "Conservation Village Model": CVM. The aim of this program is to empower local community living in and outside the national park. CVM concepts were tried to develop local economics through open access of resources of income by develop the potential of nature (such as developing ecotourism) and to help the local community by giving livestock (such as goats, catfish, and hardwood seeds), home industry, environmental service, and natural tourism (such as home stay, interpreters, and mini-hydro electric power). The GGPNP Biosphere Reserve started to implement this program in 2007 through the Design Engineering of Conservation Village Model. During the first period in the year 2007, they selected four example villages: Sukatani, Kebonpeteuy, Lemahduhur, and Sukamaju.

16.2.2 Forest Dependency Level Analysis

This study was conducted in six selected villages: Gunungputri, Sarongge, Tangsel, Pasirhantap, Pasirbuncir, and Cibeling. These villages represent the typical village in the border of GGPNP Biosphere Reserve having high dependency on forest products. Using the stratified random sampling method, demographic information and details of household economics were collected. The primary data were collected though a questionnaire to collect the data on demography and the daily economic and household activities related to forest products utilization and to identify conditions of the local people in the transition zone of GGPNP (Amelgia et al. 2009).

The interviews used the national language and a semi-structured questionnaire with open-ended questions (a type of question that has no suggested answers); the answers were classified afterward. The interviewers immediately recorded the answers on the questionnaire pages during the interviews. The questionnaire asked for data on (a) household background, such as age, income sources, education, and household size; (b) forest products utilization; and (c) opinion of forest functions.

Multiple linear regressions were used to analyze the relationship between the dependent variables (level of household dependency on forest products) and independent variables (gender, education, household size, assets, off-forest income, policy, distance of forest edge, steepness of slope, and elevation).

16.3 Results and Discussion

Some information on demographics and household economics was collected from the local people using the questionnaire. In total, 210 respondents were selected from six villages in the transition zone of GGPNP. The data revealed that 361,002 people lived within the area (Mulcahy and Mc Carthy 2010).

Table 16.3 shows general information about the six villages in the study areas. Agriculture is the main sector of occupation for the local people; most of them were mainly employed as farming labor. In Tangsel Village, most of the respondents have double jobs. Besides working as farm laborers, they also worked in the tea plantation industry, which was located in the borderline between the park and the settlement area of the local people. The tea plantation itself was established in 1921, and recently the area has been developed into an agrotourism center equipped with some sport facilities. Table 16.3 also notes that Pasirhantap Village is the nearest village to the forest edge, but it is also the farthest village from the main road in contrast with Tangsel Village. The elevation of Tangsel village is relatively high, but the steepness of the slope is not great.

Figure 16.3a shows wealth status and number of family members in the study site. Wealth status was defined by considering the ownership of land and livestock in each household. The average land size owned by each family was less than 2 ha. Thirty-five percent of the respondents were classified as poor households, meaning that they did not have any land or livestock. In several villages, the local people used to own land, but they have sold it to urban people to earn money for several reasons, such as weddings and circumcision ceremonies. Consequently, they are

Table 16.3 General information of the six villages in the study site

Village (District)	Gunung putri	Sarongge	Tangsel	Pasirhantap	Pasir Buncir	Cibeling
Total household	125	374	320	110	77	302
Occupation	Farmer labor	Farmer labor	Farmer labor	Farmer labor	Farmer labor	Farmer labor
			Labor tea company		Gum collector	Gum collector
Income (Rp/month) *10,000	50–250	30–250	45–250	30–250	45–300	45–180
$1.25 (WB poverty line) = Rp 10,091 = 1.5 kg rice						
Distance to main road (m)	7,580	7,200	1,020	13,736	4,830	5,396
Distance to forest edge (m)	12	413	3,570	2	84	286
Elevation (m)	1,221	1,558	885	558	635	593
Steepness of the slope (%)	20–30	10–20	0–10	10–20	20–30	10–20

Fig. 16.3 Wealth status (**a**) and number of family members (**b**) in the study area

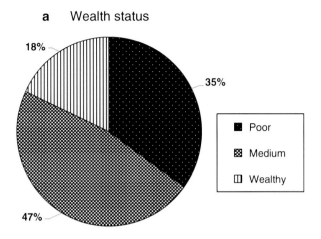

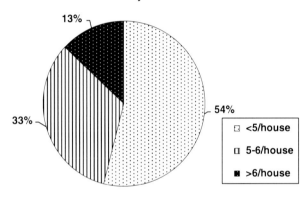

now getting used to working as farming laborers on their own previous land. Figure 16.3b shows the number of persons living in a one-unit house. In average, four or five persons occupied a one-unit house. Some households had a large number of family members because parents, adult siblings, relatives, and children lived together in one house.

Figure 16.4a describes the distance from the house to the forest edge, which is classified as (1) less than 1 km, (2) 1–2 km, and (3) more than 2 km from the forest edge. More than half of the respondents (59 %) lived close to the forest edge. The local people used footpaths to reach the forest edge, instead of roads. Some of the roads in the study site site from the main road to the village were not made of concrete or cement and almost half the roads had some potholes. Figure 16.4b shows the education level of the local community, which was very low. Most of the household leaders did not achieve elementary school or never went to school at all. Seventy-six percent of them could attain the basic education level, and only 7 % of them obtained their education at a level up to senior high school or higher.

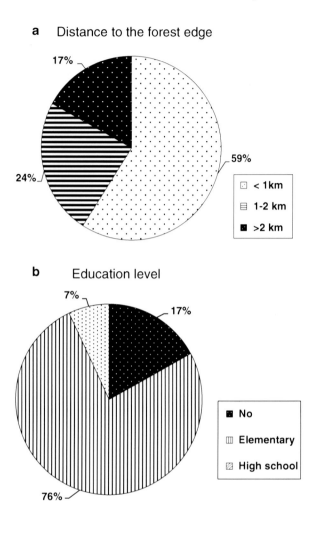

Fig. 16.4 Distance to the forest edge (**a**) and education level (**b**) in the study area

One reason for the low level of education was the distance from the schools. For example, the distance from Lembur Pasir to the closest elementary school was around 8 km. The respondents were questioned about the type of forest products utilization and the estimation of quantities for their daily consumption. The local people used the forest products either for daily consumption or for trading. The price of forest products consumed in their daily life was estimated based on the price in the village or in the closest market. The community used seven forest products, including agriculture products, fuel wood, gum, bamboo, fruits, and flavorings.

Table 16.4 indicates that each village utilized different forest products. For example, Pasirbuncir and Cibeling villages in Bogor District utilized all the forest products; Gunungputri, Sarongge, Pasirhantap, and Tangsel villages only utilized

Table 16.4 Forest products utilization and total number of respondents in six villages

Forest products	Gunung putri	Sarongge	Tangsel	Pasirhantap	Pasirbuncir	Cibeling
Agriculture	26	9	–	19	16	–
Fuel wood	19	33	29	33	34	27
Bamboo	1	12	7	7	5	7
Fruits	–	1	3	12	7	11
Gum	–	–	–	–	3	6
Flavor	–	–	–	–	12	1
Other	–	–	1	–	4	4

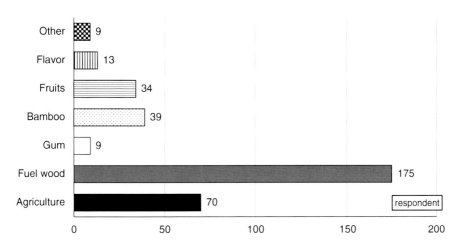

Fig. 16.5 Total respondents and forest product utilization perception

several forest products such as agricultural activities, fruits, fuel wood, and bamboo. The other two villages, Cibeling and Pasirbuncir, concentrated on collecting gum and flavorings. Fuel wood was the common forest product used by a local community for daily purposes. People collected fuel wood from dead stands and dry branches. They never cut the living trees for fuel wood (Fig. 16.5). Another forest product, bamboo, was used for construction, such as for fences or stabling for livestock. Usually, people collected wild bamboo growing in the surrounding forest area. Gum was collected every day from pine (*Pinus merkusii*) trees and sold to the markets weekly. Gum and flavors were collected only for commercial purposes.

The questionnaires could also explain the local community perception of forest function. In the questionnaire, forest function was divided into seven categories, from which each respondent could select more than one category. The functions were selected by the respondents in accordance with their forest activities. The results in Fig. 16.6 show that around 201 respondents gave their opinion, saying that the GGPNP Biosphere Reserve had as its main function the water resource. Forest as an area for conservation was the second largest perception given by the

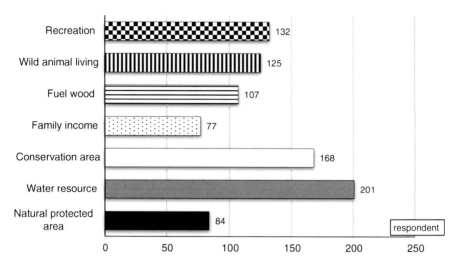

Fig. 16.6 Forest function based on local community perception

respondents concerning the forest function. No more than 77 people perceived the forest as a resource for family income. The results also showed that 95 % of respondents agreed that the GGPNP functioned as a water resource. The third function of the biosphere reserve selected by the respondents was forest for recreation. Particularly during weekends, local people go to the park located in the eastern part of the biosphere reserve to enjoy the fresh air and mountain views. Generally, the local people in the transition zone of the park gave good perceptions on the forest functions simply, because they agreed and realized that the park should exist to maintain a clean water resource.

The forest products shared in contributing to the household total income. Household cash income was generated from agricultural activities, forest products, salary, pensions, and other sources. The household dependency level on forest products was measured from the ratio between income generated from forest products and total household income. To identify the level of dependency of the local people, the income generated from forest products was classified into five levels: less than 20 %, 20–40 %, 40–60 %, 60–80 %, and more than 80 % of total income. Table 16.5 summarizes analysis of the level of dependency of the local people on forest products. About 58 % of the respondents depended on forest products, with 0–20 % of their total monthly income obtained from these products. Only 10 % of the respondents were highly dependent on the forest with more than 80 % of their total monthly income from forest products. On average, the forest products contributed about 28 % of their total income. The social conditions of communities in the transition zone of park were similar, with agriculture as the main resource of income and 76 % of education obtained below junior high school. Seven forest products mainly collected by local people were agriculture, fuel wood, gum, bamboo, fruits, flavor, and other. Eighty percent of the local community used

Table 16.5 Household level of dependency on forest products (unit: respondents)

| Level | Percentage (%) | Total respondents | | Category |
		Number	Percentage (%)	
I	0–20	122	58.10	Less dependent
II	20–40	41	19.52	Relatively dependent
III	40–60	13	6.19	Dependent
IV	60–80	13	6.19	More dependent
V	80–100	21	10.00	Highly dependent
	Total	210	100	

fuel wood for cooking. Fuel wood was one type of energy source as a substitution for kerosene and gas. Sometimes people combined use of fuel wood with that of kerosene or gas to reduce expenditure.

Agriculture was the second largest category of forest products used by the local people. They used to open forestland for agricultural activities, and therefore agriculture was considered to be the biggest threat to the biosphere reserve. The GGPNP Management Board has been trying to prevent the expansion of agriculture land by providing consultation programs to the local people. Another effort applied by the park was a lending agreement between the management board and the local people, by which the local people were able to use the forestland for their agricultural activities for a certain period. This plan did not occur in entire villages where people used the same type of forest products; it depended on the condition of each village. For example, people in most villages collect fuel wood as the main forest product, but in Gunungputri, fuel wood was the second most desirable product after agriculture products. The local people in Cibeling and Pasirbuncir mainly collected gum, because the forest edge surrounding those villages was covered by pine forest. They had started collecting gum 10 years ago when the state-owned forestry firm managed the pine forest. Starting in 2003, the management board has taken over the management of the pine forest.

The level of dependency on forest products in the transition zone of biosphere reserve was quite low. The result of statistical analysis (Fig. 16.7) shows that 58 % of respondents said that less than 20 % of the forest products contributed to the total household income. About 10 % of the respondents highly depended on forest products, which contributed about 80–100 % of their total income. Based on the field survey, households who used forestland for agricultural activities had a higher-level dependency compared to other households, because household income that is derived from agriculture depends on its location in forestland.

The level of dependency is low because surrounding the study site are several sources of occupation. For example, several industrial factories and restaurants are located, usually on the road, so that local people could work as laborers. On the other hand, the local community depends more on agriculture productivity than on forest products. The level of dependency on forest products among the local community in the GGPNP Biosphere Reserve is significantly determined by several factors. Multiple linear regression proved that from nine of independent variables

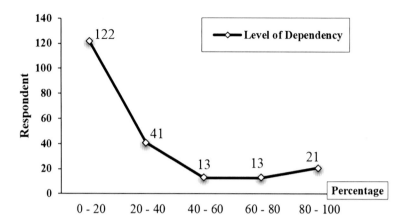

Fig. 16.7 Local community level of dependency on forest products in GGPNP Biosphere Reserve

only five variables such as gender, household size, policy, off-forest income, and elevation are mainly responsible for determining the level of dependency. Results showed that the households with low off-forest income, large household size, and located in a remote area are more dependent on forest products compared to other households. Off-forest income shows strong significant negative correlation to the level of dependency. It is implied that if the farmer could create more income from non-forest products, the dependency on forest products tends to decrease. In fact, a household with a low off-forest income is categorized as a poor family. Income relates to education level. The education level of the local people in this biosphere reserve is low. If the education level is low, the opportunity for local people to get better jobs with a good salary is small. However, agriculture activity and collecting forest products does not require a high skill level and education.

Additionally, the men collect fewer forest products than do the women. In the study site, the man is the decision maker in a family and is responsible for the family income. Women usually stay at home and take care of all family members. In daily activity, women collect forest products such as fuel wood, vegetables, and flavorings for cooking. The men usually work as agricultural laborers or as company labor. During the field survey, female respondents answered the questions in more detail than did male respondents. The level of dependency is also related to household size. It can be understood that a household of large size depends more on forest products in their livelihood. A larger household requires more fuel wood for cooking and more additional income. Furthermore, the statistical results show that the Conservation Village Model (CVM) program is not yet successful in reducing the dependency of the local community on forest products. Households who received support through the program are still dependent on forest products. Even though the sign was not appropriate, it does not mean that this program has failed to improve family welfare. The program was launched in 2007 and the local community has not yet received the benefits of the program; additional time and refinement are required to implement the program.

16 Community Aspects of Forest Ecosystems in the Gunung Gede Pangrango... 285

Another reason is the fund recipient (household leader) might lack a sense of responsibility. This program has tried to support local people by giving livestock and seedlings, but according to the field survey, several respondents who received a goat or sheep sometime sold them to obtain cash money directly. Hence, they returned to collecting forest products as additional income afterward. This research also considers the effect of natural condition related to the level of dependency on forest products. Elevation and steepness of slope are two variables that represent natural conditions.

It is statistically proven that off-forest income and elevation of the household strongly affect household forest dependency. A house located in high elevation (upper areas) is usually in a remote area and is relatively far from the market. The dwellers try to fill their needs from agricultural activities by collecting forest products. They depend more on forest products compared to local people living in a house that is located at a lower elevation. The average degree of condition of steep roads to collect forest products are around 10 % to 20 % in degree of slope. Some respondents also go to the forest area to collect forest products even though the steepness of the road is 20 % or 30 %. Gender and policy are significant at the 5 % level and household size at the 10 % level. Simply, the household with low off-forest income, large family, and located in a remote area is more dependent on forest products.

There are two kinds of dependency on forest products in the local community in the GGPNP Biosphere Reserve: dependency on forest products as economic resources and dependency on intangible forest products as functions such as water supply, fresh water, and mountain views. Although economically the value of forest products is low, unwittingly local people depend highly on intangible forest products, such as water resources, fresh air, and tourism areas. The results of the questionnaire showed that 70 % of local people are using clean water that is sourced from the forest area directly. Ninety-five percent of the local community said that the biosphere reserve functions as a water resource and 62 % of them were enjoying the forest area for recreation.

16.4 Conclusion

Forestland and forest resources are important for the local community in the transition zone of GGPNP Biosphere Reserve. Forest income constituted an average of 28 % of the local community income. The level of dependency on forest products was influenced by several factors, such as gender, household size, policy, off-forest income, and household elevation. Based on the questionnaire results, the local community in the transition zone utilized the forest for collecting several forest products (fuel wood, bamboo, gum, fruits, flavor, etc.) and used forestland for agricultural activities. Forest products served not only as a primary livelihood source but also for additional income of the local people. The forest income was more important for poor families, especially those who were located in remote areas.

The local community realized that the forest area must exist to support their life. Although not the entire local community gets direct benefit from the forest by collecting forest products, some of the local community also depend very much on the indirect benefits from this biosphere reserve such as ground water, rivers, and clean air.

16.5 Recommendations

Another study should be conducted in the future in the other villages also located surrounding the GGPNP Biosphere Reserve to obtain a comprehensive picture of forest dependency in the area. In addition, more research focusing on the effect of environmental conditions related to the level of dependency is needed. The local government should initiate some programs introducing alternative income generation through skill development and training for the local people.

By these measures, the local people can generate economic activities without damaging forest resources. Additionally, the investment in infrastructures and markets, as well as the collaboration with all involved stockholders, can stimulate the generation of new economic opportunities that in turn can create economic diversity in the villages surrounding the park.

Acknowledgments This research was supported by the Global Environmental Leaders (GELs) Education Programme of Hiroshima University, Japan and the Ministry of Forestry, Indonesia.

References

Amelgia R, Wicaksono KP, Nakagoshi N (2009) Forest product dependency excluded timber in Gede Pangrango national park in West Java. Hikobia 15:331–338

Azizia PA, Sulistyawati E (2009) Analysis of forest patches in three regencies in West Java based on satellite data. Biology Study Program, School of Life Sciences & Technology, Institut Teknologi Bandung, Bandung

GGPNP (2009) Statistics of Gunung Gede Pangrango National Park, Jakarta (in Indonesia)

Indra Exploitasia S (2010) Lesson learned and practices on tree adoption program in Mount Gede Pangrango national park in support of sustainable land use in Cibodas biosphere reserves, Indonesia. Ministry of Forestry Indonesia, Jakarta

IUCN (2008) Guidelines for applying protected area management categories. IUCN, Gland

Ministry of Forestry (2006) Strategic plan of the ministry of forestry 2005–2009 (revised). Centre of Forestry Planning and Statistics Agency for Forestry Planning, Jakarta

Ministry of Forestry (2008) Forestry statistic of Indonesia. Forestry statistic and data communication network divison. Ministry of Forestry of Indonesia, Jakarta

Ministry of Forestry (2010) Forestry statistic of Indonesia. Forestry statistic and data communication network divison. Ministry of Forestry of Indonesia, Jakarta

Mulcahy G, McCarthy J (2010) Nature based tourism. Case study: Gunung Gede Oangrango national park. Institute for Sustainability and Technology Policy, Murdoch University, Perth

Nakagoshi N, Amelgia R (2010) Forest dependency of the communities in Gunung Gede Pangrango biosphere reserves, Indonesia. In: The 4th EAFES International Congress, EAFES, Sangju, Korea, pp 140–141

Riswan S, Hartanti L (1995) Human impacts on tropical forest dynamics. J Plant Ecol 121:41–52

Siarudin M, Mile MY (2005) Evaluasi Program Pengembangan Usaha Ekonomi Pedesaan (USPED) masyarakat daerah penyangga studi kasus: resort Cisarua, Seksi Konservasi Wilayah II Bogor. Taman Nasional Gunung Gede Pangrango, Pusat Penelitian Sosial Ekonomi dan Kebijakan Kehutanan, Departemen Kehutanan, Jakarta (in Indonesian)

Wiratno K, Virza S, Harry K (2003) Valuation of Mt. Cibodas Biosphere Reserve. Ministry of Forestry Indonesia, Jakarta

Yohanes YR (2010) Forest land use by the community in Sorong Natural Tourism Park at Sorong City, West Papua Province. Faculty of Forestry, the State University of Papua, Manokwari

Chapter 17
Landscape Ecology-Based Approach for Assessing Pekarangan Condition to Preserve Protected Area in West Java

Regan Leonardus Kaswanto and Nobukazu Nakagoshi

Abstract As is widely known, the landscape ecology-based approach focuses on three aspects: structure, function, and dynamic or change. The fourth aspect, which was added recently, is culture. These four aspects were attempted to be elaborated into the homegardens condition as an ecological process in human activities. As traditional homegardens in Indonesia, *pekarangan* played an important role in building low carbon society management. To conserve protected areas, the *pekarangan* concept could be one solution to help people stay in their village. It means people will not encroach on the forest to seek more income because they could have it from *pekarangan*. Assessment of the *pekarangan* condition was conducted through interviewing and measuring the ecological, economic, and sociocultural data of 96 households located in four watersheds of West Java, Indonesia. The levels of landscape management are affected by the size of land; therefore, the scale of ownership of *pekarangan* size is considered. In each respective watershed level, four groups of *pekarangan* were defined in a purposive random village. Those four groups are G1, with a *pekarangan* size less than 120 m^2 that does not have other agricultural land (OAL), G2 ($<$120 m^2 with OAL $<$ 1,000 m^2), G3 (120–400 m^2 with no OAL), and G4 (120–400 m^2 with OAL $<$ 1,000 m^2). The results statistically proved that *pekarangan* contributed significantly to ecological conditions, household income, and also nutrition (diet) for the family members. The ecological conditions, both horizontal and vertical biodiversity, showed that *pekarangan* offer a large contribution to help the environment be more sustainable. The levels of heterogeneity, indicated by species

---

R.L. Kaswanto (✉)
Landscape Management Division, Landscape Architecture Department, Bogor Agricultural University (IPB), Kampus IPB, Dramaga-Bogor 16680, Indonesia
e-mail: kaswanto@ipb.ac.id; anto_leonardus@yahoo.com

N. Nakagoshi
Graduate School for International Development and Cooperation, Hiroshima University, Higashi-Hiroshima 739-8529, Japan

N. Nakagoshi and J.A. Mabuhay (eds.), *Designing Low Carbon Societies in Landscapes*, Ecological Research Monographs, DOI 10.1007/978-4-431-54819-5_17, © Springer Japan 2014

richness (Margalef Index), species diversity (Shannon–Wiener Index), species evenness (Simpson Index), and similarity (Sørensen Index) showed that *pekarangan* has high biodiversity. The research also proved that *pekarangan* production contributed as much as 11.5 % of a household's income and 12.9 % of household's diet in term of food expense. One recommendation is that the community should consider a suitable agroforestry practice, such as *pekarangan*, for better landscape management in the future. In addition, *pekarangan* as a small agroforestry system also can contribute significantly to a region's carbon budget while simultaneously enhancing the livelihoods of the rural population.

Keywords Agroforestry system • Environment conservation • Homegardens • Landscape management • Rural landscape

17.1 Introduction

17.1.1 Background

The landscape ecology-based approach focuses on three aspects: structure, function, and change (Forman and Godron 1986). The fourth aspect, which was added lately, is culture. These four aspects were elaborated into the homegardens concept as an ecological process in human activities. Recently, encroaching on the forest is a major problem for some local governments in Indonesia. Therefore, to reduce forest encroachment and at the same time conserve protected areas, the *pekarangan* system should be revitalized. When communities can obtain some benefits from their *pekarangan*, they may avoid destroying the forest areas. They can manage the *pekarangan* area intensively and obtain some income to fulfill their needs (Fig. 17.1).

The terminology of *pekarangan* is now used widely in much scientific research, particularly in agroforestry and environment-related areas. *Pekarangan* can be viewed as a reconstruction of a complex tropical forest ecosystem consisting of useful flora and fauna, which also has the attribute of doing minimal harm to the environment (Takeuchi 2010). As traditional homegardens in Indonesia, *pekarangan* has an important role in building a low carbon society (LCS). To conserve protected areas, *pekarangan* could be one solution to keep people in their village. It means people will not encroach on the forest to seek more income because they can obtain it from their *pekarangan*. A large number of *pekarangan* in Indonesia are a potential source of carbon sequestration and maintenance of stock. In addition, it provides additional income for the households.

In the effort to rebuild sustainable local societies, the current circumstances in which *pekarangan* systems are contributing to environmental diversity and community incomes in the face of massive urban and rural development are a concern. *Pekarangan* reflects the behavior of its ownership; it has a strong relationship to human activities. As stated by Harashina et al. (2003), human activities cause

17 Landscape Ecological Based Approach for Assessing *Pekarangan* in West Java

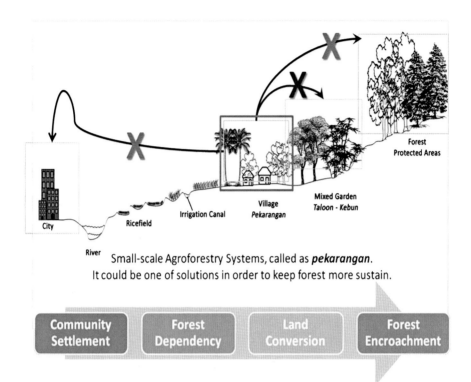

Fig. 17.1 *Pekarangan* can be one solution to keep protected areas more sustainable. People are used to immigrating to urban areas or converting land use or encroaching on forest areas when their livelihood is threatened because of reduction of income, social status, and food for consumption

significant impact on nutrient flow. Soemarwoto (1984, 1987) also said that the *pekarangan* "definition" revolves around representative, intimate, multistory combinations of various tree and crops, sometimes in association with domestic animals around the homestead. Those *pekarangan* were categorized as either survival gardens or subsistence gardens based on the Wiersum (2006) classification adapted from Niñez (1984). There are also survival gardens, that are defined as a single component of the farming system of otherwise landless rural people and a combination of production from staple food crops and complementary crops. Subsistence gardens are part of a multicomponent farming system in conjunction with permanent or shifting field production and also a complementary system to open-field staple food cultivation systems. In this type of garden, daily household provision of vegetable, herbs, spices, and fruits for immediate needs and occasional sale is found.

17.1.2 Objectives

The aims of this study are to determine the contribution of *pekarangan* in preserving protected areas and to visualize its benefit for rural communities. When *pekarangan* is well managed, it can provide real benefits in term of income, nutrition, and other needs. In addition, when communities revitalize *pekarangan* optimally, the protected areas will be more sustainable. Therefore, those communities do not find it necessary to slash-and-burn the forest to obtain additional income or food resources.

17.2 Study Site

17.2.1 Protected Areas in West Java Region

In Indonesia, there are 51 national parks: 5 of them are located in Java Island, and 4 are located in West Java (Table 17.1). The national parks of West Java are (1) Ujung Kulon, (2) Gunung Gede Pangrango, (3) Gunung Halimun Salak, and (4) Gunung Ciremai National Park. Two national parks were selected for this research, Gunung Gede Pangrango (GGP) and Gunung Halimun Salak (GHS), because they are located close to four watersheds that are being studied.

Table 17.1 General condition of four national parks in West Java

Government regulation	Name of national park			
	Gunung Gede Pangrango	Gunung Halimun-Salak	Ujung Kulon	Gunung Ciremai
Ministry of Forestry (MoF) Act	174/Kpts-II/2003	175/Kpts-II/2003	284/Kpts-II/1992	424/Menhut-II/2004
Date of act	June 10th, 2003	June 10th, 2003	February 26th, 1992	October 19th, 2004
Area (ha)	± 21.975	± 113.357	± 120.551	± 15.500
District covered	Bogor, Cianjur, and Sukabumi	Sukabumi, Bogor, and Lebak	Pandeglang	Kuningan
Coordinate location	$106°50'–107°02'E$	$106°13'–106°46'E$	$102°02'–105°37'E$	$108°19'–108°27'E$
	$06°41'–06°51'S$	$06°32'–06°55'S$	$06°30'–06°52'S$	$06°47'–06°58'S$

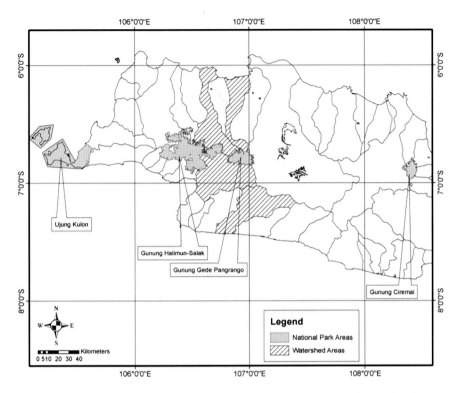

Fig. 17.2 Study areas of four watersheds within two national parks in West Java region. Gunung Halimun-Salak (GHS) and Gunung Gede Pangrango (GGP) National Park are located at high altitude

17.2.2 Four Watersheds in Two National Parks

The research sites are in four watersheds that were selected based on their location close to the national parks: Cisadane (CS), Ciliwung (CL), Cimandiri (CM), and Cibuni (CB) watersheds. In addition, we also considered their orientation, lying from south to north (Fig. 17.2). Although CB is located a little far from GGP National Park, it has actually been impacted because there are many main access roads from this area to the core of its national park.

17.3 Methods

To understand the conditions within *pekarangan*, the method for this research began with a sampling design for respondent selection. Then, interviewing by questionnaire and *pekarangan* measurements followed (Fig. 17.3).

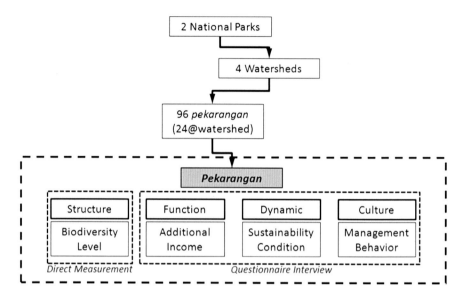

Fig. 17.3 The framework is an assessment of the basics of landscape ecology approach: structure, function, change, and culture: starting from the macroscale of two national parks and gradually reaching microscale on *pekarangan* size

17.3.1 Sampling Design for Selecting Pekarangan

The *pekarangan* plot survey samples were designed through three levels of consideration following the preceding process. The first level of consideration of this design is the different stream directions of the rivers where the watershed lies in either southern areas (SAs) and northern areas (NAs); this is called orientation level. The two sides of the flow of rivers that originate from the same mountain are expected to be almost homogeneous and have almost the same parent soil material, yet hypothetically have different climate conditions. A distinct variation of parent soil material may occur within the mountain at the water sources. In general, the NA watersheds were more humid than the SAs; because of this variation in agroclimatic zone, the NAs tend to have better agricultural activity performance than the SAs.

The second level of consideration is that of agroecological zoning, which in this study is represented by elevation stratification within a watershed where a *pekarangan* lies. There are three levels of stratification: 0–300 m, 300–700 m, and >700 m above sea level (m a.s.l.), called stream level. Ochse, as cited by (Harjadi 1989), a Dutch horticulturist working in Indonesia in the colonial era, used the 700 m a.s.l. level as a threshold between lowland and highland. Ochse said that

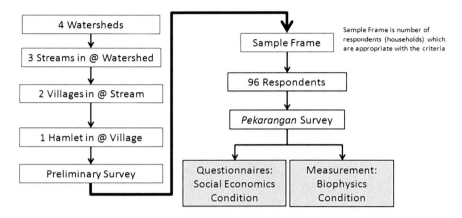

Fig. 17.4 Flowchart of the research process from preliminary to main survey. Sample frame is number of respondents who have *pekarangan* and other agriculture land (OAL) size appropriate to group criteria

above 700 m a.s.l. coconut, a plant grown widely in Indonesia, cannot grow productively. Therefore, in this survey we use this elevation threshold. The 300–700 m a.s.l. zone was used in concordance with bioclimatic conditions (Arifin et al. 2001). This zone is a transitional zone, where there is high biodiversity of plant species mixed from upper or lower zones.

The third level of consideration is the scale of ownership. At each respective stream level, four groups of *pekarangan* ownership were defined in a purposive random village. Those four groups are G1, with a *pekarangan* size less than 120 m^2 with no other agricultural land (OAL), G2 (*pekarangan* <120 m^2 with OAL < 1,000 m^2), G3 (*pekarangan* 120–400 m^2 with no OAL), and G4 (*pekarangan* 120–400 m^2 with OAL < 1,000 m^2) (Figs. 17.4, 17.5) (Arifin et al. 2008).

Following the three levels of sampling design just described, two villages were randomly selected to obtain averaged data. Then, in each village four groups were selected based on the sample frame. In total, 96 *pekarangan* that were distributed among four watersheds were investigated (Fig. 17.6).

The sampling methods were based on region, orientation, streams, villages, and location relative to the national parks. In each respective area, a sample frame consisting of group levels of *pekarangan* ownership were chosen randomly (Fig. 17.4). The consideration is the size of *pekarangan* and the ownership of other agricultural land (OAL). Representative samples were collected by (1) village selection, (2) *pekarangan* utilization, and (3) sample frame requirement. Four surveyors conducted those activities: two surveyors interviewed asked the respondents while the other two measured the *pekarangan* biophysically.

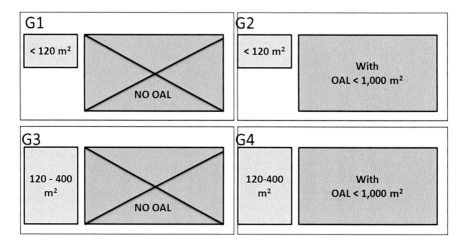

Fig. 17.5 Schematic diagram of group criteria: G1, G2, G3, G4. G1 are households with *pekarangan* less than 120 m² in size but that do not have OAL. G3 are households with *pekarangan* between 120 and 400 m² in size that have OAL but its size is less than 1,000 m². The scheme is drawn in proportional size

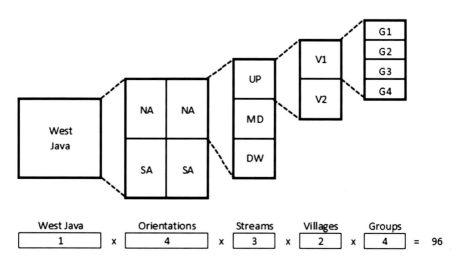

Total respondents: 1 x 4 x 3 x 2 x 4 = 96 respondents

Fig. 17.6 The sampling design shows a total of 96 respondents distributed in four watersheds. The process of sampling design is as follows: there is one region in West Java, two orientation levels (*NAs* north areas, *SAs* south areas), three stream levels (*UP* upper stream, *MD* middle stream, *DW* down stream), two villages (V1 and V2 were randomly selected), and four group levels (G1, G2, G3, and G4 based on sample frame)

17.3.2 Questionnaires

The questions asked were mainly about the utilization of vegetation in *pekarangan*, particularly plant function and its history. Before commencing interviews and data collection, the surveyors sought out the *Kepala Dusun* or *Kepala Kampung* (head of hamlet, the informal leader), who assisted in the search for *pekarangan* and helped to reassure respondents of the positive intentions of the surveyors. In average, to finish one questionnaire took at least 2 h depending on interviewing situation.

The variability and reliability tests were framed throughout the preliminary survey. This questionnaire was tested through four respondents to know the difficulties and obstacles during interviewing process. Then, a fixed questionnaire was used for interviewing the 96 respondents in our primary survey.

17.3.3 Biophysical Measurements

Two surveyors measured the biophysical data of *pekarangan*. *Pekarangan* size was measured, including open space and built-up areas. Plants and animals (livestock) were also counted. Diameter at breast height (DBH) and height of plants were measured. After that, they joined the interviewer to ask about the production of those plants and livestock. The estimation was up to 1 year of production.

The combination of direct measurement and questionnaire has been proven as the optimal approach to comprehensive investigation for estimating *pekarangan* function. Direct measurement has been conducted by some researchers, including Soemarwoto (1984, 1987), Christanty et al. (1986), Roshetko et al. (2002; 2007), Harashina et al. (2003), Abdoellah et al. (2006), Montagnini (2006), Das and Das (2010), and Kumar (2011). A comprehensive questionnaire was used, consisting of 17 pages with 228 questions, to gather detailed information on the production and utilization of *pekarangan*. Those readers with further interest in the questionnaire are invited to contact the authors directly.

17.3.4 Data Analysis

Data were gathered on the relationship between the biodiversity indices and the structures of *pekarangan*. The relationships were calculated by four indices: the Margalef, Shannon–Wienner, Simpson, and Sørensen indices. Those indices, which have different functions, show some significant results related to heterogeneity in *pekarangan*.

17.3.4.1 Species Richness (Margalef Index)

The Margalef index is a species diversity index divided into two types of species richness (how many types are there) and assessment of species evenness or dominance (how individual species are distributed among the community). The Margalef index was one of the first attempts to compensate for the effect of sample size by dividing the number of species in a sample by the natural log of the number of species collected (Death 2008).

The index is thus

$$D = (S - 1)/\ln\!N \qquad (17.1)$$

where S is the total number of species in a sample and N is the total number of individual in the *pekarangan*.

17.3.4.2 Species Diversity (Shannon–Wiener Index)

The Shannon–Wiener index is most sensitive to the number of species in a sample and considered to be biased toward measuring species richness, because determining the degree to which each factor contributes to diversity is impossible from the calculated value of H' alone (Elliott and Knoepp 2005).

The index is thus

$$H' = \sum_{i=1}^{s}(\rho_i \ln\!\rho_i) \qquad (17.2)$$

where S is the total number of species, and p_i is the frequency of the ith species (the probability that any given individual belongs to the species, hence p).

17.3.4.3 Species Evenness (Simpson Index)

The Simpson index is used to emphasize the evenness of species (Nagendra 2002). Producing values from 0 to 1, Simpson's index defines the probability that two equal-sized *pekarangan*, selected at random, belong to different *pekarangan* areas. The D also represents the uniformity of individual quantities of various species, which means the more species, the higher the uniformity and the value of D (Hong et al. 2012). The index is thus

$$D = 1 - \frac{(n_i - 1)}{N(N - 1)} \qquad (17.3)$$

where n_i = number of individuals of species i and N is the total number of organisms of all species found.

17.3.4.4 Species Similarity (Sørensen Index)

The Sørensen index is used to measure similarity of vegetation types in a *pekarangan*. The index is thus

$$QS = \frac{2C}{A + B} \tag{17.4}$$

where A and B are the number of species in samples A and B, respectively, and C is the number of species shared by the two samples. This expression is easily extended to abundance instead of presence/absence of species. An A value approaching 1.0 indicates high similarity in species present in the two landscapes (Miller and Winer 1984).

17.3.5 Data Performance

The heterogeneity of plants in *pekarangan* is calculated by biodiversity indices. The statistical analysis was conducted using statistical software (PASW). All data were analyzed statistically into three levels of consideration: (1) orientation level, southern areas (SAs) and northern areas (NAs); (2) stream level, the level of altitude inside each watershed [there are three stream levels, upper-stream (UP), middle-stream (MD), and down-stream (DW)]; and (3) group level, the group criteria for size classification of *pekarangan* plot. There are four groups: G1, G2, G3, and G4.

17.4 Results

17.4.1 The Respondents

The household number in each *pekarangan* is 4.6 persons, on average. The range of household members numbered from one to ten persons. The head of household most often was male (93.8 %); the rest of heads were women (6.3 %). The results showed that most respondents' communities were more often patriarchates. The head of household averaged 48 years of age: the youngest was 22 years old and the oldest was 90 years old. Most (93.1 %) household heads had been married, mostly at 24 years of age, although among those the youngest one married at 13 years and the oldest at 63 years of age. In general, rural villages have productive age and potential human resources.

Nine occupation types found at the survey sites. Agricultural labor (26.4 %), particularly for members of G1, and farm-owned land (25.0 %), particularly for who hold OAL (G2 and G4), were the main occupation of the head of the household.

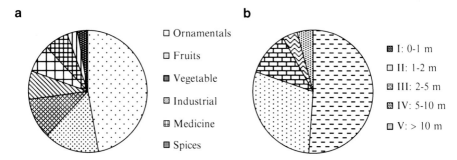

Fig. 17.7 Plant compositions in *pekarangan* were divided into eight categories and five strata. Plants in *pekarangan* are Category of Plant Utilization mostly in ornamental utilization (47.20 %) (**a**) and Strata of Plant Utilization predominantly 0–1 m in height (51.2 %) (**b**). In addition, plants with height more than 2 m occupied up to 19.85 % of total, which means *pekarangan* have potential for growth

Others had occupations in non-agriculture (13.2 %), as petty trader (4.2 %), housewife (4.2 %), driver (3.5 %), farmer of tenant land (1.4 %), and village official (0.7 %) and village elder (0.7 %), respectively. Remaining occupations (20.8 %) were a mix of many minor occupations. It can be concluded that the existence of agriculture-based occupations was prospective as initial capital for a development program.

17.4.2 Plant Heterogeneity in Pekarangan

A higher heterogeneity was predominantly found in the *pekarangan* system because of a large number of ornamental plants, meaning *pekarangan* can be more adaptable to human disturbance (Fig. 17.7). Plants are mostly ornamental in utilization (47.2 %) and predominantly 0–1 m in height (51.2 %). In addition, plants with height more than 2 m were up to 19.9 % of the total, which means *pekarangan* can support plant growth without any limitations.

The relationship between the Margalef index and the stream level shows that the middle stream has a higher index value compared to other stream levels (Fig. 17.8): this is determined to be the effect of the ecotone condition, which changed plant heterogeneity in the middle stream. Furthermore, G3 and G4 have higher index value compared to G1 and G2, but these differences are not significant for those groups at the upper stream. These conditions were caused by the different size of open area and the topography level. The middle and downstream have flat areas, which encouraged the household to plant many kinds of species.

The relationship between the Simpson index and stream level shows that the upper stream is more varied compared to other stream areas (Fig. 17.9). In addition, all respondents represent a high index value, close to 1. Thus, the uniformity of individual quantities of various species inside *pekarangan* increases by group levels

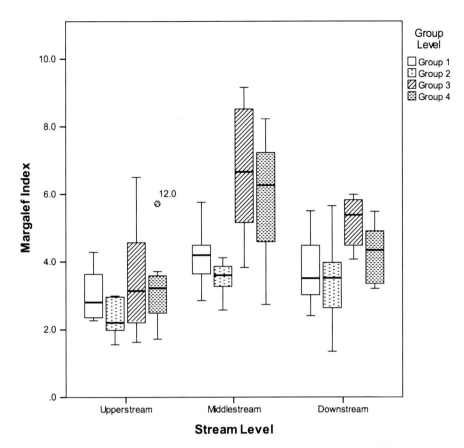

Fig. 17.8 Relationship between Margalef index and stream level shows that middle stream has a higher index than other streams, explaining the ecotone conditions among watersheds. In addition, G3 and G4 have a higher index than G1 and G2, although the difference is not significant on the upper stream

but is almost same between the stream levels. It is also proof that the more species, the higher the chance that the same species can be found.

The relationship between the Shannon–Wiener index and stream level shows that the index values range from 2.5 to 3.5, which means species diversity is high inside the *pekarangan* area (Fig. 17.10). A trend is shown that G1 and G3 have lower values compared to G2 and G4, because the ownership of OAL caused the household to also plant some crops in *pekarangan*. The Sorenson index shows that the similarity of plants among *pekarangan* is high (Fig. 17.11).

Two groups (G1 and G2) were combined to represent a small *pekarangan* group, while G3 and G4 formed a large *pekarangan* group. In addition, G1 and G3 were also combined into non-OAL groups, while G2 and G4 were crafted into OAL groups. There are no significant differences in either *pekarangan* size or OAL within the orientation and stream levels. However, the physical conditions of *pekarangan*, that

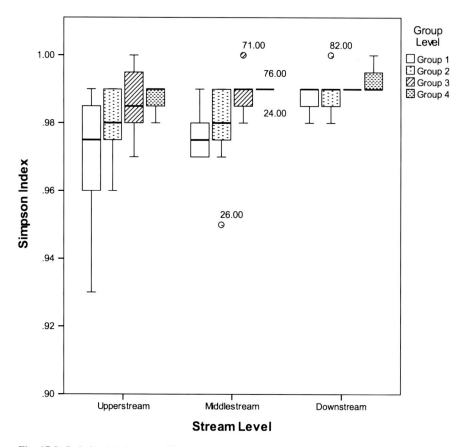

Fig. 17.9 Relationship between Simpson index and stream level

is, *pekarangan* size, building size, open area size, OAL size, number of species, and number of households, were significantly different. It was found that G1 has the lowest biodiversity, followed by G2, then G3, and finally G4 has the highest.

17.4.3 Benefits from Pekarangan

17.4.3.1 Plants and Animals for Family Health

The production of plants and animals in *pekarangan* gave much nutrition that significantly influenced the family's health. Four nutrition types were elaborated to determine the contribution of *pekarangan* production to a family's diet: calories, protein, vitamin A, and vitamin C. There were no statistically significant differences for those nutritional values between G1, G2, G3, and G4 in family

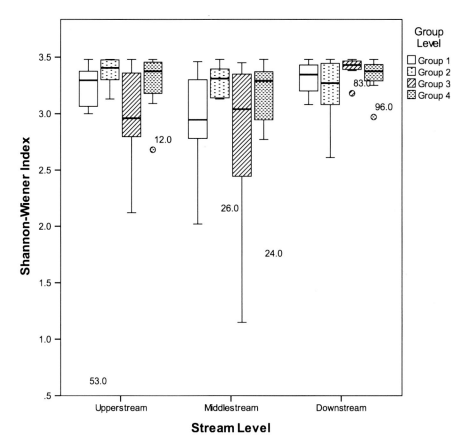

Fig. 17.10 Relationship between Shannon–Wiener index and stream level

	Upperstream	Middlestream	Downstream
Upperstream	X	0.52	0.67
Middlestream	X	X	0.63
Downstream	X	X	X

	G1	G2	G3	G4
G1	X	0.44	0.48	0.42
G2	X	X	0.69	0.58
G3	X	X	X	0.62
G4	X	X	X	X

Fig. 17.11 Similarity index among stream levels and among groups

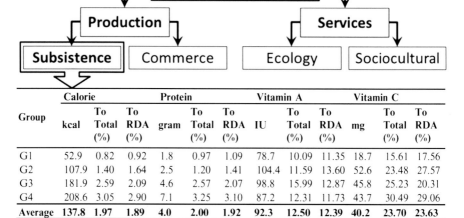

Fig. 17.12 Value of family consumption derived from *pekarangan* per day. Nutrition contribution from *pekarangan* to total consumption (%) and to recommended dietary allowance (RDA) (%). There were no significant differences between G1, G2, G3, and G4 on family consumption from *pekarangan* for all kinds of nutrition

consumption from *pekarangan*. It can be concluded that the contribution of *pekarangan* is important for all groups without exception (Fig. 17.12).

Therefore, *pekarangan* is one recommended strategy for addressing malnutrition and micronutrient deficiencies, particularly for marginal people in a forest margin area or at the buffer zone of a protected area. A number of studies have found that *pekarangan* produce a significant percentage of plants such as fruits and vegetables for household consumption. In addition, a number of studies have concluded that *pekarangan* are associated with better household nutrition (Mitchell and Hanstad 2004).

17.4.3.2 Household Income

The income derived from *pekarangan* production contributed up to 11.5 % to total income derived from the main occupations of all family members. The members of the household may sell the products of *pekarangan*, including fruits, vegetables, animal products, and other valuable materials such as bamboo or wood for construction or fuel. Some of the respondents also obtained incomes from renting the *pekarangan* area for uses such as drying harvest products, kiosks, repair shops, and producing handmade materials.

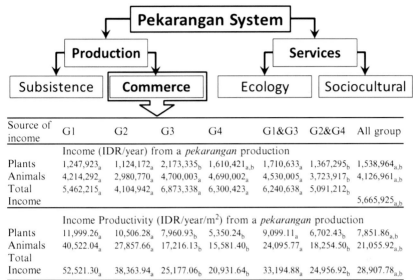

Fig. 17.13 Income derived from *pekarangan* from selling plants and animals (livestock production). Productivity is derived by calculating total income from *pekarangan* per year divided by open space area

Incomes derived from *pekarangan* could came from selling plants and livestock production (Fig. 17.13). In term of productivity, G1 and G2 are different from G3 and G4, which means small *pekarangan* are more productive. Small *pekarangan* (G1 and G2) tend to improve their *pekarangan* by cultivating more crops and caring for more livestock. However, the productivity for those who have no OAL (G1 and G3) indeed is higher compared to the others. Those of G1 and G3 cultivate their *pekarangan* intensively because they do not have OAL to manage.

17.4.3.3 Wage Security and Social Status for Household

Some respondents said that they used *pekarangan* for getting credit or loans from a formal bank or an informal lender. *Pekarangan* is positioned as a guarantee to obtain access for the bank. In addition, for those who have a main occupation with regular salary, they have an opportunity to obtain credit.

In some cases, the size of *pekarangan* also implies the social status of the household. The larger the *pekarangan* area, the more respect and honor will be received. The large *pekarangan*, particularly with a flat open area, allows conducting some neighborhood activities often. *Pekarangan* is a symbol of social status, and provides better access to trading relationships within the villages. In addition, it will increase access to government programs serving village households (Mitchell and Hanstad 2004)

17.4.3.4 Indirect Benefits from Pekarangan

There are also some benefits that could be derived indirectly from *pekarangan* other than those already mentioned: benefits to women, wealth or position in labor markets, access to obtain credit, additional income-generating activities, post-production activities, and also as storage.

The role of women in *pekarangan* is hypothetically an impact by the selection of plants and animals. This condition impacts biodiversity inside *pekarangan* indirectly, and most of them tried to introduce new species to the *pekarangan*. The sensitivity of woman to beautification is believed to increase introduction of exotic species that in the future will have an impact on native species in the protected area. *Pekarangan* also represents the woman's freedom from dependence from their neighbors. They try to show their ability to develop *pekarangan* into a beautiful place for social gathering.

The market also sometimes tends to influence the *pekarangan* condition. The households try to plant commercial trees that can be sold at market. However, in a small-scale *pekarangan*, the households only try to fulfill their subsistence needs; when they have more they will give it to their neighbors. *Pekarangan* also provided access to obtain credit from an informal lender or formal financial agency [village bank, *Koperasi Unit Desa* (*KUD*) village cooperative system]. Additional income that can be derived when someone has *pekarangan* is the opportunity to develop a small-scale home industry.

Some respondents also said that *pekarangan* also have been used for post-harvest activities such as drying and threshing. They received some money from people who used their *pekarangan* for those activities. *Pekarangan* also provide space for storing harvest production, tools, fertilizer, and other items related to farming activities.

17.4.4 Revitalizing Pekarangan Conserves Protected Area

Pekarangan systems are sustainable examples of systems represented by reliably recycling nutrients, preserving soils, and conserving protected areas indirectly. Some environmental benefits are obtained from *pekarangan*. In many cases, revitalizing *pekarangan* conserves a protected area and tends to reduce encroachment of the forest (Table 17.2).

Mainly in West Java, some forests outside the protected areas were transformed into open-field agriculture, intensive tree crop production, such as cacao (*Theobroma cacao*) systems or plantations for the beverage industry, tea (*Camellia sinensis*) and coffee (*Coffea arabica*). Thus, some abandoned lands or unsuitable lands from those

17 Landscape Ecological Based Approach for Assessing *Pekarangan* in West Java 307

Table 17.2 Differences of the *pekarangan* system compared to other utilization

Characteristics	*Pekarangan*	Commercial garden	Field agriculture
Planting and harvesting			
• Species density	High	Medium to low	Low
• Species type	Staple, vegetable, fruit (cultural)	Vegetable, fruit (market-oriented)	Staple (subsistence, agro-industrial)
• Harvest frequency	Daily, seasonal	Seasonal (short)	Seasonal (long)
• Cropping patterns	Irregular, row	Row	Row
Production and economic			
• Production objective	Home consumption (subsistence)	Market sale	Subsistence, market safe
• Economic role	Supplementary	Major economic activity	Major economic activity
• Technology needed	Simple hand tools	Hand tools or mechanized	Mechanized if possible, hand tools
• Input cost	Low	Medium to high	Medium to high
• Economic assistance	None or minor	Credit	Credit, extension
Labor and skill			
• Labor resource	Family (female, elderly, children)	Family or hired (male, female)	Family, hired (male, female)
• Labor requirements	Part-time	Full-time	Full-time
• Skills required	Garden-horticultural	Market-horticultural	Agricultural, commercial
Spatial			
• Space utilization	Horizontal, vertical	Horizontal, vertical	Horizontal
• Location	Close to dwelling	Close to urban market	Rural setting, near or distant from homestead
• Distribution	Rural and urban	Suburban	Rural
Protected area			
• Competition to local species	Medium to high	Highly competitive	Low to medium
• Disturbance	Small	Medium to large scale	Large scale
• Encroachment level	Small	Medium to high	High

Underlined entries indicate *pekarangan* benefits and unique characteristics compared to other cultivation systems

systems are gradually becoming small villages. These villages occurred sporadically within a decade. The government perceived these arising villages as an unsolved problem. The communities inside these villages then will slowly try encroaching on the forest margin to get some land for cultivating or for fuel wood, food, and other materials. Introduction of the ideas of *pekarangan* practices, which can convince those communities to maintain their own land for the long-term preservation of protected areas, is strongly to be recommended.

17.5 Discussion

Although *pekarangan* are an age-old practice in many parts of the tropics and even other parts of the world, as stated by Kumar and Nair (2004, 2006), only limited data are available about their contents and distribution, particularly in Indonesia. The number of *pekarangan* covers about 5.13 million ha of total land, of which 1.74 million ha is located in Java (BPS 2000), and then increased during the next 10 years to 8.5% of total area (BPS 2010). Research on *pekarangan* is primarily concerned with *pekarangan* in urbanization (Kaswanto et al. 2010), floristic plants (Molebatsi et al. 2010), fragmentation (Arifin and Nakagoshi 2011), single-commodity production systems (Kumar 2011), and conservation (Tabuti et al. 2011). There are few studies on species from *pekarangan* related to conservation of protected areas.

For the purpose of conserving a protected area, *pekarangan* might be one solution to help people stay within their villages. People will not encroach protected areas to seek more income or food resources, because the *pekarangan* itself could provide significant additional income and food for consumption. In some cases, population pressures and economic reasons threaten some households to resettle in forest margin areas, or even to move to protected areas. Offering *pekarangan* area to landless or marginal families can reduce pressure for encroaching forest areas. It helps both to allow families remain in their area and also to reduce forest conversion. In addition, distribution of *pekarangan* may reduce the need for land-poor families to gather fodder and fuel wood from marginal lands, contributing to the sustainability of protected areas (Mitchell and Hanstad 2004).

The species heterogeneity inside *pekarangan* is high in value, which was represented by Margalef, Shannon–Wienner, Simpson, and Sorenson indices. Invasive species that were planted in *pekarangan* related to the sustainability of protected area were also a concern. Among 214 species found in *pekarangan*, 32 of them (14.95 %) are exotic species, meaning the progress of invading species in *pekarangan* is still small. However, some exotic species were found in the upperstream area, which means the forest margin area also should be monitored.

Although the *pekarangan* systems in the tropics are claimed to sustain basic needs of communities without environmental deterioration (Schultink 2000), the ecological rationality of the harmony between humans, *pekarangan*, and the environment could be more clearly understood (Gajaseni and Gajaseni 1999). Moreover, the *pekarangan* systems also provide economic benefits for rural communities. Therefore, this research tried to analyze both economic aspects and biodiversity.

It is supported by some research that *pekarangan* in rural areas are an important factor for the economy and self-sufficiency of many households. The degree to which the *pekarangan* contribute to the provision of household food varies considerably and can only be tentatively determined (Wezel and Bender 2003). Through this research we found that *pekarangan* contribute up to 12.9 % of total income. Therefore, *pekarangan* can still has benefits for households in the future.

As stated by Soemarwoto (1987), more than 20 years ago but still relevant, *pekarangan* is a traditional agroforestry system with a promising future. Sustainable managed *pekarangan* have a significant role for the environment (Kaswanto et al. 2008; Harashina et al. 2003).

Nevertheless, we also assume that watersheds are also classified as protected areas, particularly at the upper stream area, because they basically formed as forest and conserve much biodiversity and water. Our study sites are on the watershed concept; we start to manage watershed from the microscale as well as *pekarangan*. When we preserve watershed, we also conserve the protected area, indirectly.

In term of regulation, the "segregation" of functions, however, was not complete and the boundaries remained contested. For the conservation stakeholders the primary way to achieve their goals was to increase the size and connectivity of "protected areas," while people continued to infringe on the national park (particularly on GGP and GHS) and contested the legality of forest allocation to logging concessions or conservation agencies. In part of the landscape, a more integrated and gradual transition from natural forest to human habitat survived for a number of reasons. Here, the villages, generally located below the natural forest, maintained an active interest in regularity of water flow and other ecosystem services that the forest provided. Maintaining a balance in such "integrated" landscapes depends on appropriate incentives, rather than the "command and control" approach of protected areas.

The LCS built by stakeholders in rural communities could further develop environmentally friendly activities, by passing down local wisdom for future generations. A rural LCS community, particularly those in remote areas still undisturbed by modern activity, have a way of knowledge on sustainable system management.

17.6 Conclusions and Recommendations

In conclusion, *pekarangan* can be one solution to conserve protected areas indirectly. In addition, some findings are also interesting: (1) *pekarangan* has high biodiversity even in a small area less than 400 m^2, (2) *pekarangan* gives a significant contribution to additional income, diet, and nutrition for a household, and (3) *pekarangan* is proven as one solution to keep protected areas/national parks more sustainable. The recommendations are (1) to revitalize more *pekarangan* around protected areas, and (2) to provide knowledge for marginal communities around protected areas about the benefits from *pekarangan* management.

Acknowledgments This research was supported by the Global Environmental Leaders (GELs) Education Program for Designing a Low Carbon Society (LCS) by Hiroshima University, Japan. Preliminary survey was a cooperative work between the Rural Development Institute (RDI), Seattle, USA and the Department of Landscape Architecture, Bogor Agricultural University, Indonesia. This paper has been presented in the 8th World Congress of the International Association for Landscape Ecology (IALE) on August 18–23, 2011 in Beijing, China.

References

Abdoellah O, Hadikusumah H, Takeuchi K, Okubo S, Parikesit (2006) Commercialization of homegardens in an Indonesian village: vegetation composition and functional changes. In: Kumar B, Nair P (eds) Tropical homegardens. Springer, Dordrecht, pp 233–250. doi:10.1007/978-1-4020-4948-4_13

Arifin HS, Nakagoshi N (2011) Landscape ecology and urban biodiversity in tropical Indonesian cities. Landsc Ecol Eng 7(1):33–43. doi:10.1007/s11355-010-0145-9

Arifin HS, Sakamoto K, Takeuchi T (2001) Study of rural landscape structure based on its different bio-climatic conditions in middle part of Citarum Watershed, Cianjur District, West Java, Indonesia. In: JSPS-DGHE (ed) Core university program in applied biosciences, Tokyo. The University of Tokyo, Japan, pp 99–108

Arifin HS, Munandar A, Mugnisjah WQ, Budiarti T, Arifin NHS, Pramukanto Q (2008) Homestead plot survey on Java. Rural Development Institute (RDI), Seattle

Christanty L, Abdoellah OS, Marten GG, Iskandar J (1986) Traditional agroforestry in West Java: the pekarangan (homegardens) and kebun talun (annual-perennial rotation) cropping systems. In: Marten GG (ed) Traditional agriculture in South-east Asia. Westview, Boulder, pp 132–158

Das T, Das A (2010) Litter production and decomposition in the forested areas of traditional homegardens: a case study from Barak Valley, Assam, northeast India. Agrofor Syst 79 (2):157–170. doi:10.1007/s10457-010-9284-0

Death R (2008) Margalef's index. In: Sven Erik J, Brian F (eds) Encyclopedia of ecology. Academic, Oxford, pp 2209–2210

Elliott KJ, Knoepp JD (2005) The effects of three regeneration harvest methods on plant diversity and soil characteristics in the southern Appalachians. For Ecol Manag 211(3):296–317. doi:10.1016/j.foreco.2005.02.064

Forman RTT, Godron M (1986) Landscape ecology. Wiley, New York

Gajaseni J, Gajaseni N (1999) Ecological rationalities of the traditional homegarden system in the Chao Phraya Basin, Thailand. Agrofor Syst 46(1):3–23. doi:10.1023/a:1006188504677

Harashina K, Takeuchi K, Tsunekawa A, Arifin HS (2003) Nitrogen flows due to human activities in the Cianjur-Cisokan watershed area in the middle Citarum drainage basin, West Java, Indonesia: a case study at hamlet scale. Agric Ecosyst Environ 100(1):75–90. doi:10.1016/S0167-8809(03)00173-7

Harjadi S (1989) Introduction to horticulture. Life Science Inter University Center, Bogor Agricultural University, Bogor

Hong J, Liu S, Shi G, Zhang Y (2012) Soil seed bank techniques for restoring wetland vegetation diversity in Yeyahu Wetland, Beijing. Ecol Eng 42:192–202. doi:10.1016/j.ecoleng.2012.01.004

Kaswanto, Arifin HS, Munandar A, Iiyama K (2008) Sustainable water management in the rural landscape of Cianjur Watershed, Cianjur District, West Java, Indonesia. J Int Soc Southeast Asian Agric Sci 14(1):33–45

Kaswanto, Nakagoshi N, Arifin HS (2010) Impact of land use changes on spatial pattern of landscape during two decades (1989–2009) in West Java region. Hikobia 15:363–376

Kumar BM (2011) Species richness and aboveground carbon stocks in the homegardens of central Kerala, India. Agric Ecosyst Environ 140(3/4):430–440. doi:10.1016/j.agee.2011.01.006

Kumar BM, Nair PKR (2004) The enigma of tropical homegardens. Agrofor Syst 61–62(1):135–152. doi:10.1023/B:AGFO.0000028995.13227.ca

Miller PR, Winer AM (1984) Composition and dominance in Los Angeles Basin urban vegetation. Urban Ecol 8(1-2):29–54. doi:10.1016/0304-4009(84)90005-6

Mitchell R, Hanstad T (2004) Small homegarden plots and sustainable livelihoods for the poor. FAO LSP Working Paper 11. Access to Natural Resources Sub-Programme. Rural Development Institue (RDI), Seattle, USA

Molebatsi LY, Siebert SJ, Cilliers SS, Struwig M, Kruger A (2010) A comparative floristic analysis of peri-urban and rural homegardens in North-West, South Africa. S Afr J Bot 76 (2):414. doi:10.1016/j.sajb.2010.02.076

Montagnini F (2006) Homegardens of Mesoamerica: biodiversity, food security, and nutrient management. In: Kumar B, Nair P (eds) Tropical homegardens. Springer, Dordrecht, pp 61–84. doi:10.1007/978-1-4020-4948-4_5

Nagendra H (2002) Opposite trends in response for the Shannon and Simpson indices of landscape diversity. Appl Geogr 22(2):175–186. doi:10.1016/s0143-6228(02)00002-4

Nair PKR, Kumar BM (2006) Introduction. In: Kumar B, Nair P (eds) Tropical homegardens. Springer, Dordrecht, pp 1–10. doi:10.1007/978-1-4020-4948-4_1

Niñez V (1984) Household gardens: theoretical considerations on an old survival strategy. Potatoes in Food Systems Research series report No. 1. International Potato Center, Lima, Peru

Roshetko JM, Delaney M, Hairiah K, Purnomosidhi P (2002) Carbon stocks in Indonesian homegarden systems: can smallholder systems be targeted for increased carbon storage? Am J Altern Agric 17(3):138–148

Roshetko J, Lasco R, Angeles M (2007) Smallholder agroforestry systems for carbon storage. Mitig Adapt Strat Glob Change 12(2):219–242. doi:10.1007/s11027-005-9010-9

Schultink G (2000) Critical environmental indicators: performance indices and assessment models for sustainable rural development planning. Ecol Model 130(1-3):47–58. doi:10.1016/s0304-3800(00)00212-x

Soemarwoto O (1984) The talun-kebun system, a modified shifting cultivation, in West Java. Environmentalist 4:96–98. doi:10.1007/bf01907300

Soemarwoto O (1987) Homegardens: a traditional agroforestry system with promising future. In: Steppler HA, Nair PKR (eds) A decade of development. ICRAF, Nairobi, pp 157–170

Tabuti JRS, Muwanika VB, Arinaitwe MZ, Ticktin T (2011) Conservation of priority woody species on farmlands: a case study from Nawaikoke sub-county, Uganda. Appl Geogr 31 (2):456–462. doi:10.1016/j.apgeog.2010.10.006

Takeuchi K (2010) Rebuilding the relationship between people and nature: the Satoyama initiative. Ecol Res 25(5):7. doi:10.1007/s11284-010-0745-8

Wezel A, Bender S (2003) Plant species diversity of homegardens of Cuba and its significance for household food supply. Agrofor Syst 57(1):39–49. doi:10.1023/a:1022973912195

Wiersum K (2006) Diversity and change in homegarden cultivation in Indonesia. In: Kumar B, Nair P (eds) Tropical homegardens. Springer, Dordrecht, pp 13–24. doi:10.1007/978-1-4020-4948-4_2

Chapter 18
Integrating the Aerial Photos and DTM to Estimate the Area and Niche of *Arundo formosana* in Jiou-Jiou Peaks Natural Reserve of Taiwan

Jeng-I Tsai and Fong-Long Feng

Abstract Earthquakes and typhoons have affected land use and land cover (LU/LC) in Taiwan, but an endemic grass, *Arundo formosana*, remains widely distributed. However, we lack knowledge about the niche of *A. formosana*. The purpose of this study was to estimate the area of *A. formosana* distribution by using scientific evidence and to describe its niche. In 2000, the Jiou-Jiou Peaks Natural Reserve was reported to be used to protect the unusual topography and complex biodiversity in the region. Several vegetation types can no longer grow in this region because of natural disturbances. However, *A. formosana* is able to grow. Because of the abundant roots and foliage of *A. formosana*, erosion is reduced. *A. formosana* can hang downward and thrive in crevices and cliffs, so its niche area might be underestimated if researchers ignored this characteristic. Today, most remote sensing images are two dimensional (2D), and sometimes 2D spatial information is insufficient to explain all natural phenomena. Therefore, we integrated ortho-aerial photographs and the digital terrain model (DTM) to estimate and analyze the niche area of *A. formosana*. The results indicated that the niche area of the species increased from 26.34 % to 32.86 % after the slope factor was considered. The results also exhibited that surfaces facing northeast, east, south, and southeast from 22.5° to 202.5° were more suitable for the growth of *A. formosana*. The slopes of the surfaces with *A. formosana* growth ranged from 0.00° to 81.82°, with an average of 53.99° ± 13.45°; slopes of 74° to 78° were the most suitable for *A. formosana* growth. The steeper peaks

---

J.-I Tsai
Department of Forestry, Graduate Student, National Chung Hsing University,
250, Kuo Kuang Rd, Taichung 402, Taiwan, R.O.C
e-mail: d9833003@mail.nchu.edu.tw

F.-L. Feng (✉)
Professor, Department of Forestry, National Chung Hsing University,
250, Kuo Kuang Rd, Taichung 402, Taiwan, R.O.C.
e-mail: flfeng@nchu.edu.tw

N. Nakagoshi and J.A. Mabuhay (eds.), *Designing Low Carbon Societies in Landscapes*, Ecological Research Monographs,
DOI 10.1007/978-4-431-54819-5_18, © Springer Japan 2014

(steeper than 79°) were still covered with bare soil. This niche information could provide land managers with valuable information for water and soil conservation and eco-technology.

Keywords *Arundo formosana* • DTM • Ortho-aerial photographs • Jiou-Jiou Peaks Natural Reserve • Niche

18.1 Introduction

The 921 (Chi-Chi) earthquake occurred in the Chi-Chi and the Sun Moon Lake area on the morning of 21 September 1999. It was a serious disturbance, with a magnitude of 7.3 on the Richter scale, and caused severe casualties and damage in Taiwan. The Jiou-Jiou Peaks area was one of the most seriously affected areas. Lin et al. (2001) indicated that the loose, unstable ground and the stones that had accumulated on the surface of the slopes were responsible for the occurrence of expanding mudflows and landslides during rainy seasons. Moreover, the amount of water retained in the soil for plants was critical for the recovery of vegetation in this area (Lin et al. 2005). The landscape still changes at the inner sides and outer edges of terraces every year (Chang 2000). According to data of Nantou Forest District Office in 2004, the vegetational composition differed before and after earthquake. Distinct types of pioneer species cover the landslides area of the Jiou-Jiou peaks, and *Arundo formosana* grows best in steeper slopes (Nantou Forest District Office 2004).

Arundo formosana (Formosan giant reed, pendent reed, Tagai, and Gaogan; called *A. formosana* later) is a grass endemic to Taiwan and an evergreen perennial herb. *A. formosana* is also considered a rock plant that can hang downward and thrive in rock crevices and cliffs. Because its seeds are small, light, and can float in the wind, the species is widely distributed in low altitudes in Taiwan [below 1,800 m mean sea level (MSL)], Lanyu, and Lutao. The reed is a high-pulp material, and the roots and stems are often used in the manufacture of handicrafts and baskets. The soil conservation capacity of *A. formosana* is an important function in preservation of the ecosystem and topography. We lacked information about the niche (ecological site) of this species and have tried to understand it. A researcher might underestimate the niche area of a rock plant because the plant can grow on steep slopes. For this reason, ortho-aerial photographs (spatial resolution, 0.25 m $\times$ 0.25 m) and a digital terrain model (DTM; 5 m $\times$ 5 m) were integrated in this study to analyze the distribution and niche of *A. formosana*. Field survey data were also collected for accurate assessment of image classification and identification.

18.2 Materials and Methods

18.2.1 Study Area

The Jiou-Jiou peaks area is situated on the north shore of Wu River of the middle course of Dadu River. Its steep slopes reduce human disturbance, so several species of wildlife could be protected, which increased ecosystem complexity and biodiversity. The 921 earthquake, Richter magnitude of 7.3, caused serious casualties and damage in Taiwan, and the Jiou-Jiou Peaks area was one of the most seriously affected areas. Taiwan Forestry Bureau announced the establishment of "Jiou-Jiou Peaks Natural Reserve" in accordance with Article 49 of the Cultural Heritage Preservation Act and Article 72 of the Enforcement Rules of the same law on 22 May 2000 to protect the unusual topography and complex biodiversity. This is the first natural reserve being designated for the purpose of preserving a collapsed cliff and ecological succession. This reserve, 1,198.4466 ha, 200–800 m altitude, is located between the 8th and 20th compartments of Puli Working Circle. Its range is from 225,420.000, 2,661,408.000 to 230,510.000, 2,652,100.000 m in the TWD97 coordinate system (Fig. 18.1). The rock formation is the Tou-Ke-Shan (Toukemountain) stratification, mostly formed by gravel, minor sandstone, and rock.

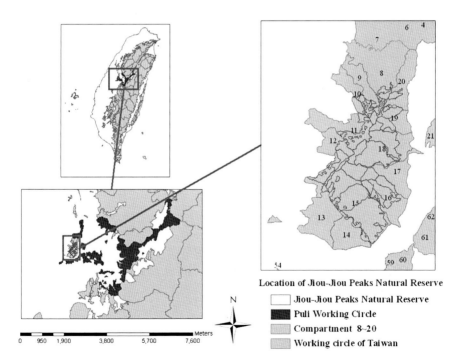

Fig. 18.1 Location of Jiou-Jiou Peaks Natural Reserve, 8th–20th compartments of Puli Working Circle, Taiwan

According to the Shuang Dung weather station, the average precipitation is 2,259.2 mm/year, mainly concentrated from May to August. Rainfall type is convective precipitation as thunderstorms and typhoons (Nantou Forest District Office 2004; Taiwan Forestry Bureau 2009).

18.2.2 *Arundo formosana*

Before the 921 earthquake, the forest vegetation zones were mainly classified into *Ficus–Machilus* forests, *Machilus–Castanopsis* forests, and forests dominated by some pioneer species in this reserve. The steeper slopes and dry environment reduced human disturbances, and precious species such as *Zelkova serrata*, *Schima superba*, *Liquidambar formosana*, *Ulmus parvifolia*, *Firmiana simplex*, *Cyclobalanopsis glauca*, and *Carpinus kawakamii* grow on the lower slopes (Liao 1992). The 921 earthquake changed the LU/LC completely. The area affected by landslides was covered by pioneer species such as *Trema orientalis*, *Sapium discolor*, *Mallotus paniculatus*, *Mallotus japonicus*, *Rhus chinensis*, *Fagara ailanthoides*, and *Macaranga tanarius*. *A. formosana* grows well in this area.

Arundo formosana can be classified into the category "grass" and the family Poaceae (subfamily Gramineae). It can hang downward and thrive in crevices and cliffs and often grows and forms a secondary grassland. The seeds of *A. formosana* are small, light, and can float in the wind. The species can be found at low altitudes in Taiwan, Orchid Island, and Green Island. The leaves are alternate, long-lanceolate, and papyraceous, with basal filamentous hairs. Leaf length is 10–20 cm and width is 1–2 cm. Flowering occurs from June to November as a panicle, terminal, light yellow spikelet, with three flowers about 0.6 cm in size. *A. formosana* is a high-pulp material, and the roots and stems are used in the manufacture of handicrafts and baskets in Taiwan (Endemic Species Research Institute 2003). Almost all the steep slopes of Jiou-Jiou Peaks Natural Reserve were covered by this species (Fig. 18.2).

18.2.3 *Image Classification and Accuracy Assessment*

Interpreters classify (LU/LC) easily by using image tone (color) and texture (Campbell 2002; Lillesand et al. 2007). The color of *A. formosana* was similar to that of other vegetation in some regions, so the texture was used to extract data for this species more effectively. Many speckle reduction filters (e.g., Lee, Lee-Sigma, Frost) assume a Gaussian distribution for speckle noise (Lee 1981; Frost et al. 1982). The maximum a posteriori (MAP) filter is used to attempt to estimate the original pixel digital number (DN), which is assumed to lie between the local average and the degraded (actual) pixel DN. MAP logic maximizes the a posteriori probability density function with respect to the original image (Frost et al. 1982). However, a recent study has shown this to be an invalid assumption.

Fig. 18.2 The particular distribution of *Arundo formosana*

Natural vegetation areas have been shown to be more properly modeled as having a gamma-distributed cross section. A textural image was created using 3 × 3 pixels and a moving window gamma filter. We established the distribution of *A. formosana* with color and texture.

Image classification is the major task of remote sensing (RS), including image analysis and pattern recognition. The maximum-likelihood classifier (MLC) of supervised classification provides higher accuracy assessment than minimum distance to means classifier, parallelepiped classifier, and unsupervised classification methods (Giannetti et al. 2001; Boles et al. 2004; Lillesand et al. 2007). Classification error matrix is one of the most common ways to express the classification accuracy of an image. Its overall accuracy can be assessed with the kappa coefficient (K), which is a calculation of agreement or accuracy for LU/LC classification (Congalton 1991; Congalton and Green 1999). The result of a kappa analysis is computed as follows:

$$\hat{k} = \frac{N \sum_{i=1}^{r} n_{ii} - \sum_{i=1}^{r} (n_{i+} \times n_{+i})}{N^2 - \sum_{i=1}^{r} (n_{i+} \times n_{+i})} \quad (18.1)$$

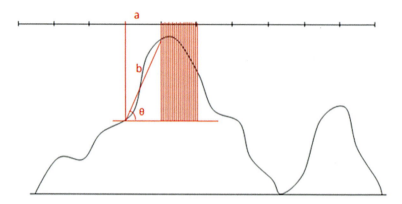

Fig. 18.3 Illustration of the hypotenuses in slope surfaces of different spatial scales: a = adjacent (read from RS data), b = hypotenuse, and θ = slope degree (obtained from raster image)

where r is the number of rows in the matrix, n_{ii} is the number of observations in row i and column i, n_{i+} and n_{+i} are the marginal totals for row i and column i, respectively, and N is the total number of observations (Bishop et al. 1975). This accurate assessment has been recognized as a useful tool in judging the fitness of remote sensing data. Janssen and Vanderwel (1994) pointed out that it should be higher than 0.7 to ensure a good classification result.

18.2.4 Estimation of Slope Area

An ortho-aerial photograph is helpful and effective in obtaining the status and estimation of (LU/LC) area in flat regions. However, *A. formosana* grows well in steep regions where trees cannot grow, and the terrain is extremely steep in this reserve. Therefore, the area of *A. formosana* growth might be underestimated if we use conventional methods to estimate growth area and if we consider the terrain in this study.

The slope degree was calculated with DTM by using Horn's algorithm (Horn 1981), which is the best estimation method (Skidmore 1989) in ArcGIS. The hypotenuses of all the pixels were calculated using the normal RS data and slope degrees. Figure 18.3 is an exaggerated example to explain the concept of the trigonometric function. The raster data of the hypotenuses were calculated pixel by pixel by using Eq. 18.2:

$$b = a \times \sec \theta \qquad (18.2)$$

where a = adjacent, b = hypotenuse, and θ = slope degree (slope degree $\div$ 180 $\times$ π).

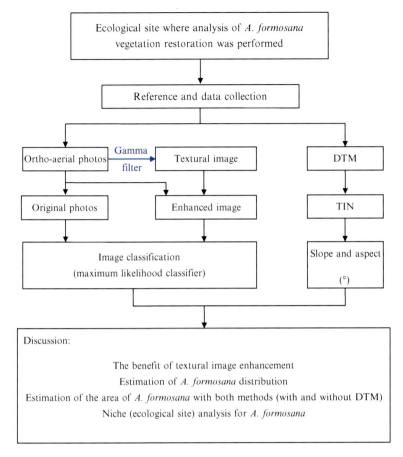

Fig. 18.4 Flowchart for the analysis procedure for *A. formosana* distribution, area, and niche

18.2.5 Niche Analysis for A. formosana

Although *A. formosana* is distributed widely below 1,800 m MSL in Taiwan, we still lacked niche information; therefore, we performed this study to gain an understanding of the niche. Slope and aspect were computed using Horn's algorithm. The aspect values were grouped into eight principal directions (north, northeast, east, southeast, south, southwest, west, and northwest). Data of the slope and aspect helped us to measure the terrain spatial distribution of *A. formosana*. The relationship between environmental factors (slopes and aspects) and (LU/LC) was evaluated using two-way analysis of variance (ANOVA) and cluster analysis. Above the synthesis states, we integrated the ortho-aerial photographs and DTM after considering the characteristics of *A. formosana*. A textural image was created with the gamma filter to help us to estimate the distribution accurately. The niche area of the species was also calculated using both methods [normal and three-dimensional (3D) methods], and we discussed the benefits of texture image enhancement and the 3D method. Then, the niche information was also obtained. Figure 18.4 is a flowchart that describes all these concepts.

18.3 Results and Discussion

18.3.1 Effect of Textural Enhancement on Image Classification

Figure 18.5 shows the original ortho-aerial photograph and enhanced images (the original photograph combined with the textural images). The MLC of supervised classification was adopted to discuss the benefit of this approach. Some environmental effects, such as shadow, aerosol, and terrain, create spectral variance and impact the accuracy of image classification. The spectral variance can not be corrected and reduced completely, even with the rapid development of the RS technique. Our results demonstrated that the enhanced image improved the classification results in shadowy regions; that is, the effect of image tone combined with texture is reduced shadow and terrain effect, although this approach did not work on nonreflective regions. The improvement in classification is shown in Fig. 18.6. Figure 18.7 illustrates the field survey data.

A total of 256 random points were evaluated, and the overall kappa value was 0.85 (Table 18.1). The kappa coefficients for forest, *A. formosana*, basin, and bare soil were 0.85, 0.81, 0.92, and 0.82, respectively. According to the study by Janssen and Vanderwel in 1994, the kappa value should be higher than 0.7. Therefore, the classification results could be applied to the analysis of environmental factors, allowing us to gain niche information.

18.3.2 Estimation of Niche Area for A. formosana

Because *A. formosana* is a rock plant, the DTM was integrated to estimate the niche area more accurately. The raster data of slope degrees and slope areas were calculated. The slopes ranged from $0.00°$ to $81.82°$; this means that the increases in the hypotenuse and area ranged from 1 to 8.73 fold (with trigonometric function method). The distribution of *A. formosana* was extracted from the original photographs and enhanced image. The RS data (0.25 m $\times$ 0.25 m) have a dissimilar spatial resolution compared to that obtained by DTM (5 m $\times$ 5 m). A pixel of the DTM image was separated into 400 subpixels to overlay raster data with two different resolutions appropriate. This system could be used to estimate the target area with the 3D method more easily.

18 Integrating the Aerial Photos and DTM to Estimate the Area and Niche... 321

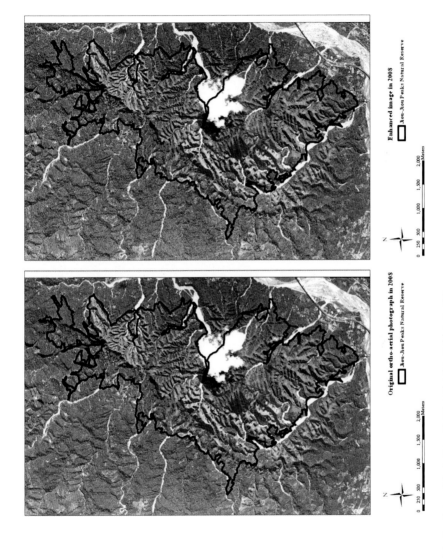

Fig. 18.5 Original ortho-aerial photograph (**a**) and enhanced image (**b**)

Fig. 18.6 Estimation results of original photographs and enhanced image. (*Punctiform polygons* are classification results with image tone; *Slash polygons* are results with image tone and texture)

Because of cloud cover, the total study area was recalculated using ArcGIS (from 1,198.45 ha to 1,138.79 ha). The niche area for *A. formosana* was estimated using normal and 3D methods (combined with DTM). The results, including two classification results (classification by only color and by both color and texture) and the 3D method results, are listed in Table 18.2. The texture could help us extract another 15.62 ha (1.37%) of the target. The niche areas for *A. formosana* were estimated to be

Fig. 18.7 Field survey data in part of the Jiou-Jiou Peaks Natural Reserve: specific terrain in study area (**a**) and land cover of different directions in parts of study area (**b**)

324 J.-I Tsai and F.-L. Feng

Table 18.1 Error matrix of land use image in Jiou-Jiou Peaks Natural Reserve

Classified data	Reference data				
	Forest	*Arundo formosana*	Basin	Bare soil	Row total
Forest	57	7	0	0	64
A. formosana	7	55	0	2	64
Basin	0	0	60	4	64
Bare soil	1	2	6	55	64
Column total	65	64	66	61	256
Overall classification accuracy = 88.67 %					

Table 18.2 Estimation of areas for *A. formosana* in Jiou-Jiou Peaks Natural Reserve

	A. formosana area (ha)	*A. formosana* area (%)	Total area (ha)
Image tone	284.38	24.97	1,138.79
Image tone and texture	300.00	26.34	1,138.79
3D method	628.07	32.86	1,911.26[a]

[a]Estimation of three-dimensional (3D) total area with the same 3D method

26.34 % and 32.86 % of the total investigated area by using the normal and 3D methods, respectively. This species can grow in steep regions where trees cannot. The (LU/LC) percentage of the species increased if the slope was considered; this was the reason why the area extended about 6.52 % with the 3D method.

18.3.3 Environmental Factors and Niche Analysis

We lacked niche information for *A. formosana*, so the environmental factors (slopes and aspects) were evaluated in this study. This information is valuable for land management decisions regarding soil and water conservation and eco-technology. The slope was measured in degrees in the image. The aspect values were grouped into eight principal directions (north, northeast, east, southeast, south, southwest, west, and northwest). We used two-way ANOVA and cluster analysis in SAS to analyze and recognize the niche of *A. formosana*. The results indicated that slope and aspect have significant impacts on the growth of this species ($P_r < 0.0001$).

A. *formosana* grew on slopes ranging from 0.00° to 81.82°, with an average of 53.99° ± 13.45° (mean ± standard deviation). Slopes of 74–78° are the most suitable for this species. The steeper peaks, where slopes were steeper than 79°, were still covered with bare soil (Fig. 18.7a). Cluster analysis demonstrated that northeast, east, south, and southeast directions from 22.5° to 202.5° were more suitable for *A. formosana* growth than the other directions. According to Liu and Su (1983), the south aspect is drier than the north in the Northern Hemisphere because of the radiance effect, and this might explain the directionality of *A. formosana* growth (Fig. 18.7b).

18.4 Conclusion

Earthquake and multitemporal typhoons caused serious landslides in the Jiou-Jiou Peaks Natural Reserve. Despite these natural disasters, *A. formosana* can still be found growing well in rock crevices. After observing its growth characteristics, we integrated ortho-aerial photographs and DTM to estimate its niche area accurately. A textural image was created using a gamma filter, and the original photographs were enhanced with this image. It is difficult to distinguish the variance between these images and observe the benefit of the enhancement via visual observation. However, our results showed that the textural image enhancement reduced the shadow effect of the image classification. Because of the ability of *A. formosana* to grow in steeply sloped areas, we estimated the growth distribution and niche area of *A. formosana* by using normal and 3D methods. The areas of *A. formosana* growth were estimated to be 26.34 % and 32.86 % of the investigated area by using normal and 3D methods, respectively. The reason for this discrepancy is that *A. formosana* thrives in steep regions where trees cannot generate and grow. Our results also suggest that slopes ranging from 74° to 78° and southeast to southwest aspects were the most suitable places for *A. formosana* growth. Several kinds of rock plants grow, generate, and thrive in Taiwan; thus, the methodology that we have described here can be applied to other similar species or regions. Moreover, our niche analysis provided valuable information for land managers making decisions regarding soil and water conservation and eco-technology.

Acknowledgments We thank the Nantou Forest District Office and Taichung Working Station for providing the field survey data.

References

Bishop YM, Fienberg SE, Holland PW (1975) Discrete multivariate analysis: theory and practice. MIT, Cambridge, p 575

Boles SH, Xiao X, Liu J, Zhang Q, Munkhtuya S, Chen S, Ojima D (2004) Land cover characterization of Temperate East Asia using multi-temporal VEGETATION sensor data. Remote Sens Environ 90:389–477

Chang SC (2000) The survey and designation of potentially landslide hazardous settlements after Chi-Chi earthquake. Proceedings of the second national conference on landslide stabilization and disaster prevention research in Taiwan, Taipei, Taiwan (in Chinese)

Campbell BJ (2002) Introduction to remote sensing, 3rd edn. Taylor and Francis, London, pp 124–127, 319–351

Congalton RG (1991) A review of assessing the accuracy of classifications of remotely sensed data. Remote Sens Environ 37(1):35–46

Congalton RG, Green K (1999) Assessing the accuracy of remotely sensed data: principles and practices. Lewis, Boca Raton

Endemic Species Research Institute (2003) The legend of Jiou-Jiou Peaks. Taiwan Endemic Species Research Institute (in Chinese)

Frost VS, Stiles JA, Shanmugan KS, Holtzman JC (1982) A model for radar images and its application to adaptive digital filtering of multiplicative noise. IEEE Trans Pattern Anal Mach Intell 4(2):157–166

Giannetti F, Montanarella L, Salandin R (2001) Integrated use of satellite images, DEMs, soil and substrate data in studying mountainous lands in supporting their sustainable development. Earth Observ Geoinf 3:25–29

Horn BP (1981) Hill shading and the reflectance map. Proc IEEE 69:14–47

Janssen LLF, Vanderwel JM (1994) Accuracy assessment of satellite derived land-cover data: a review. Photogram Eng Remote Sens 60(4):419–426

Lee JS (1981) Speckle analysis and smoothing of synthetic aperture radar images. Comput Graphics Image Process 17(1):24–32

Liao CC (1992) The studies of vegetation ecology and flora analysis in the area of Houyenshan in Shuandong of Nantou County in Taiwan. Bull Exp For Natl Chung Hsing Univ 14(1):1–60 (in Chinese)

Lillesand TM, Kiefer RW, Chipman JW (2007) Remote sensing and image interpretation, 6th edn. Taylor and Francis, London, pp 482–500

Lin CY, Wu JP, Lin WT (2001) The priority of revegetation for the landslides caused by the catastrophic Chi-Chi earthquake at Ninety-Nine peaks in NantouAera. J Chin Soil Water Conserv 32:59–66 (in Chinese)

Lin WT, Chou WC, Lin CY, Huang PH, Tsai JS (2005) Vegetation recovery monitoring and assessment at landslides caused by earthquake in central Taiwan. For Ecol Manag 210:55–66

Liu TS, Su HJ (1983) Forest ecology. Sangwu, Taipei, pp 47–50

Nantou Forest District Office (2004) Gorgenous mountains and water. Introduction to the protection areas, pp 12–17 (in Chinese)

Skidmore AK (1989) A comparison of techniques for calculating gradient and aspect from a grid digital elevation model. Int J Geogr Inf Syst 3(4):323–334

Taiwan Forestry Bureau (2009) http://www.forest.gov.tw/mp.asp?mp=1

Chapter 19
REDD+ Readiness Through Selected Project Activities, Financial Mechanisms, and Provincial Perspectives in Indonesia

Ima Yudin Rayaningtyas and Nobukazu Nakagoshi

Abstract Climate change has become the international issue of the moment. Because most of Indonesia's emissions result from deforestation, land degradation, inappropriate land uses, and land use conversion, reducing emissions from deforestation and forest degradation (REDD)+ has become an eminent priority for Indonesia. This study primarily draws on insights from national and subnational experiences on REDD+ readiness in Indonesia. A literature review, web-based searches, a questionnaire, and in-depth interviews were applied to compile data. Comprehending the REDD+ current status and the challenge of a REDD+ financial mechanism were obtained from reviewing secondary data. The provincial level perspective on REDD+ was acquired through assessment of the questionnaire and in-depth interview results. This study showed that REDD+ issues are evolving quickly and several established regulations are as yet inadequate to cope with REDD+ uncertainties. Hence, remaining issues related to strategy, capacity building, methodology, regulation, and financing schemes need to be developed and strengthened to succeed in REDD+ implementation. In addition, this study revealed that the knowledge level on REDD+ issues between stakeholders at the province level was comparable. However, the stakeholders at the province level had a tendency to have different opinions related to REDD+ issues.

Keywords Deforestation • Knowledge level • Opinion • Readiness • REDD+ working group • REDD+

I.Y. Rayaningtyas (✉) • N. Nakagoshi
Graduate School for International Development and Cooperation, Hiroshima University, 1-5-1 Kagamiyama, Higashi-Hiroshima 739-8529, Japan
e-mail: rayaningtyas@gmail.com

N. Nakagoshi and J.A. Mabuhay (eds.), *Designing Low Carbon Societies in Landscapes*, Ecological Research Monographs, DOI 10.1007/978-4-431-54819-5_19, © Springer Japan 2014

19.1 Introduction

19.1.1 Background

Approximately 15–20 % of global greenhouse gas (GHG) emissions are ascribed to deforestation and forest degradation. In addressing this problem, the United Nations Framework Convention on Climate Change (UNFCCC) established an international mechanism using carbon market or financial incentives to reduce CO_2 emissions from deforestation and forest degradation, which is referred to as REDD+. REDD+ is becoming one of the main pillars of a post-2012 international climate regime (Corbera and Schroeder 2011). REDD has also the potential to reduce emissions generated by the forest sector and simultaneously creating sustainable development benefits for communities in tropical countries (Cerbu et al. 2011).

At the moment, REDD+ is considered as the most promising mechanism for conserving tropical forest, even though after 6 years of discussions and negotiations the REDD+ mechanisms is yet to be endorsed (Venter and Koh 2012). REDD+ projects currently face uncertainty over future demand for carbon credits, the potential for inconsistent donor support in the long term, carbon market volatility, investor preference for low-cost emissions mitigation over co-benefits, and the possibility of a short-lived REDD+ mechanism (Phelps et al. 2011). The success of REDD will ultimately depend on the existence of national arrangements that can deliver emission reductions at scale (Streck 2010).

As mentioned by UN-REDD Programme, Indonesia's major emissions result from deforestation, land degradation, inappropriate land uses, and land use conversion. REDD+ is a good start to tackling all these issues, but it is not a short-term solution; it will most likely involve massive political trade-offs, and it will not be implemented on a large scale in the near future (Lederer 2012). The highest deforestation rate in Indonesia accounted for approximately 2.83 million ha/year and occurred during 1997–2000 because of forest fire. Afterward, the deforestation rate dropped off to 1.08 million ha/year throughout 2001–2003. Unfortunately, for the period of 2004–2006, the deforestation rate increased to 1.17 million ha/year. Hence, REDD+ has become an eminent priority for Indonesia.

Because the REDD+ initiatives are positively noticed by many stakeholders, Indonesia must encounter many challenges to enable REDD+ implementation (Indonesia Task Force 2012). During the readiness phase, 2009–2012, Indonesia needs to prepare methodologies and policies to facilitate REDD+ implementation in Indonesia. However, it is still insufficient, and there are many remaining elements that need to be developed, strengthened, and synchronized to confront REDD+ full implementation post 2012. Up to now, the scientific reference on REDD+ is still focusing on international and national issues. Discussion on subnational or provincial experience on REDD+ readiness is still limited. The establishment of appropriate and complementary REDD mechanisms at the international and national level as well as the establishment of appropriate and complementary projects at the subnational level will determine REDD success (Blom et al. 2010). Therefore, this study examines REDD+ activities at the national and provincial level in Indonesia.

19.1.2 Objectives

This study primarily recognize national and subnational experiences on REDD+ readiness with the following objectives:

a. To acknowledge the current status of REDD+ in Indonesia
b. To comprehend the challenge of a REDD+ financial mechanism in Indonesia
c. To assess the knowledge level and realize the opinion on the REDD+ program in the provincial level: case study of Central Sulawesi Province REDD+ Working Group

19.2 Study Area

There are 46 REDD+ projects in Indonesia. Indonesia is one of the pilot countries, undertaking the UN-REDD Programme Indonesia. This project is jointly managed by the United Nations Environment Programme (UNEP), the United Nations Development Programme (UNDP), the Food and Agriculture Organization (FAO), and the Ministry of Forestry of Indonesia.

After public discussions involving stakeholders at both national and provincial levels, the Ministry of Forestry designated Central Sulawesi Province as an UN-REDD Programme Indonesia project site in 2010. Central Sulawesi Province was selected for the following reasons: a 60 % forest cover with a 2 % rate of deforestation and forest degradation, a relatively high carbon density, a supportive local government, and because Sulawesi is still underrepresented in REDD projects compared to Kalimantan and Sumatera (UNREDD Programme 2010).

The objective of the UN-REDD Programme Indonesia is "to support Indonesia in attaining REDD-Readiness." More specifically, this implies strengthening multi-stakeholder participation and consensus at the national level, to succeed in demonstration of establishing a REL, MRV, and fair payment systems based on the national REDD+ strategy, to provide a toolkit for priority setting toward carbon benefits and co-benefits, and to develop capacity building to implement REDD+ at local levels.

The UN-REDD Indonesia Programme was selected as a case study because this project was the only project in Indonesia that is managed by the United Nations organizations (UNDP, UNEP, FAO) and this project has established a REDD+ working group. Therefore, this project has become a role model of REDD+ implementation in Indonesia

19.3 Methods

The data and information were collected from October 2010 to July 2012. Through desk review, the institution, regulation/policy, strategy, and capacity building related to REDD+ were identified and analyzed. Comprehending the REDD+ current status and the challenge of a REDD+ financial mechanism were obtained from reviewing secondary data. The provincial level perspectives on REDD+ were acquired through assessment of the questionnaire and in-depth interview results.

The first survey in March 2011 was accomplished over 3 weeks. There were several discussions and consultations with multiple stakeholders, namely the Chief of Forestry Services for Central Sulawesi Province, the Chief of Natural Resource Conservation Agency (NRCA), the Chief of Lore Lindu National Park, the Chief of Forestry Planning Agency Regional XVI, and REDD+ working group members.

The second survey was carried out in September 2011. Over 4 weeks, a deliberately purposive random sampling questionnaire was given to the REDD+ working group members. A supplementary interview to confirm answers and opinions was also conducted. The questionnaire had three parts: knowledge, opinions, and open questions. Thirty-six questions of the knowledge section and 21 questions of the opinions part were examined and analyzed. There were four open questions related to their observations of the REDD+ programme in Central Sulawesi Province. Seventy questionnaires were disseminated and 30 questionnaires were returned. Of these, 26 questionnaires were analyzed, which consisted of responses from government, the academic, local community, local NGOs, and the private sector. Four questionnaires were discarded because of invalidity. Descriptive statistics, multiple linear regression, and analysis of variance (ANOVA) were generated using XLSTAT and Predictive Analytics Software (PASW) 18. To distinguish between the opinions of stakeholders, chi-square analysis was used with the following formula:

$$X^2 = \sum \frac{(O - E)^2}{E} \tag{19.1}$$

where O is the observed frequency, E is the expected frequency, and χ^2 is the chi square.

To classify which part of REDD+ issues the stakeholders acknowledged, the following classification has been proposed:

a. *Red level*: if the average percentage of yes answers to the knowledge questions is less than 25 %, it reveals that the respondents have very poor knowledge.
b. *Yellow level*: if the average percentage is in the range 25 % $\leq n$ <50 %, it shows that the respondents have poor knowledge.
c. *Green level*: if the average percentage is in the range 50 % $\leq n$ < 75 %, it indicates that the respondents have sufficient knowledge.
d. *Blue level*: if the average percentage is greater than 75 %, it indicates that the respondents have good knowledge.

With the purpose of recognizing the REDD+ working group knowledge level, age, gender, education, group of stakeholders, and level of participation were analyzed. Because the REDD+ concerns evolving issues and international negotiations on REDD+ still continue, some collected information and project activities may have changed from their main objectives and activities. In spite of these deficiencies, this study provides current development of REDD+ readiness in Indonesia at both the national level and the provincial level.

19.4 Results

19.4.1 REDD+ Current Status in Indonesia

The current status of REDD+ in Indonesia was obtained from reviewing secondary data. This section recorded the REDD+ development during the past 7 years and its strategy in Indonesia.

The emergence of REDD+ and the establishment of institutions, regulations, and information related to REDD+ from 2005 to 2012 are presented in Fig. 19.1.

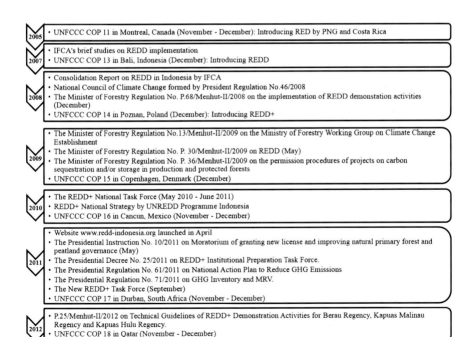

Fig. 19.1 REDD+ emergence in Indonesia

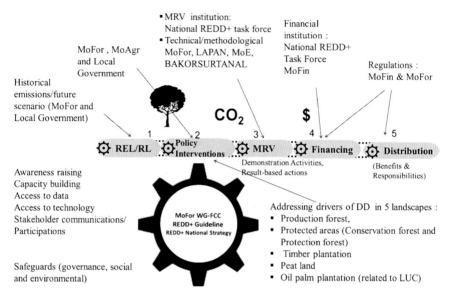

Fig. 19.2 REDD+ strategy in Indonesia. [From IFCA Study (Ministry of Forestry 2008)]

The Indonesia REDD+ strategy comprises five key elements: reference emission level (REL) or reference level (RL), policy intervention, MRV mechanism, financing, and distribution (Fig. 19.2).

19.4.2 Challenge of REDD+ Financial Status in Indonesia

At this time, REDD+ projects are funded by international and national funds. The Government of Indonesia obtained international funds through bilateral cooperation (Australia, Germany, Japan, Norway, etc.) and multi-donors [the World Bank, the International Tropical Timber Organization (ITTO), the United Nations Environment Programme (UNEP), the United Nations Development Programme (UNDP), the Food and Agriculture Organization (FAO), and The Nature Conservancy (TNC)]. The overview of climate change funds in term of REDD+ is shown in Fig. 19.3.

19.4.3 Provincial Perspectives on REDD+

The provincial level perspective on REDD+ was acquired through assessment of the questionnaire and in-depth interview results. This section explains the status of the REDD+ working group in Central Sulawesi Province, and also their knowledge level and opinion on REDD+.

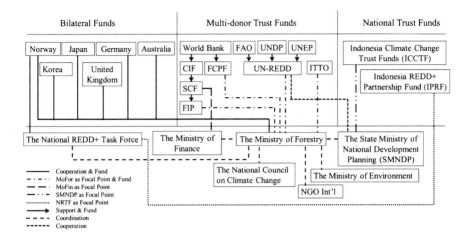

Fig. 19.3 Overview of climate change funds in terms of REDD+

19.4.3.1 REDD+ Working Group of Central Sulawesi Province

According to Central Sulawesi Province (2011), on February 18, 2011, there were 77 REDD+ working group members in total, 4 members of the advisory board, 8 coordinators, and 65 members of the working group. However, just over 50 % of members were actively involved in the UN-REDD Programme Indonesia. Each sub-working group was represented by multi-stakeholders, composed of government, academic, local NGOs, local community, and private sectors. This composition enabled the working group members to share their knowledge and concerns from different perspectives. The sub-working groups were in authority of the following tasks:

a. Sub-working group I is responsible for formulating the local strategy to support REDD+ implementation.
b. Sub-working group II prepares the institutional and methodological strategy.
c. Sub-working group III is in charge of developing criteria and indicators for demonstration activities.
d. Sub-working group IV is accountable for enabling capacity building to attain consensus on the basis of free, prior, and information consent (FPIC).

19.4.3.2 Knowledge Level of REDD+ Working Group

The 36 questions in the knowledge section of the questionnaire encompassed knowledge of terminology, regulations, and activities related to the REDD+ program. The questions were used to assess the knowledge level of REDD+ working group members. The number of 'yes' answers were counted and averaged as a percentage to obtain a score. According to each type of stakeholder, the results in

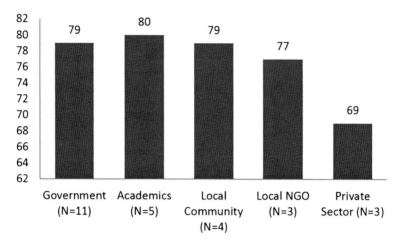

Fig. 19.4 Level of knowledge between stakeholders

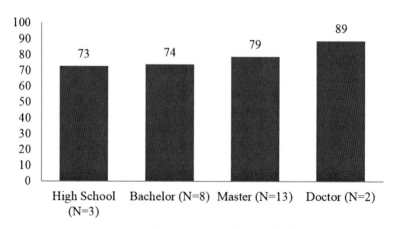

Fig. 19.5 Level of knowledge based on education

Fig. 19.4 show that the academics had the highest knowledge of REDD+, attaining 80 %, followed by the government and the local community, who obtained 79 %. The local NGO achieved 77 % and the private sector had the least knowledge, scoring only 69 %.

In relationship to education level, the knowledge level of holders of doctor's, master's, and bachelor's degrees and high school graduates were 89 %, 79 %, 74 %, and 73 %, respectively (Fig. 19.5).

19 REDD+ Readiness Through Selected Project Activities, Financial Mechanisms... 335

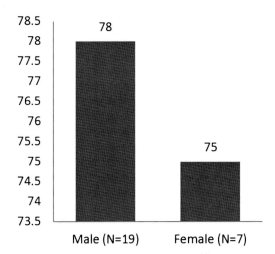

Fig. 19.6 Level of knowledge based on gender

Regarding gender, men achieved 78 % and women gained 75 % (Fig. 19.6).

After categorizing the questions into four parts, the results are as follows (Fig. 19.7):

a. There are two questions in the red level (6 %): Project Idea Note (PIN) or Concept Note; and Have you ever read/known the standard of project development and/or carbon market mechanism outside REDD+?
b. There are eight questions in the yellow level (22 %): Project Design Document (PDD); Voluntary Carbon Standard (VCS); Certified Emission Reduction (CER); Leakage; Business as Usual (BAU); Structure of REDD+ Financial Mechanism; Ministry of Forestry (2009a); and Measurement above and below ground carbon.
c. There are five questions in the green level (14 %): Carbon Offset; Minister of Forestry Regulation No. P. 68/Menhut-II/2008 on demonstration activities (DA); Ministry of Forestry (2009b); National Action Plan on Climate Change; and Total of carbon emissions from forestry sector.
d. The 21 questions in the blue level (58 %): Have you ever heard of REDD+ mechanisms/program?; Do you follow REDD+ progress?; REDD/REDD+ meaning; The differences between REDD and REDD+; Reference Emissions Level (REL); Measureable Reportable Verifiable (MRV); Demonstration Activities (DA); REDD+ Working Group; Co-benefit of REDD+; United Nations Framework Convention on Climate Change (UNFCCC); UNREDD Programme Indonesia; Cleaning Development Mechanism (CDM); Safeguards; Free Prior and Informed Consent (FPIC); Carbon stock; Kyoto Protocol; Climate change mitigation; Climate change adaptation; Greenhouse gas emissions; REDD+

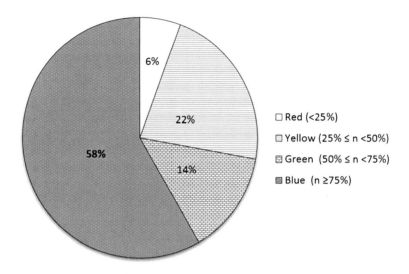

Fig. 19.7 Categorization of knowledge questions

Table 19.1 Model summary of knowledge level

				SE of the estimate	Change statistics					
Model	R	R^2	Adjusted R squared		R squared change	F change	df 1	df 2	Significant F change	Durbin–Watson
1	0.704[a]	0.496	0.370	8.12751	0.496	3.938	5	20	0.012	2.169

[a]Predictors: (constant), participation, education, gender, age, stakeholder
Dependent variable: knowledge

Table 19.2 Analysis of variance (ANOVA) of knowledge level

Model	Sum of squares	df	Mean square	F	Significance
Regression	1,300.536	5	260.107	3.938	0.012[a]
Residual	1,321.128	20	66.056		
Total	2,621.664	25			

[a]Predictors: (constant), participation, education, gender, age, stakeholder
Dependent variable: knowledge

National Strategy; and President Soesilo Bambang Yudhoyono's statement that Indonesia will reduce emissions 26 % by 2020 and by 41 % with international support.

Statistical analysis using the ANOVA calculation in Tables 19.1 and 19.2 showed that knowledge as a dependent variable could be explained by participation, education, gender, age, and stakeholder as independent variables with a significance of 0.012 and an R^2 of 0.496.

19 REDD+ Readiness Through Selected Project Activities, Financial Mechanisms... 337

The model of knowledge level was as follows:

$$LK = 60.138 - (0.365 * SH) + (1.213 * EDU) - (0.411 * AGE) \\ + (3.089 * GDR) + (1.053 * PAR)$$

where LK is level of knowledge; SH is stakeholder; EDU is education; AGE is age; GDR is gender; and PAR is participation (Table 19.3).

REDD+ working group members obtained REDD+ issues information from a range of sources. Ninety-two percent of members obtained REDD+ information from seminars, workshops, and focus group discussions, 62 % from books and the Internet, 46 % from leaflets or brochures, 42 % from newspapers, and 21 % from television. This result revealed that knowledge on REDD+ issues was highly affected by participation in seminars, workshops, and focus group discussions. In addition, 54 % of working group members were of the opinion that the amount of REDD+ information provided was sufficient.

19.4.3.3 Opinion of REDD+ Working Group

The opinions of REDD+ working group members were assessed by asking 21 questions using a Likert scale from 1 (strongly disagree) to 5 (strongly agree). Data were analyzed using ANOVA (Table 19.4), chi-square test, and multiple comparisons least-square distance (LSD) to distinguish whether all stakeholders had the same opinion on particular REDD+ issues (Table 19.5). The statistical analysis of the chi-square test was 2.5×10^{-13} and the significance was 0.034 (Table 19.4). The results confirmed that the respondents had varying opinions. The variations between stakeholders are summarized in Table 19.5.

19.5 Discussion

19.5.1 REDD+ Current Status in Indonesia

Indonesia is now in the second phase or readiness phase of REDD+. In this phase, Indonesia is preparing REDD+ methodologies and policies to facilitate REDD+ implementation. Romijn et al. (2012) stated that Indonesia, one country at the forefront of REDD+ framework concerns, positively has a high proportion of forest area with high soil carbon content, high national engagement in the UNFCCC REDD+ process, very good monitoring capacities in forest area change, good forest inventory capacity, and low remote sensing technical challenges. However, the completeness of greenhouse gas inventory and forest area that is affected by fire is still at a medium level. Furthermore, carbon pool reporting capacity is limited.

An effective REDD+ mechanism for Indonesia must address the decision-making processes and the categorization of conversion and production forest.

Table 19.3 Coefficients of knowledge level

Model		Unstandardized coefficients		Standardized coefficients			Correlations			Collinearity statistics	
		B	SE	Beta	t	Significance	Zero-order	Partial	Part	Tolerance	VIF
1	(Constant)	60.138	18.103		3.322	0.003					
	Stakeholder	−0.365	1.610	−0.051	−0.226	−0.823	−0.240	−0.051	−0.036	0.494	2.023
	Education	1.213	0.887	0.268	1.367	0.187	0.354	0.292	0.217	0.656	1.524
	Age	−0.411	−0.232	−0.324	−1.771	0.092	−0.185	−0.368	−0.281	0.753	1.328
	Gender	3.089	4.076	0.136	0.758	0.457	0.143	0.167	0.120	0.777	1.287
	Participation	1.053	0.294	0.606	3.582	0.002	0.511	0.625	0.569	0.881	1.135

Dependent variable: knowledge

19 REDD+ Readiness Through Selected Project Activities, Financial Mechanisms... 339

Table 19.4 Analysis of variance (ANOVA) of opinion

Model	Sum of squares	df	Mean square	F	Significance
Between groups	7.560	4	1.890	2.626	0.034
Within groups	388.697	540	0.720		
Total	396.257	544			

Table 19.5 Multiple comparisons: least square distances (LSD) of opinion

$(I)1$	$(J)1$	Mean difference $(I-J)$	Standard error	Significance	95 % confidence interval	
					Lower bound	Upper bound
1	2	−0.112	0.100	0.262	−0.31	0.08
	3	0.223	0.108	0.039	0.01	0.44
	4	−0.074	0.121	0.539	−0.31	0.16
	5	0.180	0.121	0.137	−0.06	0.42
2	1	0.112	0.100	0.262	−0.08	0.31
	3	0.336	0.124	0.007	0.09	0.58
	4	0.038	0.135	0.778	−0.23	0.30
	5	0.292	0.135	0.031	0.03	0.56
3	1	−0.223	0.108	0.039	−0.44	−0.0.1
	2	−0.336	0.124	0.007	−0.58	−0.09
	4	−0.298	0.141	0.036	−0.58	−0.02
	5	−0.044	0.141	0.758	−0.32	0.23
4	1	0.074	0.121	0.539	−0.16	0.31
	2	−0.038	0.135	0.778	−0.30	0.23
	3	0.298	0.141	0.036	0.02	0.58
	5	0.254	0.151	0.094	−0.04	0.55
5	1	−0.180	0.121	0.137	−0.42	0.06
	2	−0.292	0.135	0.031	−0.56	−0.03
	3	0.044	0.141	0.758	−0.23	0.32
	4	−0.254	0.151	0.094	−0.55	0.04

Note: (1) the government; (2) the academics; (3) the local communities; (4) the local nongovernmental organization; and (5) the private sector

The deficiency of clarity in land allocation procedures, especially over conversion forest, may cause high transaction costs and consequently limited attractiveness for potential investors (Brockhaus et al. 2012). If REDD+ is to be successful in supporting the alignment of policies across state entities, it will require the development of coordination and conflict negotiation mechanisms under top-level political leadership (Larson and Petkova 2011).

19.5.1.1 REDD+ Development in Indonesia

Papua New Guinea and Costa Rica called for the inclusion of reducing emissions from deforestation in developing countries (RED) at UNFCCC COP 11, 2005, in

Montreal, Canada. Afterward, the Indonesia Forest Climate Alliance (IFCA) carried out brief studies on Indonesia's preparedness to implement REDD methodologically and institutionally. At UNFCCC COP13, 2007, in Bali, Indonesia, the additional "D," which stands for forest degradation was added; consequently, it was renamed REDD.

In 2008, IFCA published a Consolidation Report on REDD in Indonesia, following the Ministry of Forestry, who developed a REDD+ strategy and readiness strategy from 2009 to 2012. In addition, the Ministry of Forestry also formulated some REDD regulations and established a climate change working group that was amended three times. Furthermore, the National Council of Climate Change was established based on President Regulation No. 46/2008.

During 2008 to 2009, at UNFCCC COP 14 and COP 15, respectively, discussions on REDD continued in a wider scope, which was reflected in the acronym REDD+. The "plus" symbolizes the consideration of the role of conservation, sustainable management of forests, and enhancement of forest carbon stocks. Indonesia started to implement the readiness strategy of REDD+ in 2009.

In 2010, the REDD+ National Task Force established by the President and the Ministry of Forestry developed the National REDD+ Strategy. Indonesia's official REDD+ program website www.redd-indonesia.org was launched in April 2011. Currently, there are more than 46 REDD+ projects in Indonesia. Thus far several regulations have been formulated to support the implementation of REDD+.

19.5.1.2 REDD+ Strategy

Although the design of frameworks, mechanisms, and arrangements to implement REDD programs have received significant attention, it is not yet clear how REDD+ will function on the ground or how the participation of local populations will be assured (Cronkleton et al. 2011). Indonesia's REDD+ strategy consists of five elements, being the following:

a. Reference emission level (REL) or reference level (RL)

The international perspective of REL/RL is the amount of gross emissions from a geographic area estimated within a reference time period: reducing emission from deforestation and forest degradation (REL) and conservation, sustainable forest management, and enhancing forest carbon stock (RL). REL/RL should be based on historical data of land uses, GHGs emission/removals, and socioeconomic variables, and once set, it cannot be changed during the implementation period (UNFCCC 2009).

Referring to the Ministry of Forestry (2009a), REDD+ in Indonesia is being implemented through a national approach that is carried out at the subnational level (provincial, district, or management unit). Thus, REL is set at the national, subnational, and project site levels. REL at the national level will be established by the Ministry of Forestry and at the subnational level will first be settled by local governments (provincial or district) before finally being confirmed with national REL. REL at the project site location will be determined and then confirmed with the national and subnational REL.

b. Policy intervention

From the time REDD mechanisms were introduced, the government of Indonesia has been actively involved in international discussions and negotiations. To support REDD+ implementation, the government has formulated some regulations, such as The Minister of Forestry Regulation No. P.68/Menhut-II/2008 on the implementation of REDD demonstration activities; Ministry of Forestry (2009a); Ministry of Forestry (2009b); The Minister of Forestry Regulation No.13/Menhut-II/2009 on the Establishment of the Ministry of Forestry Working Group on Climate Change; The Presidential Instruction No. 10/2011 on Moratorium of granting new license and improving natural primary forest and peat land governance; The Presidential Regulation No. 61/2011 on National Action Plan to Reduce Greenhouse Gas Emissions; The Presidential Regulation No. 71/2011 on Greenhouse Gas Inventory and MRV; The Presidential Decree No. 25/2011 on REDD+ Institutional Preparation Task Force; The Minister of Forestry Regulation No.25/Menhut-II/2012 on Technical Guidelines of REDD+ Demonstration Activities for Berau Regency, Kapuas Malinau Regency and Kapuas Hulu Regency.

Nonetheless, those regulations are still in general and are not comparable with the rapidly evolving REDD+ issues. Hence, those regulations are insufficient in answering the REDD+ challenge.

c. Measurable, reportable, verifiable (MRV)

MRV was introduced during COP 13 UNFCCC in Bali, 2007 and outlined in the UNFCCC (2007) paragraph 1b (i) and (ii) as follows:

(i) is measurable, reportable and verifiable nationally appropriate mitigation commitments or action, including quantified emission limitation and reduction objectives by all developed country parties, while ensuring the comparability of efforts among them, taking into account differences in their national circumstances;

(ii) is nationally appropriate mitigation actions by developing country parties in the context of sustainable development, supported and enabled by technology, financing and capacity-building in a measurable, reportable and verifiable manner.

Since 2007, MRV discussions have been held on what is MRV, who should conduct it, and how it will be applied in both developed and developing countries. For Indonesia, MRV will be carried out by the National GHG Inventory System Center (NGISC) under the supervision of the Ministry of Environment. The NGISC will coordinate with all stakeholders to formulate a document on GHG emissions. The stakeholders will be supported at both the provincial and district level, as coordinated by the Governor of each province. Meanwhile, all stakeholders will submit a climate change mitigation plan to the National Planning Agency, which will then be validated by an expert panel under the coordination of the Ministry of Environment. Each plan will then be

Table 19.6 Benefit distribution of carbon credit based on forest type

	Distribution (%)		
Permit holder/developer	Government	Community	Developer
IUPHHK-HA (Timber Forest Product for Natural Forest)	20	20	60
IUPHHK-HT (Timber Forest Product for Plantation Forest)	20	20	60
IUPHHK-RE (Timber Forest Product for Ecosystem Restoration)	20	20	60
IUPHHK-HTR (Timber Forest Product for People Plantation Forest)	20	50	30
Hutan Rakyat (People Forest)	10	70	20
Hutan Kemasyarakatan (Community Forest)	20	50	30
Hutan Adat (Customary Forest)	10	70	20
Hutan Desa (Village Forest)	20	50	30
KPH (Forest Management Unit)	30	20	50
KHDTK (Special Purpose Forest Area)	50	20	30
Hutan Lindung (Protection Forest)	50	20	30

Note: The government shares will be divided proportionally between central, provincial and district level about 40 %, 20 % and 40 % respectively
Source: Ministry of Forestry (2009b)

incorporated into the National Medium-term Development Plan, the National Long-term Development Plan, and the National Report on the Implementation of Mitigation Actions and Reduction Targets as part of national communication on climate change (Ministry of Environment 2009). In addition, greenhouse gas inventory and MRV were outlined in Presidential Regulation No. 71/2011.

d. Financing

The proposed financial scheme and REDD+ payment scheme reflect the multi-source fund management system (with multiple outlets/windows), multi-recipients (private, community, local government), and multiple purposes (input, output/investment, and to the success of policy or strategic initiatives), to achieve REDD+ objectives. This fund will be used to reduce emissions from deforestation and forest degradation, in addition to the generation of additional benefits (co-benefits) such as conservation of biodiversity (UNREDD Programme Indonesia 2011).

The categories of financing comprise national REDD+ strategies, REDD+ action plans and capacity building, development of MRV systems and regulatory reforms, demonstration activities, and performance-based payments for emissions reductions. A REDD+ financial mechanism is offered through carbon market and financial incentives.

e. Benefit distribution

Related to Annex III of the Ministry of Forestry (2009b), the benefit distribution of carbon credit based on forest type is shown in Table 19.6.

19.5.2 Overview of Climate Change Funds in Terms of REDD+

Focusing on the REDD+ financing mechanism, there are three main funds, namely, bilateral funds (Norway, Japan, Germany, Australia, Korea, United Kingdom, etc.), multi-donor trust funds (the World Bank and United Nations bodies), and national trust funds [Indonesia REDD+ Partnership Fund (IRPF) and Indonesia Climate Change Trust Fund (ICCTF)]. These funds are managed by government institutions, namely, the Ministry of Finance, the Ministry of Forestry, the State Ministry of National Development Planning, and the National REDD+ Task Force.

As a follow-up on the Letter of Intent (LOI) between Indonesia and Norway, the government has established the IRPF system to facilitate REDD+ implementation programs in Indonesia through financing services and fund distribution. Previously, the Indonesia Climate Change Trust Fund (ICCTF) has been launched to support a wider scope of climate change issues, constituting energy, transportation, agriculture, etc.

Concerning financial institutions and funding instruments, the Indonesian REDD+ Partnership Fund (IRPF) and its implementation system was set up to facilitate REDD + implementation programs in Indonesia permanently through financing services and distributing funds. IRPF includes both public and private funds, both nationally and internationally, which reach REDD+ beneficiaries at the community level within and surrounding forest areas. IRPF will start with a managing grant from Norway and will develop with additional funds from multiple sources (UNREDD Programme Indonesia 2011).

19.5.3 Provincial Perspectives on REDD+

This section discusses the status of the REDD+ working group in Central Sulawesi Province, and also their knowledge level and opinions on REDD+.

19.5.3.1 REDD+ Working Group of Central Sulawesi Province

The REDD+ working group is responsible for developing REDD+ infrastructure and social, economic, and environmental safeguard mechanisms to enable Central Sulawesi to be 'ready' for REDD+ implementation. By the end of 2011, sub-working group 1 had formulated the draft of REDD+ local strategy. It is a guideline for REDD+ implementation at the provincial level. Sub-working group 2 is developing the institutional planning and methodology to undertake REL/RL and MRV mechanisms at the provincial level. Sub-working group 3 has developed the criteria and indicators for demonstration activities. Five areas have been nominated, namely, Dampelas region in Donggala district, Tinombo region in

Parigi Moutong district, Lore Lindu National Park in Poso district, Tojo Una-una, and Toli-toli (Forest People Programme 2011). Meanwhile, sub-working group 4 is developing and testing methods of UNREDD FPIC implementation in Central Sulawesi Province. During a seminar and workshop, the REDD+ working group members identified their capacity, capability, competence, and constraints in developing REDD+ infrastructure. Also, they developed a social, economic, and environmental safeguards mechanism to facilitate Central Sulawesi Province as 'ready' for REDD+ implementation.

19.5.3.2 Knowledge Level of REDD+ Working Group

The knowledge level of the academics, the government, the local community, the local NGOs, and the private sector are 80 %, 79 %, 79 %, 77 %, and 69 %, respectively. According to the ANOVA, the significance was 0.012 with a 0.05 threshold of significance, which means there was no significance. Hence, the knowledge level between stakeholders was identical or similar.

Variables with a significance value less than 0.05 would have a significant influence on the knowledge level. From five independent variables, namely, stakeholder, education, age, gender, and participation, only the participation variable had a value less than the minimum 0.05, that is 0.002. This result revealed that the participation of REDD+ working group members was the most significant factor affecting their knowledge level and that the influence of other variables was not significant.

This circumstance seems to indicate that the UNREDD Programme Indonesia successfully achieved their objective to enable the stakeholders' capacity in acknowledging REDD+ issues and the stakeholders were also actively involved in REDD+ discussions. Of interest was that the stakeholders (the government, the academics, the local communities, the local NGO, and the private sector) and education (high school, undergraduate, master's, and doctoral) variables had no significant impact. Thus, the origin of institutions and education level do not enhance knowledge level.

In regard to the local communities, it was observed that the representatives of local communities in the REDD+ working group were not from the communities who actually live in the surrounding forest area. However, they came from the Indigenous Peoples Alliance of the Archipelago organization called Aliansi Masyarakat Adat Nusantara (AMAN) in Bahasa Indonesia. The members of this organization already have good knowledge of forest issues. They were also having experiences participating in the national and international conferences. Therefore, their level of knowledge was equal to the other groups. As education level does not significantly decide the knowledge level, it indicates that the REDD+ scheme can be learned and understood through intensive participation and discussion.

19.5.3.3 Opinion of REDD+ Working Group

According to the statistical analysis, the variation in opinions between stakeholders can be summarized as follows.

a. The government shares similar opinions with academics, local NGOs, and the private sector, yet has opinions different from those of the local communities.
b. The view of academics is shared with the government and local NGOs; however, it differs from the opinions of the local communities and the private sector.
c. The beliefs of the local communities are shared only with the private sector; all other groups disagree with the local communities.
d. The local NGOs are in agreement with the government and the academics, yet they are in disagreement with the local communities and the private sector.
e. The private sectors only share consensus with the government and the local communities, but differ in opinions with the academics and the local NGO.

In regard to the responses, all stakeholders agreed that the deforestation and forest degradation in Central Sulawesi Province is relatively high and that it is caused by land conversion and the mismanagement of forest. They also were of the opinion that the REDD+ program could help to reduce deforestation and forest degradation. However, they believe that REDD+ has many uncertainties in addition to expecting REDD+ implementation to take a significant amount of time and expense. Furthermore, they agreed that commitment and cooperation from all stakeholders is needed for the REDD+ program to succeed, even though there are still different levels of knowledge between central and local levels, which need to be addressed for REDD+ to succeed. As for indigenous communities to be able to participate in REDD+, it remains clear that they will support the projects that are effective and bring more benefit than harm and that indigenous people are only willing to consider such projects if they clearly see preconditions in place that would safeguard their cultures, territories, and autonomy (Reed 2011).

19.6 Conclusion

The Government of Indonesia has been conducting numerous activities and established a number of regulations related to REDD+. However, as these issues are evolving rapidly, it is still inadequate to face current REDD+ uncertainties. Hence, strategy, methodology, regulation, development of capacity building, MRV, financial mechanisms, and the benefit distribution system still need to be strengthened to allow REDD+ projects to succeed.

The funding mechanisms from both bilateral and multi-donors are divergent. The Indonesian Government must identify each target of these funds to avoid overlapping activities in the implementation stage. As a consequence of many institution being involved, there is a risk related to operational, transaction, and opportunity costs of REDD+ implementation. Thus, the government should be cautious and act quickly to synchronize the roles of all institutions.

The knowledge level of the academics, the government, the local community, the local NGOs, and the private sector are 80 %, 79 %, 79 %, 77 %, and 69 %, respectively. According to the statistical analysis, it was identical. From five independent variables, this study revealed that only the participation was the most significant factor affecting the REDD+ working group knowledge level. In addition, the REDD+ working group tended to have different opinions between stakeholders related to REDD+ issues.

Acknowledgments Our deep gratitude to the Joint Japan/World Bank Graduate Scholarship Program, the Global Environment Leaders Program of IDEC Hiroshima University, the Ministry of Forestry of Indonesia, the UNREDD Programme Indonesia, the Forestry Service of Central Sulawesi Province, and the REDD+ Working Group Members of Central Sulawesi Province, which supported me to conduct this research.

References

Blom B, Sunderland T, Murdiyarso D (2010) Getting REDD to work locally: lessons learned from integrated conservation and development projects. Environ Sci Policy 13:164–172

Brockhaus M, Obidzinski K, Dermawan A, Laumonier Y, Luttrell C (2012) An overview of forest and land allocation policies in Indonesia: Is the current framework sufficient to meet the needs of REDD+? For Policy Econ 18:30–37

Central Sulawesi Province (2011) Governor Decree No.522/84/DISHUTDA-G.ST/2011 on establishment of working group on reducing emissions from deforestation and forest degradation plus (REDD+ WG) of Central Sulawesi Province 2011. Palu

Cerbu GA, Swallow BM, Thompson DY (2011) Locating REDD: a global survey and analysis of REDD readiness and demonstration activities. Environ Sci Policy 14:168–180

Corbera E, Schroeder H (2011) Governing and implementing REDD+. Environ Sci Policy 14:89–99

Cronkleton P, Bray DB, Medina G (2011) Community forest management and the emergence of multi-scale governance institutions: lessons for REDD+ development from Mexico, Brazil and Bolivia. Forests 2:451–473

Forest People Programme (2011) CENTRAL SULAWESI: UN-REDD Indonesia's Pilot Province. Indonesia

Indonesia Task Force (2012) Refining REDD+ in Indonesia: policy recommendations for increasing effectiveness, efficiency, and equity. University of Indonesia, University of Washington, and the Henry M. Jackson School of International Studies

Larson AM, Petkova E (2011) An introduction to forest governance, people and REDD+ in Latin America: obstacles and opportunities. Forests 2:86–111

Lederer M (2012) REDD+ governance. Wiley Interdiscipl Rev Clim Change 3:107–113

Ministry of Environment (2009) Indonesia's climate change action plan and MRV. Jakarta

Ministry of Forestry (2008) IFCA compilation report: reducing emission from the deforestation and forest degradation in Indonesia. Indonesia

Ministry of Forestry (2009a) Minister of Forestry Regulation No. P.30/Menhut-II/2009 on REDD. Jakarta (in Indonesian)

19 REDD+ Readiness Through Selected Project Activities, Financial Mechanisms. . . 347

Ministry of Forestry (2009b) Minister of Forestry Regulation No.P.36/Menhut-II/2009 on the permission procedures of projects on carbon sequestration and/or storage in production and protected forests. Jakarta (in Indonesian)

Phelps J, Webb EL, Koh LP (2011) Risky business: an uncertain future for biodiversity conservation finance through REDD+. Conserv Lett 4:88–94

Reed P (2011) REDD+ and the indigenous question: a case study from Ecuador. Forests 2:525–549

Romijn E, Herold M, Kooistra L, Murdiyarso D, Verchot L (2012) Assessing capacities of non-Annex I countries for national forest monitoring in the context of REDD+. Environ Sci Policy 19–20:33–48

Streck C (2010) Reducing emissions from deforestation and forest degradation: national implementation of REDD schemes. Clim Change 100:389–394

UNFCCC (2007) Decision -/CP.13Bali Action Plan

UNFCCC (2009) PNG views on Reference Emission Levels and Reference Levels for REDD

UNREDD Programme (2010) Indonesia Joint Programme: Pilot province selection. Jakarta

UNREDD Programme Indonesia (2011) REDD+ National strategy. UNREDD Programme Indonesia, Jakarta. (in Indonesian)

Venter O, Koh LP (2012) Reducing emissions from deforestation and forest degradation (REDD+): game changer or just another quick fix? Ann N Y Acad Sci 1249:137–150

Index

A
Abundance, 194
Accessible environment, 81
Accuracy, 244
Accuracy assessment, 231
Accurate assessment, 318
Agricultural expansion, 167
Agricultural population, 94
Agriculture, 282, 300
Agroecological zoning, 294
Agroforestry, 139, 290
Agrotourism, 278
Air conditioners, 17
Altingia axalsa, 274
Amazilia
    A. candida, 194
    A. yucatanensis, 194
Amazona albifrons, 194
AMeDAS, 16
Anaphalis javanica, 274
Anthropogenic heat emissions, 20
Aratinga nana, 194
Arundo formosana, 314, 316
Asian elephants, 167
Avian susceptibility, 195
Avifauna, 181

B
Back-casting approach, 92
Batang Merao watershed, 241
Beijing, 59, 60, 69
Bioclimatic, 295
Biomass, 7, 8
Biosphere Reserve, 146, 271, 275, 277, 282, 285
Bird feeding guild(s), 179, 183, 185

Bos
    B. gaurus, 211
    B. javanicus, 275
Buffer zone, 148, 275

C
Callenbach, 35–41, 53–55
Callitris, 65
Camellia sinensis, 306
Cameron Highlands, 227
Camili Basin, 255
Camili Biosphere Reserve, 255, 257
Canopy frugivores (CF), 194
Carbon sinks, 6
Cardinalis cardinalis, 195
Carpaticum occidentale, 120
Cassia siamea, 169
C. cinnamomeus, 195
Central Sulawesi Province, 330
Chinese pine, 65, 67, 69
CO_2, 16
Coefficient of ecological stability, 122–123
Coexistence, 171
Coffea arabica, 306
Colonial period, 208–210
Color digital orthophoto data, 17
Composting, 50–52
Comprehensive management system (CMS), 161–162
Connectivity, 212
Conservation, 277
Cooling effects, 16
Cooling potential, 17
Core zone, 147
Correlation analysis, 22, 28
Crypturellus cinnamomeus, 194

Cultural approach, 165
Cultural diversity, 157
Cultural landscapes, 75
Cultural services, 99, 106
Cumulative effects, 125

D

Dactylortyx thoracicus, 195
Dadohae Marine National Park, 147
Decline and aging of rural populations, 94
Deforestation, 4, 249, 328
Degazettement/regazettement, 208
Degree of human impact, 217
Dendrocincla, 179, 193
Development trends, 207–211
Dicerorhinus sumatrensis, 208, 211
Diurnal, 66
Diversity, 192
Dynamic changes, 250

E

Ecological function, 132
Ecological landscape stability, 121
Ecological network, 84
Ecosystems, 84
  management plan, 146
  services, 99
Ecotopia, 35–53, 55
Ecotopian, 39
Ecotourism, 253–256, 262–263, 266, 267
Ecovillage, 130
Educational status, 259, 260
Egretta eulophote, 150
The elephant culture, 170
Elephant-related problems, 164
Elephus maximus, 211
Emergence of REDD+, 331
Energy Policy of the Slovak Republic, 111
Environmental benefits, 306
Environmental change, 227
Environmental Impact Assessment (EIA), 113
Environmental impacts, 250
Environmental parameters, 61
Environmental protection, 269
Environment-friendly, 174
Eucalyptus, 65
Eurasian oystercatcher, 150
European strategies, plans and programs, 111
Evaluated sustainable energy facilities, 125
Ex-situ, 211

F

Fagus sylvatica L, 63
Family health, 302, 304
Feeding guild(s), 183, 190–191
Financial mechanism, 330
Financial status, 332
Floor area, 21
Forest
  cover, 225
  dependency, 286
  edge, 279
  fragmentation and habitat
    degradation, 221
  management, 272
  products, 280, 284, 285
Forestland, 283
The forest priority bioculture, 168
Fuel wood, 265–266, 281

G

GIS, 17
Global warming, 30
The Great Tohoku Earthquake, 49
The Great Tohoku Earthquake
    Disaster, 94
Greenery, 130
Greenhouse gas (GHG), 16, 30
Greenhouse gas emissions, 90
Green roofs, 18
Green space, 18, 31, 75
Growing season, 63
Guild Species Composition, 186–191
Gunung Gede Pangrango, 273

H

Haliptilon, 150
Heat island, 16
Heterogeneity, 299
Highway net, 174
Historical and cultural sites, 206
Home gardens, 141
Home place, 37, 40, 53, 55
Household(s), 258, 278, 284
Household income, 304–305
Houzuki Sennari Ichi, 48–50
Human behavior, 175
Human disturbances, 172
Human–elephant conflicts, 171
Human impact, 218
Human land use, 212, 221

Index

I

IALE. *See* International Association for Landscape Ecology (IALE)
Image classification, 317, 320
Impact assessment, 114
Indigenous knowledge, 154
Indirect benefits, 306
Indonesia, 329
Infrared orthophoto data, 17
Infrastructure, 18
Intercarpaticum, 120
International Association for Landscape Ecology (IALE), 4, 8
Isachne pangrangensis, 274

J

Japan Low-Carbon Society Project, 90, 92
Jiou-Jiou Peaks Natural Reserve, 315
Job opportunities, 268
Juniper bush, 79

K

Kappa coefficients, 320
Kitakyushu City, 75
Kougenn-ji, 48–50
Kyoto protocol, 110

L

Land cover, 231
Land degradation, 245
Landscapes, 115, 229
    changes, 212–217
    composition, 236
    degradation, 6
    ecological stability coefficients, 123
    ecology, 290
    elements, 212
    management, 240
    matrix, 193
    metric, 235
    pattern, 225, 232
    units, 180
Land use, 7, 265
Land use/land cover (LULC), 240, 241, 245
Large-scale rubber plantations, 208
Liquid petroleum gas (LPG) usage, 266–267
Literature analysis, 166
Local community, 279
Local NGO, 334
Low-carbon, 157

Low carbon society(ies), 4, 7
LPG usage. *See* Liquid petroleum gas (LPG) usage
LULC. *See* Land use/land cover (LULC)
LULC classification, 243

M

Mean daily maximum temperature, 17
Meleagris ocellata, 195
Meteorological, 68
Migration, 261–262
Minato-ku, 17
Ministry of Land, 18
Ministry of the environment, 18
Motomachi park, 41–48, 50, 53, 54
Multicollinearity, 29
Multifunctional landscape planning (MFLP), 78, 81
Multiple regression analysis, 22

N

National Action Plan on Climate Change, 335
National carbon sequestration strategy, 196
National park, 211, 276, 309
Natural disasters, 325
Naturalness of protected areas, 217
Natural reserve, 314
Natural space, 74
Negative impact, 263–264
Niche, 319–320, 322
Niche information, 320
NPO bank of green resources, 51–53

O

Omniscape, 84
Ornamental plants, 300
Overexploitation of commercial hunting, 208

P

Pannonicum, 120
Panthera tigris sondaica, 274–275
Participation, 263
Participatory Rural Appraisal (PRA), 166
Patagioenas flavirostris, 194
Pekarangan, 130, 132, 290, 297, 308
Peninsular Malaysia, 206, 208–209
Phenomenon, 16
Photosynthetically, 61

352   Index

Physical activity, 74
Pine trees, 60, 65
Pinus
    P. merkusii, 281
    P. tabulaeformis, 59
Plant life, 79
Poaching, 171, 208
Podocarpus imbricatus, 274
Potential soil erosion, 243
Process planning, 76, 81
Protected area(s), 207, 211, 212, 221, 253, 272, 309
Provincial perspectives, 332–337
Provisioning services, 99

Q
Quality of landscape, 115
Quercus ilex, 65
Quercus spp., 274

R
REDD+
    development, 339–340
    financing mechanism, 343
    implementation, 337
    information, 337
    issues, 341
    payment, 342
    program, 345
    projects, 345
    strategies, 342
    uncertainties, 345
    working group, 335, 337, 343
REDD mechanisms, 341
Regulating services, 99
Rehabilitation of rural areas, 96
Remote sensing, 224
Residential area, 76
Rock plant, 320
Rubber and oil palm, 212
Rubber plantation, 173
Rural landscape, 107

S
Sapium discolor, 316
Satochi-Satoyama, 96
Satoumi, 97
Satoyama, 90
Schima
    S. superba, 316
    S. walichii, 274

Selfish consciousness, 171
Shinan Dadohae Biosphere Reserve (SDBR), 146, 155
Site elevation, 21
Slope degree, 318
Socioeconomic, 259–261
Socioeconomic and political scenarios, 221
Soil moisture content, 69
Species diversity, 298
Species–environment relationships, 183
Species even, 298
Species guild composition, 192–196
Species richness, 191, 194, 298
Species similarity, 299
SPOT 5, 229
Staying periods, 264
Stomatal sensitivity, 67
Superstitions, 171
Supervised classification, 233
Survey, 105–106
Sustainability and sustainable development, 221
Sustainable, 156
    energy, 110
    forestry management, 258
    land management, 224

T
Taiwan, 325
Tapirus indicus, 211
Text analysis, 97
Theobroma cacao, 306
Tityra semifasciata, 194
Tokyo, 17
Towards a Low-Carbon Society Japanese Society, 91
Traditional ecological knowledge (TEK), 16
Transition zone, 148, 275
Transpiration, 66, 69
Transpiration patterns, 58, 62–65
Transport and tourism, 18
Tropical dry forest (TDF), 178, 196
Tropical forests, 272
Tropical mountain forest, 236

U
The unique bioculture landscape, 169
The unique TEK, 168
UN-REDD, 328
UNREDD Programme, 344
Urban conurbation, 218
Urban ecology, 40, 54
Urban environment, 58

Index

Urban green ecosystem, 58
Urbanization, 58
USLE method, 243

V
Vegetation structure, 134
Vegetation variables, 181–182

W
Water stress, 62
West Java, 292
Wildlife sanctuaries, 211
Woodlands, 17, 18

X
Xiphorhynchus, 179, 193
Xishuangbanna region, 165

Y
Yellow River Basin, 8
Yucatan Peninsula, 180

Z
Zelkova serrata, 316
Zenaida asiatica, 195
Zoobenthic, 150

Printed by Publishers' Graphics LLC
LMO140522.23.33.47